Mathematical Analyses of Decisions, Voting and Games

CONTEMPORARY MATHEMATICS

795

Mathematical Analyses of Decisions, Voting and Games

AMS Special Session
Mathematics of Decisions, Elections and Games
April 8, 2022
Virtual

Michael A. Jones
David McCune
Jennifer M. Wilson
Editors

AMERICAN MATHEMATICAL SOCIETY

2020 *Mathematics Subject Classification*. Primary 91-06, 90B06, 91A05, 91A18, 91A46, 91B06, 91B12, 91B14, 91B32.

Library of Congress Cataloging-in-Publication Data

Names: AMS Special Session on Mathematics of Decisions, Elections and Games (2022 : Online), author. | Jones, Michael A., 1967- editor. | McCune, David, 1984- editor. | Wilson, Jennifer Mia, 1965- editor.

Title: Mathematical analyses of decisions, voting and games : Virtual AMS Special Session on Mathematics of Decisions, Elections and Games, April 8, 2022 / Michael A. Jones, David McCune, Jennifer M. Wilson, editors.

Description: Providence, Rhode Island : American Mathematical Society, 2024. | Series: Contemporary mathematics, 0271-4132 ; 795 | Includes bibliographical references.

Identifiers: LCCN 2023042077 | ISBN 9781470469788 (paperback) | 9781470476083 (ebook)

Subjects: LCSH: Decision making–Mathematical models–Congresses. | Voting–Mathematical models–Congresses. | Game theory–Congresses. | AMS: Game theory, economics, social and behavioral sciences – Proceedings, conferences, collections, etc.. | Operations research, mathematical programming – Operations research and management science – Transportation, logistics. | Game theory, economics, social and behavioral sciences – Game theory – 2-person games. | Game theory, economics, social and behavioral sciences – Game theory – Games in extensive form. | Game theory, economics, social and behavioral sciences – Game theory – Combinatorial games. | Game theory, economics, social and behavioral sciences – Mathematical economics – Decision theory. | Game theory, economics, social and behavioral sciences – Mathematical economics – Voting theory. | Game theory, economics, social and behavioral sciences – Mathematical economics – Social choice. | Game theory, economics, social and behavioral sciences – Mathematical economics – Resource and cost allocation.

Classification: LCC QA279.4 .A64 2022 | DDC 519.3–dc23/eng/20240105

LC record available at https://lccn.loc.gov/2023042077

Visit the AMS home page at `https://www.ams.org/`

10 9 8 7 6 5 4 3 2 1 29 28 27 26 25 24

Contents

Preface

The papers included in this volume are based on talks given at the AMS Special Sections on the Mathematics of Decisions, Elections, and Games at the 2022 Joint Mathematics Meetings. The session was co-organized by Michael A. Jones (Mathematical Reviews), David McCune (William Jewell College), and Jennifer M. Wilson (The New School). Decision theory, voting theory and game theory are three intertwined areas in the mathematical social sciences that involve making optimal decisions in different contexts.

Decision theory studies instances in which individuals make choices, often under uncertainty, while game theory analyzes optimal decisions when outcomes are affected by all the players' actions. Voting theory and its application to elections, considers the outcomes when individual decisions on submitting ballots for candidates are aggregated to determine a social choice. These areas share a characteristic diversity of mathematical techniques encompassing both continuous and discrete analysis, algebraic and graph theoretic approaches, and probabilistic and deterministic reasoning. The papers in this collection span this diversity. We describe the papers below.

When to stop consulting, by Kilgour and Brams, analyzes optimal strategies for a decision maker who must make a binary choice but has the option to consult experts a fixed number of times. Similar to the Secretary Problem, in which an individual must select a job candidate after a sequence of interviews, the Consultation Problem concerns a situation when a sequence of interactions should terminate a process and trigger a decision. The decision maker's decision to either stop or to consult results in a Bayesian decision binary tree in which optimizing the expected utility depends on the cost of the consultation.

Connecting Arrow's theorem, voting theory, and the Traveling Salesperson Problem, by Saari, shows how the analysis of pairwise voting methods and the traveling salesperson problem can be unified and simplified by treating the corresponding weighted graph as a vector space and decomposing it into portions that contain cycles and those that do not. Saari shows that finding a winner from a given profile reduces to analyzing the non-cyclic component of the graph while finding an optimal solution for the traveling salesperson problem can be simplified by analyzing the graph's cyclic component.

Graph theory is also used in *Piercing numbers in circular societies*, by Mazur, Sondjaja, Wright, and Yarnall. In their work, the authors extend the analysis of approval voting to situations when individuals express approval for intervals on a circle. Previous work has analyzed bounds on the number of voters who approve of a single candidate when preferences live on a fixed interval or on a circle. This paper asks a different question: what is the minimum number of candidates needed to satisfy all voters? The authors analyze the situation when individual approval sets are of equal length and consider both deterministic and probabilistic models.

Impossibility theorems involving weakenings of expansion consistency and resoluteness in voting, by Holliday, Norman, Pacuit, and Zahedian, examines how versions of Arrow's axiom of expansion consistency are incompatible with other fundamental axioms in voting theory including anonymity, neutrality and resoluteness. Specifically, the authors consider binary expansion consistency, which states that a candidate who is the social choice from an initial field of candidates, and who beats a new candidate in a pairwise comparison, should remain the social choice when the new candidate is added to the field. Using a SAT solver, the authors show that even weakened versions of resoluteness are incompatible with binary expansion consistency. The proof makes use of a novel method for extending impossibility results on pairwise and majority voting methods to impossibility results on all voting methods, which may have application in future analyses.

Voting on cyclic orders, group theory, and ballots, by Crisman, Holleran, Martin, and Noonan, uses representation theory to analyze both the ballots for voting on cyclic orders and points-based procedures that aggregate such ballots. In this setting, voters provide rankings of the cyclic orders of n candidates. The authors provide a full characterization of points-based voting procedures for two ballot types when $n = 4$ and offer observations for $n = 5$.

Conditions for fairness anomalies in instant-runoff voting, by Graham-Squire, extends previous work that analyzed necessary and sufficient conditions for instant-runoff voting to demonstrate monotonicity paradoxes for three-candidate when ballots are fully-ranked. In this paper, he determines necessary and sufficient conditions for monotonicity paradoxes for three-candidate elections when ballots are not fully ranked, and necessary conditions for elections with four or more candidates. Additionally, he examines no-show anomalies in three-candidates elections for both fully-ranked and not fully-ranked ballots.

In *An iterative procedure for apportionment and its use in the 2016 Georgia Republican primary*, by Jones, McCune and Wilson, the authors analyze an apportionment method used to allocate at-large delegates in Georgia to presidential candidates during the 2016 Republican presidential primary. This method, the Iterated Lower Quota method, is based on a sequence of rounds in which each candidate receive a succession of lower quotas. The authors analyze the properties of this method, including the maximum number of rounds that may occur, and compare it to Hamilton's method. They also discuss its implementation in the 2016 primaries.

Double moves by each player in chess openings make the game fairer, by Brams and Ismail, presents a high-level analysis of work also presented at the 2001 IEEE Conference on Games in which the authors make the case that the apparent advantage in chess given to white (as the first player) can be mitigated by changing the usual sequence of moves, WB/WB/WB/WB, to the sequence WB/BW/WB/WB....

How lies induced cooperation in Golden Balls: A game-theoretic analysis, by Brams and Mor, analyzes a well-known episode from the British TV game show *Golden Balls*, in which one of the two contestants lies about his strategy. This incident transformed a situation which can be modeled by a standard Prisoners'

Dilemma into a game in which the contestants were able to improve their outcomes from the expected one when both players defect.

We would like to thank the authors of these papers, as well as all the presenters in the 2022 AMS Special Section on the Mathematics of Decisions, Elections, and Games, for their interesting contributions and their enthusiastic participation, which was especially appreciated as the 2022 Joint Meetings were moved online. We hope this volume and the session contribute to strengthening the interdisciplinary community of mathematicians and others who work in these areas.

Contemporary Mathematics
Volume **795**, 2024
https://doi.org/10.1090/conm/795/15963

When to stop consulting

D. Marc Kilgour and Steven J. Brams

ABSTRACT. Consider a decision-maker whose objective is to maximize the probability of making a correct binary (Yes-No) decision. To gain information, the decision-maker can consult independent experts. But how many? When should consultation stop?

We model the decision-maker's current state of knowledge using the number of recommendations for each alternative. After every consultation, the decision-maker decides whether to make a decision immediately, based on the advice already received, or to consult again. We impose a ceiling on the number of consultations and, possibly, a cost for each consultation.

Surprisingly, we find that even when consultation is free, it can be optimal *not* to consult. Less surprising, consultation may be uniquely optimal when evidence is scarce, especially if it is approximately balanced between Yes and No. For a decision-maker who, initially, is maximally uncertain, we obtain complete optimal strategies when consultation is free, and we show that there is a fundamental limit on expected utility ($\frac{3}{4}$, where the minimum utility is 0 and the maximum is 1), even when consultation is unlimited. We then explore the strategic effects of a positive cost, linking them to the zero-cost case. We also compare our results on consultation to strategies in the Secretary Problem, another decision problem in which information is collected sequentially until a decision is made.

Individuals facing difficult decisions often ask for advice from family, friends, or colleagues. New points of view and different expertise can provide useful facts and relevant perspectives. But consulting is not without costs, including the cost of delaying the eventual decision. Our concern is whether and when consultation stops helping. At what point does consulting lose its effectiveness as a strategy? How much should a decision-maker consult?

The decision to stop consulting can be crucial, whether the decision-maker is a powerful political leader or a private individual with limited scope for action. For example, as the Cuban Missile Crisis developed, U.S. President John F. Kennedy consulted individually with members of EXCOMM, the Executive Committee of the National Security Council, before deciding on a response [**2, 5, 6**]. Anecdotal evidence suggests that the committee 'hawks' dominated the early recommendations, but Kennedy kept on consulting. A prominent early 'dove,' U.N. Ambassador Adlai Stevenson, was later joined by others; eventually, opposition to escalation prevailed.

2020 *Mathematics Subject Classification.* Primary 90B50; Secondary 62C05, 91B06.
The first author was supported in part by NSERC Discovery Grant #8588.

We model the decision-maker's problem, focusing on strategies for deciding when to stop consulting. We also ask how limitations on the amount of consultation affect the quality of the (eventual) decision, and how much consultants can help. Our objective is to provide advice to decision-makers about when to stop consulting and get on with the decision.

1. Introduction

This paper concerns a decision-maker (DM) who must make a binary (Yes/No) decision. One of these two alternatives is correct, but DM does not know which one. DM's objective is to maximize the probability of making the correct choice. We do not specify the details, but instead assume that, shortly after the choice is made, it becomes clear whether or not it was correct. We model DM as receiving 1 unit of utility for a correct choice, and 0 otherwise; we ask whether there is a consultation strategy that maximizes expected utility.

We assume that DM has beliefs about which choice is correct, which we capture in DM's (subjective) probability, p, that Yes is correct. DM will use any advice in favor of Yes or No to revise appropriately the value of p.

We model consultation as an iterative process in which, at any time, DM can choose to Stop—make a choice of Yes or No based on the current value of p—or to Consult, obtaining advice from a new "expert." After consultation, DM adjusts the value of p to reflect the advice received, and then chooses whether to Stop or to Consult again.

A DM who chooses Stop makes a Yes/No decision immediately. It is clear that, to maximize DM's expected utility, the best choice X_{op} depends on the current value of p; specifically,

$$X_{op} = \begin{cases} \text{Yes}, & \text{if } p \geq \frac{1}{2} \\ \text{No}, & \text{if } p \leq \frac{1}{2} \end{cases}.$$

We propose two restrictions on the amount of consultation:

- DM may Consult a maximum of n times; and
- DM's final utility is reduced by c for each consultation.

Thus our model contains two parameters: $n \geq 0$, an integer representing the ceiling on the number of consultations; and $c \geq 0$, representing the cost (including delay cost) of each consultation.

Before we describe the details of our model, we compare it to another sequential decision problem, the Secretary Problem [**3**, **4**, **7**]. Both the Secretary Problem and the Consultation Problem involve sequences of information; the problem ends when a decision is made, which may occur at any point in the sequence. In the Secretary Problem, there is a sequence of job candidates, and the decision-maker must select one; the only available information about a candidate is their ordinal ranking relative to previous candidates. The decision to hire a candidate must be made before the next candidate is interviewed. Traditionally, success is measured by the probability that the candidate chosen is, in fact, the overall best candidate. In the Consultation Problem, the alternatives are fixed but estimates of their utilities change as evidence is accumulated (utilities are also diminished by the cost of consultation if $c > 0$). The Secretary Problem and the Consultation Problem are different, but both concern if and when a finite sequence of random events should trigger a decision.

2. Model

We assume that (1) initially, DM is maximally uncertain, and (2) DM sees each possible consultant as equally credible. (Possible variations in these assumptions are discussed in the Conclusions.) We implement our assumptions by modeling DM's beliefs about p as DM's **state**, (h, k), where h and k are positive integers. In state (h, k), DM's probability that Yes is correct is $p = \frac{h}{h+k}$.

We assume that DM's initial state is $(h, k) = (1, 1)$, so DM's initial probability that Yes is correct is $\frac{1}{2}$. Let $h - 1$ be the number of Yes recommendations that DM has received so far, and let $k - 1$ be the number of No recommendations. We further assume that, in state (h, k), DM believes that the probability that the next consultant will recommend Yes is $\frac{h}{h+k}$, which is identical to DM's current probability that Yes is correct. Thus, a consultation implies a state transition

$$(h,k) \longrightarrow \begin{cases} (h+1,k) & \text{with probability } \frac{h}{h+k} \\ (h,k+1) & \text{with probability } \frac{k}{h+k} \end{cases}.$$

For example, suppose that DM consults 3 times and receives 1 Yes and 2 No's. Then after these consultations, DM's state is $(h, k) = (2, 3)$, DM's subjective probability that Yes is correct is $\frac{2}{5}$, which is also DM's subjective probability that the next consultant will recommend Yes, in which case the state would transition to $(3, 3)$.

Given a decision to Stop in state (h, k), DM maximizes utility by choosing

$$X_{op} = \begin{cases} \text{Yes} & \text{if } h \geq k \\ \text{No} & \text{if } h \leq k \end{cases}.$$

Thus, at state (h, k), DM's immediate expected utility for choosing Stop is

$$u(X_{op}) = u(h,k) = \frac{\max\{h,k\}}{h+k}. \tag{2.1}$$

But in state (h, k), there have been $h+k-2$ consultations, so DM's net utility will equal $u(X_{op}) - (h + k - 2)c$. For instance, in state $(h, k) = (2, 3)$, DM can obtain expected utility of $\frac{3}{5} - 3c$ by choosing No.

DM's current state (h, k) reflects the recommendations so far received. The set of possible states forms a simple lattice, shown in Figure 1, which also shows the transition probabilities moving to the right, starting at state $(h, k) = (1, 1)$. Figure 1 also shows DM's immediate expected utility for choosing Stop at each state (h, k), as determined by (2.1).

In fact, assuming consultation cost $c = 0$, our model is a standard Bayesian decision model in which, progressing through the decision tree, DM's current probabilities and expected payoffs reflect DM's experience so far. Thus, DM's probability for p is Beta-distributed [**1**], with probability density function

$$f(x) = \frac{\Gamma(h+k)}{\Gamma(h)\Gamma(k)} x^{h-1}(1-x)^{k-1} \quad (0 < x < 1).$$

(If ℓ is a positive integer, then $\Gamma(\ell) = (\ell - 1)!$.) In state (h, k), the expected value of p is

$$E[p] = \frac{h}{h+k}$$

so that DM's current belief about p is equal to the mean of the current distribution of p.

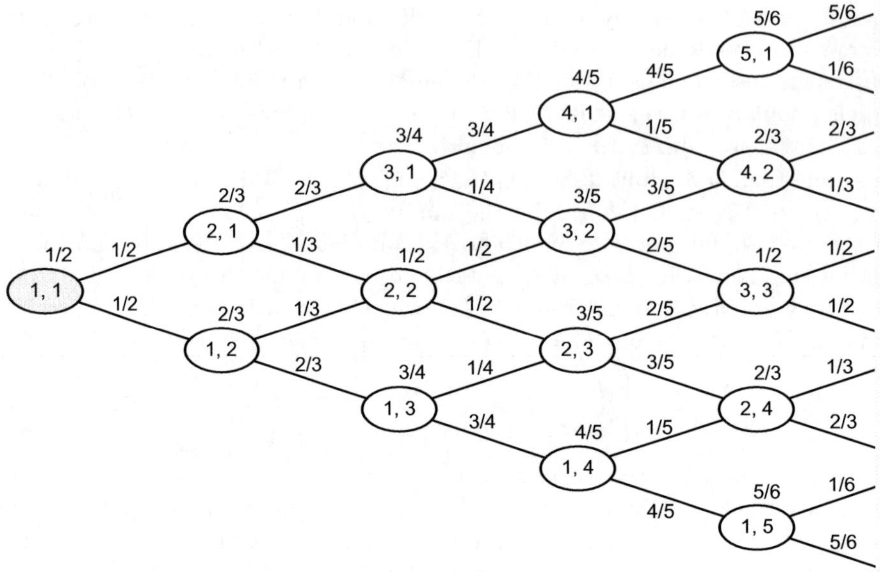

FIGURE 1. States, transition probabilities, and immediate utilities

3. No-cost models

We begin by solving all No-Cost Models, in which $c = 0$ and $n = 0, 1, \ldots$. If the process arrives at a state (shown below as a node), DM must decide whether to Stop (S) or Consult (C)—except, of course, that S is the only option if Consult has already been chosen n times. DM's expected utility for choosing optimally at node $(h, k) = (1, 1)$ and, if appropriate, all subsequent nodes, is denoted $U_c(n)$. We first study $U_0(n)$.

3.1. No-cost model: $n = 0$. If no consultations are allowed, DM must Stop at $(h, k) = (1, 1)$. There are no decisions to make, and by (2.1) DM's expected utility is $U_0(0) = \frac{1}{2}$.

3.2. No-cost model: $n = 1$. If exactly one consultation is allowed, DM may stop at $(h, k) = (1, 1)$, which would produce expected utility $U_0(0) = \frac{1}{2}$, or consult once, ending up at either $(2, 1)$ or $(1, 2)$. Using (2.1), DM's decision problem is as shown in Figure 2, where nodes labelled '0' are random (i.e., choices by Nature) and thicker (darker) branches indicate choices consistent with backward induction from the endpoints. It is easy to see that DM's optimal choice at $(1, 1)$ is C, and DM's expected utility for choosing optimally throughout is $U_0(1) = \frac{2}{3}$.

3.3. No-cost model: $n = 2$. If exactly two consultations are allowed, DM has a decision to make at $(h, k) = (1, 1)$; if that decision is Consult, then there is another decision at $(2, 1)$ or $(1, 2)$. Using (2.1), DM's decision problem is as shown in Figure 3. Note that the figure is not a conventional decision tree, as the node at

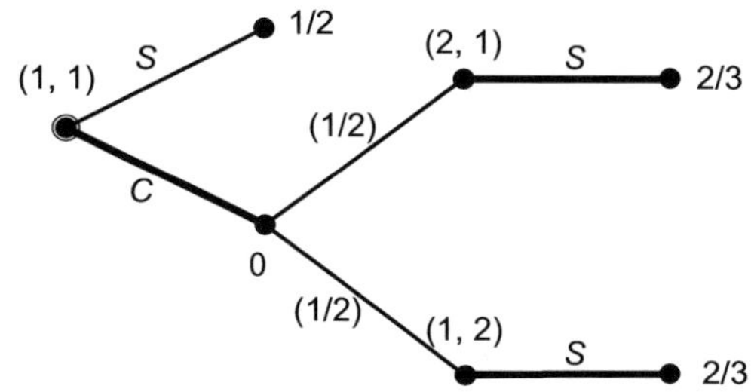

FIGURE 2. Decision Problem, $c = 0$, $n = 1$

$(2, 2)$ has indegree 2. But it is easy to see that this simplification does not change the analysis.

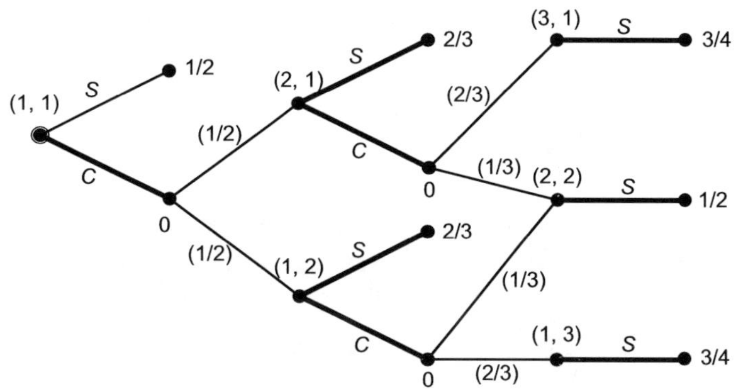

FIGURE 3. Decision Problem, $c = 0$, $n = 2$

Again, the darker branches in Figure 3 indicate rational choices. It is clear that at $(1, 1)$ DM's optimal choice is C, and at $(2, 1)$ and $(1, 2)$ it is C or S. DM's expected utility starting at $(1, 1)$ is $U_0(2) = \frac{2}{3}$.

3.4. No-cost model: $n = 3$. By a similar analysis, it can be shown that, if $c = 0$ and $n = 3$, DM's optimal choice at $(1, 1), (2, 1), (1, 2)$ and $(2, 2)$ is C, whereas at $(3, 1)$ and $(1, 3)$ it is either C or S. The expected utility of the optimal strategy is $U_0(3) = \frac{7}{10}$.

3.5. No-cost model: General analysis. Assuming that the current state is (h, k), we compare two options, (S) Stop Immediately versus (CS) Consult and then Stop Immediately. First note that, at state (h, k), there have already been $h + k - 2$ consultations, so CS is infeasible unless $n > h + k - 2$.

LEMMA 3.1. *Assume $c = 0$ and $n > h + k - 2$. If $h \neq k$, then at (h, k) DM's expected utilities for option S and option CS are equal.*

PROOF. Assume $h > k$. At (h, k), S yields expected utility $\frac{h}{h+k}$. Option C produces state $(h + 1, k)$ with probability $\frac{h}{h+k}$ and state $(h, k + 1)$ with probability $\frac{k}{h+k}$. Using (2.1), CS yields expected utility

$$\frac{h}{h+k} \cdot \frac{h+1}{h+k+1} + \frac{k}{h+k} \cdot \frac{h}{h+k+1} = \frac{h^2 + h + hk}{(h+k)(h+k+1)} = \frac{h}{h+k}.$$

The proof is analogous if $h < k$. □

The message of Lemma 3.1 is that it is never an advantage to Consult if you plan to Stop immediately after consulting. Crucial to this result is that the state, (h, k), is not located on the centre-line of Figure 1—in other words, that $h \neq k$.

Along the centre-line of Figure 1, the situation is different, as shown by the next lemma.

LEMMA 3.2. *Assume $c = 0$ and $n > 2h - 2$. At state (h, h), option CS yields greater expected utility than option S.*

PROOF. At (h, h), S yields expected utility $\frac{1}{2}$ by (2.1), while option C produces state $(h + 1, h)$ with probability $\frac{1}{2}$ and state $(h, h + 1)$ with probability $\frac{1}{2}$. Again using (2.1), CS at (h, h) yields expected utility

$$\frac{1}{2} \cdot \frac{h+1}{2h+1} + \frac{1}{2} \cdot \frac{h+1}{2h+1} = \frac{h+1}{2h+1} > \frac{1}{2}.$$

□

Lemma 3.2 provides an important contrast with Lemma 3.1. If it happens that $h = k$, then choosing CS, Consult and then Stop, is an improvement on choosing S, Stop immediately.

LEMMA 3.3. *Assume $c = 0$ and $n > h + k - 2$. If either S does not maximize DM's expected utility at $(h + 1, k)$ or S does not maximize DM's expected utility at $(h, k + 1)$, then only C maximizes DM's expected utility at (h, k).*

PROOF. If $h = k$, Lemma 3.2 shows that Stop at (h, k) cannot maximize expected utility. If $h \neq k$, Lemma 3.1 shows that S and CS produce equal expected utility. Suppose that, at $(h + 1, k)$, DM has a choice that begins with C and maximizes expected utility. Then at (h, k) the choice of C must maximize expected utility. The argument is similar if there is a utility-maximizing choice beginning with C at $(h, k + 1)$. □

The contribution of Lemma 3.3 is to link the CS versus S comparison to backward induction. If either S is not an optimal choice at $(h + 1, k)$ or S is not an optimal choice at $(h, k+1)$, then C is uniquely optimal at (h, k). But, for example, if $n = h + k - 1$, then S is mandatory at $(h + 1, k)$ or $(h, k + 1)$ since no more consultations are allowed, so both S and C are optimal choices at (h, k).

To study backward induction, it is convenient to separate cases according to the parity of n.

THEOREM 3.4. *If $c = 0$ and $n = 2m - 2$ consultations are permitted, where $m \geq 1$, DM's unique optimal choice is C at any state (h, k) satisfying $\max\{h, k\} \leq m - 1$. At every other state, both C and S are optimal.*

PROOF. If C is always chosen, then the final state (h, k) satisfies $h + k = 2m$. At state (m, m), C is not permitted, so the expected utility is $\frac{1}{2}$. By Lemma 3.2, if $1 \leq t < m$, DM's utility is increased by choosing C at state (t, t). By Lemma 3.3, the expected utility increases by choosing C at any state $(t-1, t)$, where $2 \leq t < m$. By induction, the expected utility increases by choosing C at any state (r, s), where $1 \leq s < r < m$ or $1 \leq r < s < m$. It is easy to see that such states (h, k) are exactly those that satisfy $\max\{h, k\} \leq m-1$. At all other states, both C and S are optimal, by Lemma 3.1. □

Note that DM is indifferent between S and C exactly when $\max\{h, k\} \geq m$.

Figure 4 shows DM's optimal strategy and expected utilities when $n = 4$. Since $m = 3$, Theorem 3.4 shows that the optimal choice is to consult at all states (h, k) where $\max\{h, k\} \leq 2$, that is, at (1, 1), (1, 2), (2, 1), and (2, 2). At all other states either C or S may be chosen. Each state is labelled with DM's expected utility at that state should it be attained when the optimal strategy is played. Notice that whether Consult is uniquely optimal depends on both the balance of evidence and the amount of evidence; the evidence favours Yes at (2, 1), and the unique optimal choice is Consult, whereas at (3, 2) the evidence still favours Yes, but now DM is indifferent between Consult and Stop.

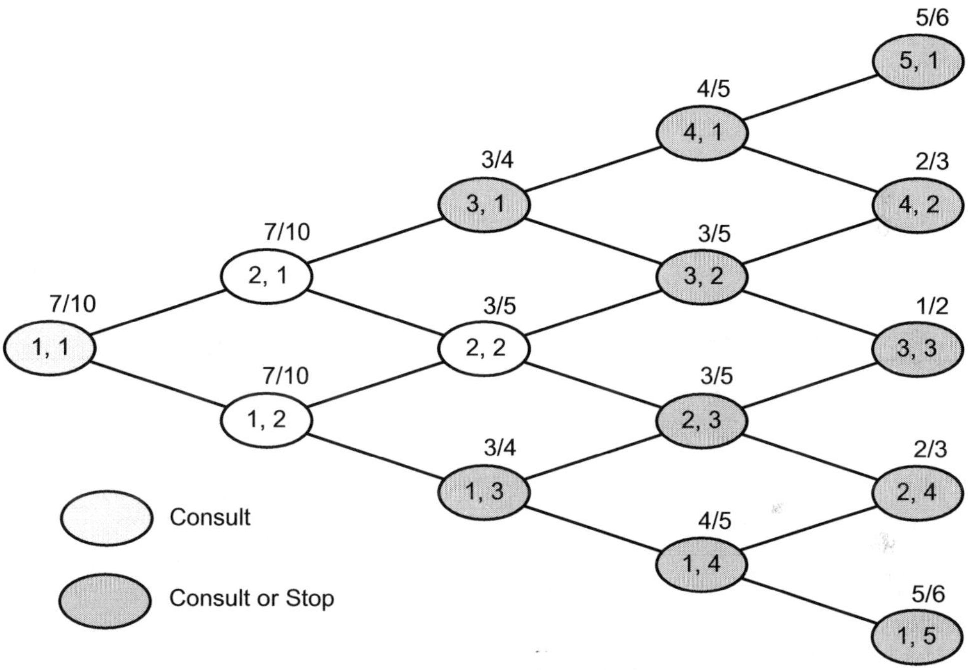

FIGURE 4. Optimal strategies and expected utilities when $n = 4$ and $c = 0$

THEOREM 3.5. *If $c = 0$ and $n = 2m-1$ consultations are allowed, where $m \geq 1$, DM's optimal strategy is C at any state (h, k) such that $\max\{h, k\} \leq m$. At every other state, both C and S are optimal.*

PROOF. If C is always chosen, then the final state (h, k) must satisfy $h + k = 2m + 1$. At state (m, m), one more consultation is permitted. By Lemma 3.2, DM strictly prefers C at state (m, m), and at any state (t, t) where $1 \leq t \leq m$. By Lemma 3.3, the expected utility increases by choosing C at any state $(t - 1, t)$, where $2 \leq t \leq m$. By induction, expected utility increases by choosing Consult at any state (r, s), where $1 \leq s < r \leq m$ or $1 \leq r < s \leq m$. At all other states, both C and S are optimal, by Lemma 3.1. □

Now it is the case that DM is indifferent between S and C exactly when $\max\{h, k\} > m$. Figure 5 shows DM's optimal strategy and expected utilities when $n = 4$. Since $m = 3$, Theorem 3.5 shows that the optimal choice is to consult at all states (h, k) where $\max\{h, k\} \leq 3$, that is, at (1, 1), (1, 2), (2, 1), (3, 1), (2, 2), and (1, 3). At all other states either C or S may be chosen.

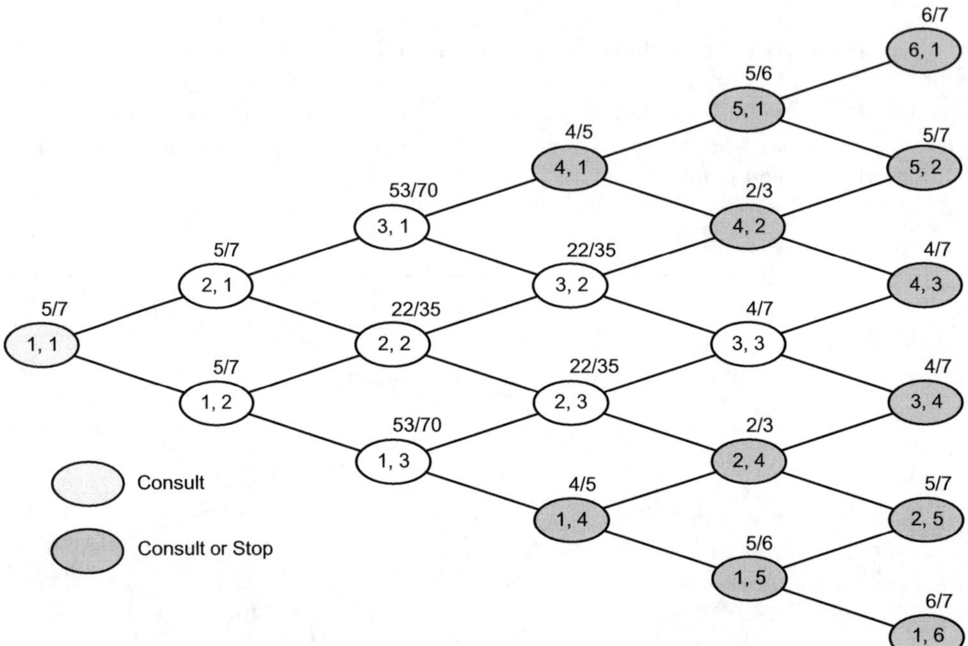

FIGURE 5. Optimal strategies and expected utilities when $n = 5$ and $c = 0$

To state the optimal strategy formally, define the set of *Consultation Nodes* to be

$$\mathcal{C}(n) = \left\{(h,k) : h \geq 1, k \geq 1, \max\{h,k\} = \left\lfloor \frac{n+1}{2} \right\rfloor \right\}.$$

For example, $(h, k) = (3, 2) \in \mathcal{C}(5)$, because $\max\{h, k\} = 3 = \frac{5+1}{2}$, but $(h, k) = (4, 2) \notin \mathcal{C}(5)$.

If $c = 0$ and the maximum number of consultations is n, a strategy is optimal if and only if it specifies C at all nodes of $\mathcal{C}(n)$. At all other nodes, either C or S may be chosen. Notice that, if n is odd, then $\mathcal{C}(n) = \mathcal{C}(n + 1)$.

When n consultations are permitted and the cost per consultation is c, denote DM's expected utility using the optimal strategy by $U_c^*(n)$. We can now specify $U_0^*(n)$.

THEOREM 3.6.

$$U_0^*(n) = \begin{cases} \frac{3n+2}{4n+4} & \textit{if } n \textit{ even} \\ \frac{3n+5}{4n+8} & \textit{if } n \textit{ odd} \end{cases}.$$

PROOF. The strategy to always Consult is optimal. By induction, after n consultations, each possible state $(n + 1, 1)$, $(n, 2)$, ..., $(1, n + 1)$ has probability $\frac{1}{n+1}$. Suppose that $n > 0$ is even, and set $n = 2m - 2$ so $m \geq 1$. All possible final states, $(j, 2m - j)$ where $j = 1, 2, \ldots, 2m - 1$, have probability $\frac{1}{2m-1}$. It follows that

$$U_0^*(n) = \frac{1}{2m-1} \sum_{j=1}^{2m-1} \frac{\max\{j, 2m-j\}}{2m} = \frac{3m-2}{4m-2}.$$

Substitution of $m = \frac{n+2}{2}$ now shows that $U_0^*(n) = \frac{3n+2}{4n+4}$ as required. Similarly, if n is odd, set $n = 2m - 1$ with $m \geq 1$. Then a similar calculation shows that

$$U_o^*(n) = \frac{1}{2m} \sum_{j=1}^{2m} \frac{\max\{j, 2m+1-j\}}{2m} = \frac{3m+1}{4m+2}$$

and substitution for $m = \frac{n+1}{2}$ now produces $U_0^*(n) = \frac{3n+5}{4n+8}$. □

Note that, if n is odd, so that $U_0^*(n) = \frac{3n+5}{4n+8}$, then $n+1$ is even, and $U_0^*(n+1) = \frac{3(n+1)+2}{4(n+1)+4} = \frac{3n+5}{4n+8} = U_0^*(n)$. Thus we have that, for example, $U_0^*(0) < U_0^*(1) = U_0^*(2) < U_0^*(3) = U_0^*(4) < U_0^*(5) \ldots$.

It is immediate from Theorem 3.6 that

$$U_0^*(n) \longrightarrow \frac{3}{4} \text{ as } n \longrightarrow \infty.$$

For some intuition into this limit, recall that after n consultations, each possible state $(n+1, 1), (n, 2), \ldots, (1, n+1)$ has equal probability. In the limit, $\frac{h}{h+k}$ can be thought of as a uniform random variable Y on $[0, 1]$; the limiting value of $\frac{\max\{h,k\}}{h+k}$ is $M(Y) = \max\{Y, 1 - Y\}$. It is easy to check that the expected value of $M(Y)$ is $\frac{3}{4}$, confirming the limiting value obtained above.

As $U_0^*(n)$ is non-decreasing, it follows that, as more consultations are permitted, the expected utility (following the optimal strategy) increases toward an upper bound of $\frac{3}{4}$. No amount of consultation can possibly increase expected utility beyond this bound, making it a fundamental limit on the value of consultation in our model.

4. Cost models

Now we turn to Cost Models, in which $c > 0$ and $n = 0, 1, \ldots$. Expected utility is denoted $U_c(n)$; if an optimal strategy is followed, DM's expected utility is denoted $U_c^*(n)$.

4.1. Cost model: $n = 0$. If no consultations are permitted, there are no decisions, and the expected utility is $U_c^*(0) = \frac{1}{2}$.

4.2. Cost model: $n = 1$. If 1 consultation is permitted, the only choice is C or S at (1, 1). DM's expected utility then equals

$$\begin{cases} \frac{1}{2}, & \text{if S} \\ \frac{2}{3} - c, & \text{if C} \end{cases}.$$

It is clear that the optimal strategy is C if $c \leq \frac{1}{6}$ and S if $c \geq \frac{1}{6}$, so that

$$U_c^*(1) = \begin{cases} \frac{2}{3} - c, & \text{if } c \leq \frac{1}{6} \\ \frac{1}{2}, & \text{if } c \geq \frac{1}{6} \end{cases}.$$

Thus, if $c > 0$ is small enough, the optimal strategy is C, exactly the same as at $c = 0$, but if c is large enough, the optimal strategy changes from C to S. Note also that $U_c^*(1)$ is a continuous function of c.

We now compare the choices S and CS at a typical node (h, k).

LEMMA 4.1. *Assume $c > 0$ and $n > h + k - 2$. If $h \neq k$ then, at (h, k), S gives greater expected utility than CS.*

PROOF. At (h, k) there have been $h + k - 2$ consultations, so S yields expected utility $\frac{h}{h+k} - (h + k - 2)c$, whereas CS yields expected utility

$$\frac{h}{h+k}\left(\frac{h+1}{h+k+1} - (h+k-1)c\right) + \frac{k}{h+k}\left(\frac{h}{h+k+1} - (h+k-1)c\right),$$

which is easily shown to equal $\frac{h}{h+k} - (h+k-1)c$. The utility of S therefore exceeds the utility of CS by $c > 0$. The proof is analogous if $h < k$. □

LEMMA 4.2. *Assume $n > 2h - 4$. At state (h, h), CS yields greater expected utility than S if and only if $c < \frac{1}{2(2h+1)}$.*

PROOF. The expected utility for S is $\frac{1}{2} - (2h - 2)c$, and the expected utility for CS is $\frac{h+1}{2h+1} - (2h - 1)c$. It is easily shown that the expected utility for CS is greater if and only if the specified condition is satisfied. □

Thus, C is better at (h, h) provided c is small enough. In particular, our analysis of the $h = 1, n = 1$ case is confirmed, and it also applies to $n = 2$. Note that if it is optimal to choose C when $c > 0$, then it must be optimal to choose C when $c = 0$. Therefore, the nodes where it is optimal to choose C are always a subset of $\mathcal{C}(n)$.

Starting at a state (h, k) where $h \neq k$, let G be the conditional strategy of continuing to choose C until the state returns to the centre line, and then choosing C once more. (If $h > k$, proceeding directly to the centre line will produce the state (h, h); strategy G calls for one more choice of C when this state is attained.)

LEMMA 4.3. *Assume $c > 0$ and $n > 2h - 2$. At state $(h, h - 1)$, G yields greater expected utility than S if and only if $c < \frac{h-1}{2(2h+1)(3h-2)}$.*

PROOF. At state (h, k), the expected utility for S is

$$\frac{h}{2h-1} - (2h-3)c.$$

The expected utility for G is

$$\frac{h}{2h-1}\left(\frac{h+1}{2h} - (2h-2)c\right) + \frac{h-1}{2h+1}\left(\frac{h+1}{2h+1} - (2h-1)c\right),$$

which can be shown to equal

$$\frac{4h-1}{2h-1}\left(\frac{h+1}{2(2h+1)} - (h-1)c\right).$$

After some manipulation, it can be seen that the expected utility for G is greater than for S if and only if

$$c < \frac{h-1}{2(2h+1)(3h-2)}.$$

□

It follows from Lemmas 4.2 and 4.3 that the optimal strategy when $n = 3$ is given by

- At (1, 1), choose C if $c \leq \frac{1}{6}$ and S otherwise;
- At (2, 1) and (1, 2), choose C if $c \leq \frac{1}{40}$ and S otherwise;
- At (2, 2), choose C; and
- At all other nodes, choose S.

The resulting utility is given by

$$U_c^*(3) = \begin{cases} \frac{7}{10} - 3c, & \text{if } c \leq \frac{1}{40} \\ \frac{2}{3} - c, & \text{if } \frac{1}{40} \leq c \leq \frac{1}{6} \\ \frac{1}{2}, & \text{if } \frac{1}{6} \leq c \end{cases}.$$

Moreover, the same strategy is optimal when $n = 4$, and $U_c^*(4) = U_c^*(3)$. Again, note that $U_c^*(3)$ is a continuous function of c for $c \geq 0$.

This optimal strategy is consistent with the optimal choices shown (for $n = 3$) in Figure 6. Note that these choices are "subgame perfect" in the sense that they prescribe the optimal choice on arriving at each node. Should DM arrive at node (2, 2), the optimal choice is to consult as long as $c < \frac{1}{10}$. However, DM can reach (2, 2) only by consulting at (2, 1) or (1, 2), and at these two nodes the optimal choice is Consult only if $c < \frac{1}{40}$. Thus, if the optimal strategy (starting at (1, 1)) is followed, testing whether $c < \frac{1}{10}$ at (2, 2) is irrelevant; this node cannot be attained unless $c < \frac{1}{40}$. But if, perhaps as a result of error, (2, 2) is reached even though $c > \frac{1}{40}$, the optimal choice is Consult if $c < \frac{1}{10}$ and Stop if $c > \frac{1}{10}$.

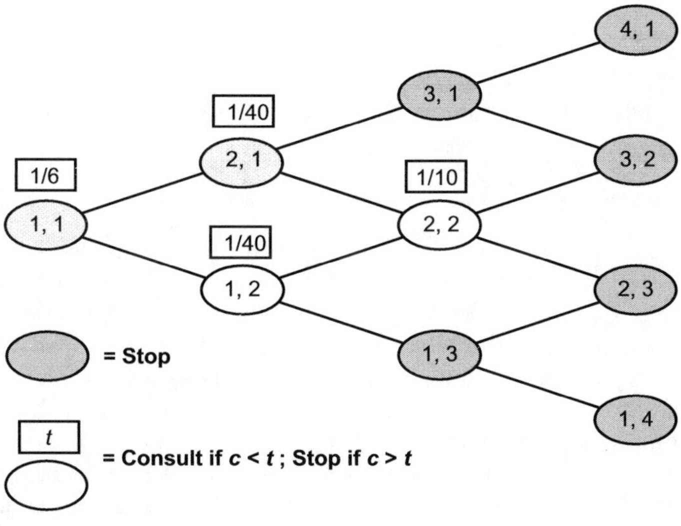

FIGURE 6. Optimal strategies and expected utilities when $n = 3$ and $c \geq 0$

When $n = 5$ and $c \geq 0$, a similar calculation yields Figure 7. In view of Figure 5, we anticipate that when $c \longrightarrow 0$, $U_c^*(5) \longrightarrow \frac{5}{7}$. This is the case, but the strategy to Consult at every node of $\mathcal{C}(5)$ is implemented only if $c < \frac{1}{189}$.

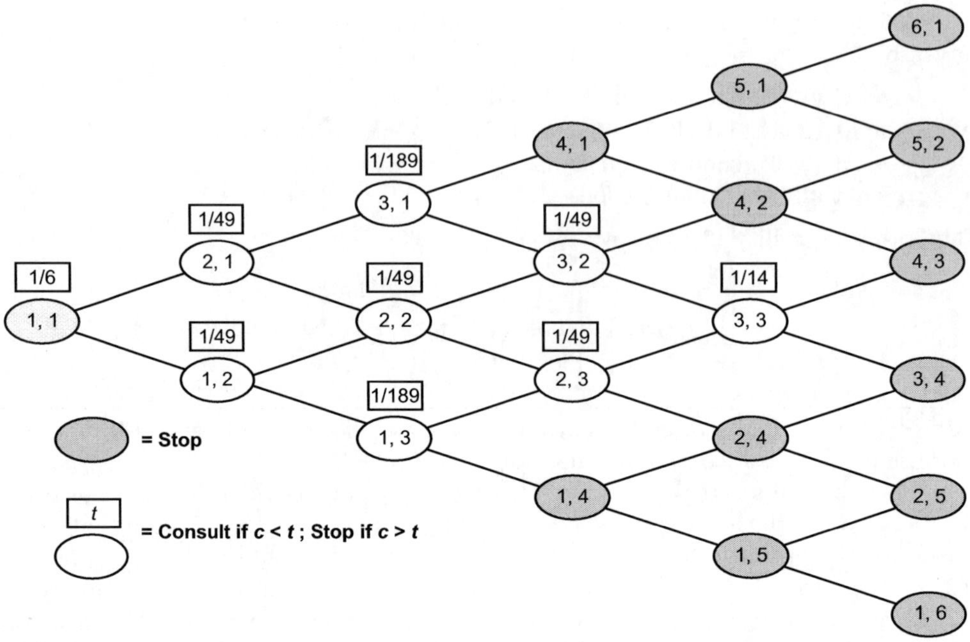

FIGURE 7. Optimal strategies and expected utilities when $n = 5$ and $c \geq 0$

5. Conclusions

Our model suggests some general rules of thumb for decision-makers who may consult to gain evidence to support a decision.

- Consult when there is relatively little evidence, or when the evidence is approximately balanced.
- Consult less as consultation becomes costlier, particularly when the evidence is less balanced.

The deliberations over the U.S. response to the Soviet missiles in Cuba demonstrate that continued consultation can be important, and the choice to keep consulting or to act can be crucial. At the outset of the Cuban missile crisis, EXCOMM had 13 members, mostly cabinet secretaries, but it then expanded to include 13 so-called advisers, who were brought in to offer counsel on various matters. (Because these advisors came and went, they do not follow strictly our model that allows consultants only to be added.) The consensus that the committee reached was that if a blockade of Cuba failed to induce Soviet withdrawal of the missiles, it could be followed by more aggressive action—specifically, an air strike—to try to eliminate the threat that the missiles posed, but at the possible cost of provoking a nuclear war. After the United States secretly agreed to withdraw its comparable missiles from Turkey if the Soviet Union withdrew its missiles from Cuba, the crisis subsided, illustrating how consultations, which occurred over 13 harrowing days, may lead to a propitious solution to a crisis.

One important conclusion from our model is that there is a fundamental upper limit on the utility of a decision-maker who is initially maximally uncertain (i.e., who begins at $(h,k) = (1,1)$), no matter how much consultation is undertaken. This limit applies even if consultation is free ($c = 0$).

By being specific, our model makes these observations explicit. For example, when consultation is costless ($c = 0$) and n consultations are permitted, the rule of thumb indicates that a DM who begins at $(h,k) = (1,1)$ should consult at any state (h,k) where

$$1 \leq h, k \leq \frac{n+1}{2}.$$

To determine whether to consult at a state (h,k), a simple test is whether strategy G is feasible; if $h > k$, say, the question is whether, if there were an unbroken sequence of No's resulting in the state (h,h), would one more consultation be permitted? If the answer to this question is positive, then G is feasible, and—if c is low enough—the decision-maker should consult at (h,k).

Simply guessing initially gives a maximally uncertain decision-maker an expected utility of $\frac{1}{2}$. Consultation can raise this expected utility, but even with unlimited costless consultation it cannot exceed the fundamental limit of $\frac{3}{4}$. Consultation can increase expected utility if costs are zero, but it does so at a decreasing rate. Even following an optimal strategy, the upper bound on the expected utility approaches $\frac{3}{4}$, so perfection—always making the right decision—is unattainable through consultation because of the inherent uncertainties in the model.

We conclude with some comments on possible variations in our model, which we leave for a future project. First, a decision-maker who is not maximally uncertain may start at initial state (h_0, k_0) for any $h_0 > 0$ and $k_0 > 0$. Moreover, consultation may cause a state transition

$$(h,k) \longrightarrow \begin{cases} (h+\alpha, k) & \text{with probability } \frac{h}{h+k} \\ (h, k+\alpha) & \text{with probability } \frac{k}{h+k} \end{cases}.$$

for some fixed $\alpha > 0$. To add more complexity, consultant ℓ may have cost c_ℓ and transition the state by an amount α_ℓ. But distinguishing consultants in this way may raise another decision problem, namely, how DM should choose a consultant to get the most information, α_ℓ, at the least cost, c_ℓ.

We have assumed additive costs; another variation on our model could be multiplicative costs, so the net utility of Stop at (h,k) would be $u(X_{op}) \cdot (1+\gamma)^{h+k-2}$ where $\gamma > 0$ is a new parameter. A hybrid model comprising both additive and multiplicative costs is also possible. We do not explore these variant of our model, but we conjecture that they have similar properties to the additive model.

Acknowledgment

The authors are grateful to the editors and an anonymous referee for comments that helped to improve the manuscript.

References

[1] Beta-binomial Distribution, Wikipedia, `https://en.wikipedia.org/wiki/Beta-binomial_distribution`, accessed 2022-04-28.

[2] Steven J. Brams, *Theory of Moves*, Cambridge University Press, 1993, pp. 130–38.

[3] Steven J. Brams and D. Marc Kilgour, *Revisiting the Secretary Problem*, Preprint, New York University.

[4] F. Thomas Bruss, *A unified approach to a class of best choice problems with an unknown number of options*, Ann. Probab. **12** (1984), no. 3, 882–889. MR744243

[5] Robert F. Kennedy, *Thirteen Days*, W. W. Norton, 1969.
[6] Jeanne Pavy, *The Role of the Executive Committee in the Cuban Missile Crisis*, http://people.loyno.edu/~history/journal/1983-4/1983-4.htm/, accessed 2022-04-02.
[7] Secretary Problem, Wikipedia, https://en.wikipedia.org/wiki/Secretary_problem, accessed 2022-04-28.

Department of Mathematics, Wilfrid Laurier University, Waterloo, Ontario N2L 3C5, Canada
Email address: mkilgour@wlu.ca

Department of Politics, New York University, New York, New York 10012
Email address: steven.brams@nyu.edu

Contemporary Mathematics
Volume **795**, 2024
https://doi.org/10.1090/conm/795/15964

How lies induced cooperation in *Golden Balls*: A game-theoretic analysis

Steven J. Brams and Ben D. Mor

ABSTRACT. We analyze a well-known episode of a popular British TV game show, *Golden Balls*, in which one of the two contestants lied about what he intended to do, which had the salutary effect of inducing both contestants to cooperate in what is normally a Prisoners' Dilemma (PD), wherein one or both contestants usually defected. This "solution" to PD assumes that the liar desired to be honorable in fulfilling his pledge to split the jackpot if he won but, surprisingly, he achieved this end without having to do so, astonishing the audience and receiving its acclaim. We note that this action has a biblical precedent in King Solomon's decision to cut a baby in two and, more generally, in resolving international conflicts, including the 1962 Cuban missile crisis.

1. Introduction

Beginning in June 2007 and running for more than two years, a popular British daytime game show called *Golden Balls* led to substantial frustration for many contestants, who were lied to and often betrayed by the other contestants. This is not surprising, because in the final round of *Golden Balls*, two contestants play a game called "Split or Steal" that, we argue, is a Prisoners' Dilemma (PD), wherein the cooperative strategy is to Split, but it is dominated by the noncooperative strategy of Steal (we give more details on the game in the next section). In 69% of 287 episodes of the show, either one (44%) or both (25%) players chose Steal, leaving only 31% in which both players chose the cooperative strategy of splitting the jackpot [**13**].

In one episode (`https://www.youtube.com/watch?v=S0qjK3TWZE8`), however, one of the contestants announced in advance how he would play and then reneged on both his promises (in fact, we never learn whether he would have kept his second promise). Paradoxically, these lies led both players to cooperate in the original game.

In this article, we show that the player who made the announcement (A) complicated the game by pledging that he would choose a particular option in the original game and, if feasible, then make a later choice outside the original game. If the payoff to this player depends only on money and status—how did I do in comparison to my opponent?—we show in the next section that (i) the original

2020 *Mathematics Subject Classification.* Primary 91A05; Secondary 91A80.
Key words and phrases. 2-person game theory; rationality; Nash equilibrium; Prisoners' Dilemma.

game is a 2×2 PD, in which cooperation is strongly dominated for both players; and (ii) A's announcement induces a 3×2 game, in which cooperation is weakly dominated for A and thus still of no help in fostering cooperation. But if A's payoffs also depend on his honor in fulfilling his second promise of splitting later if B splits in the original game, the game is transformed into another 3×2 game, but one of incomplete information in which there was misperception by B of A's preferences.

How can this happen if the players' strategies do not change in the second 3×2 game? First, A's announcement signaled that his goal might not be only to maximize his monetary payoff. Instead, it suggested that he might wish to honor his pledge about what he said he would do (split his winnings) after play of the original game.

But to get to this point, A had to prevent B from choosing noncooperation, which A did by credibly promising not to cooperate himself. Then A, correctly surmising that B would be forced to cooperate, could afford to act honorably and cooperate himself.

We begin our analysis by describing the rules of *Golden Balls* and then discuss why this show and, in particular, one episode has attracted so much attention. Among other reasons, the game is thought to simulate real-life situations more validly than laboratory settings, cf. [**7**], [**13**], [**5**], and [**12**]. Also, *Golden Balls* is an example of a one-shot PD, whereas in most experiments with PD, repeated play is assumed and some cooperation is expected. Cooperation is also more likely when players think ahead, as postulated in the theory of moves, cf. [**1**] and [**3**].

Although A's announcement in the aforementioned episode of *Golden Balls* was only cheap talk, it had the salutary effect of inducing B to believe A's lie not to cooperate and cooperate himself to try to salvage some chance of winning part of the jackpot. Indeed, A's lie laid the groundwork for cooperation in a game of incomplete information, whereby B incorrectly perceived A's preferences that—fortunately for both players—redounded to their mutual benefit.

2. Rules of Golden Balls

After a series of preliminary rounds, four contestants are reduced to two. We focus on this final stage of the game (called "Split or Steal"), which is governed by the following rules:

- Each contestant is given two balls. When opened, one indicates Split and the other Steal.
- Each contestant secretly opens his or her ball—to determine which is Split and which is Steal—and chooses one. Before making a choice, however, the contestants may speak to each other and also ask for advice from the host.
- If both choose Split, they each receive half the jackpot.
- If both choose Steal, neither contestant wins anything.
- If one contestant chooses Steal and the other Split, the Steal contestant wins the entire jackpot and the Split contestant nothing.

We assume that the best outcome for A and the other contestant, B, is to win everything (payoff of 4), next best to win half (payoff of 3), next worst to win nothing when the opponent also wins nothing (payoff of 2), and worst to win nothing when the opponent wins everything (payoff of 1), where the payoffs are rankings and not cardinal utilities. We rank the last outcome worst because of the anger, humiliation, or shame a player would feel if he or she were betrayed into thinking an opponent would Split—but chose instead Steal—when he or she Split. This game has some similarities to "divide the dollar"—in that if the two players are too greedy, they receive nothing—but there are significant differences [**4**].

Later we provide evidence for this preference ranking in a post-game quote from B. In extant studies of Split or Steal, where this assumption is not made, the game is referred to as a variant of PD or "weak" PD, because if only monetary payoffs underlie players' preferences, then being betrayed when choosing Split is as bad as mutual defection. In this case, unlike PD, the Steal strategy is weakly dominant, and there are three Nash equilibria.

As noted earlier, the payoffs indicate only an ordinal ranking of outcomes from best to worst. As in PD, the noncooperative strategy of Steal for each player strongly dominates the cooperative strategy of Split, rendering $(2, 2)$ the unique Nash-equilibrium outcome (denoted by $*$ in Game 1 in Figure 1), at which both players obtain nothing.

		B	
		Split	Steal
A	Split	$(3, 3)$	$(1, 4)$
	Steal	$(4, 1)$	$(2, 2)^*$

Game 1

FIGURE 1. The game Split or Steal (Game 1) with ordinal payoffs; the Nash equilibrium is for each player to play Steal, resulting in a payoff of $(2, 2)^*$.

In the actual play of *Golden Balls* on TV, each contestant usually tries to persuade his or her opponent to Split, promising to reciprocate so that both obtain half the jackpot. But this strategy is not convincing, often leading, as we indicated earlier, for one or both players to Steal and forgo the cooperative (3,3) outcome.

3. A's announcement

In the aforementioned episode of the game, which was between two men, A emphatically said that he would choose Steal. B chose Split, presumably because A had said that upon completion of the original game, he would give B half the jackpot if he had won, keeping the other half for himself.

This, of course, is cheap talk, because B has no assurance that A will keep his promise and give him half the jackpot. Because it is possible that A will do so, however, we give him a choice of keeping or not keeping his word after he Steals (as he said he would do) and B Splits, which yields the 3×2 game (Game 2) shown in Figure 2.

		B Split	*B* Steal
A	Split	$(3,3)$	$(1,4)$
A	Steal, then Split (if possible)	$(3,3)$	$(2,2)$
A	Steal, then don't Split (if possible)	$(4,1)$	$(2,2)^*$

Game 2

FIGURE 2. The game Split or Steal, with the option of player A keeping his promise of giving B half the jackpot (Game 2), with ordinal payoffs; the Nash equilibrium is for each player to play Steal, resulting in a payoff of $(2,2)^*$.

"If possible" becomes possible when A Steals and B Splits in the original game, in which case A can then either keep his promise and Split later or not. We know of no other game-theoretic treatment of this episode of *Golden Balls* that recognizes that A, after his announcement, has three, not two, strategies. For other game-theoretic models of this episode of *Golden Balls*, see [**8**], [**11**], and [**6**].

Game 2 in Figure 2 could also be written in extensive form (i.e., as a game tree), whereby A makes a decision after learning the outcome of the 2×2 game. But the normal form (i.e., as a payoff matrix) makes it easier to compare with the earlier 2×2 game and the final game we discuss in the next section. Notice that, as in Game 1 from Figure 1, each player does best when he wins the entire jackpot (4), next best (3) when there is a split, next worst (2) when both players Steal and there is nothing to split, and worst (1) when one wins the entire jackpot and the other nothing.

Observe that A has a weakly dominant strategy of "Steal, then don't Split (if possible)," whereas both of B's strategies are undominated. But given A's weakly dominant strategy, $(2,2)$ in the lower right is the unique Nash-equilibrium outcome (starred) in Game 2, echoing the Nash-equilibrium outcome in the 2×2 game in which both players Steal and, consequently, walk away empty-handed.

These strategies, however, were not the choices of the players in the TV game—quite the opposite: Both Split—suggesting that the payoffs in Game 2 are not an accurate reflection of the players' preferences. Instead, we believe, A had another goal in mind besides maximizing his winnings.

Notice that we do not prescribe what the players should do but work backwards from their actual choices to infer what their goals must have been to act in the way that they did. In effect, we reconstruct a game in order to try to offer a coherent explanation, through "revealed preference," of why the players' choices are consistent with their actions (i.e., are rational). While this reasoning may appear tautologous, it is the foundation of all science, including mathematics, in which nonobvious theorems are derived from assumptions. Here the rational choices of A and B in a game provide an explanation of their behavior that, on first blush, seems inexplicable.

4. The players' preferences

We suggest that the effect of A's announcement was not only to increase A's strategies in Game 1 from two to three in Game 2 but also to alter B's perception of the game. A wanted to appear sincere in declaring his intention to choose Steal in the original game, but he also wanted to say that if B Split—so A would win the entire jackpot—he would honor his pledge to split the jackpot later.

But being honorable for A, we think, does not simply mean that he privately takes pride in "doing the right thing" by keeping his pledge to Split. He also wants to demonstrate publicly—in front of the studio audience as well as 2 million TV viewers—that he acted honorably. We postulate that A cared about the public perception of his honorability, which was immediately manifest in the astonished reactions of the host and the studio audience. Viewers at that time, and later on YouTube, generally applauded his daring choice (there have been some 10 million views of the video and 10 thousand comments on it).

It was daring because it was risky and might well have backfired. So how does A prevent this when B has a strong incentive to Steal and possibly win the entire jackpot? He accomplished this by (i) promising that he would Steal so that B does not think that he can Steal himself and win everything, and (ii) credibly promising to Split after play of the original game, giving B the hope that he might obtain half the jackpot, even though A's promise to Split later is cheap talk.

If A is able to dissuade B from choosing Steal at the outset, then he will be in a position to honor his promise to Split if he wins the jackpot. But rather than Splitting privately after play of the original game, A can choose Split at the outset and gain acclaim not only for his beneficence but also for his brilliance in telling a forgivable lie to undermine the dominance of his strategy of Steal in Game 2, which induces B to choose Steal himself.

We assume that B's preferences remain the same as in Game 2, evidenced by B's (actually, "Ibrahim's" in the episode) response in a Radiolab interview to a question about whether he would have chosen to Split, knowing that A ("Nick") was also going to Split: "No, never:" Indeed, "It was Steal or nothing," because, Ibrahim said, "I would rather that both of them walk away without money than be duped into choosing Split, only to have all the money taken by the other contestant" [**10**]. Clearly, Ibrahim's social standing as being nobody's fool was as important to him as the money.

How does A's interest in keeping his promise and acting honorably—especially publicly—change his preferences in Game 2? We give these preferences in Game 3 (Figure 3). First, the Split-Split outcome becomes $(4,3)$ rather than $(3,3)$, making it better for A than the other $(3,3)$ outcome in Game 2, because A demonstrates his willingness to Split at the outset rather than privately after play of the original game. The second change in A's preferences is more dramatic: $(4,1)$ in Game 2 becomes $(1,1)$ in Game 3, because A thoroughly dishonors himself by breaking his promise to Split when B does. To be sure, A wins the entire jackpot, but he evinced no interest in doing so at the price of reneging on his promise, which he never had to do by choosing Split at the outset.

		B	
		Split	Steal
	Split	$(4,3)$	$(1,4)$
A	Steal, then Split (if possible)	$(3,3)$	$(2,2)$
	Steal, then don't Split (if possible)	$(1,1)$	$(2,2)^*$

Game 3

FIGURE 3. The game Split or Steal, with the option of player A keeping his promise of giving B half the jackpot (Game 3), where the ordinal payoffs are adjusted to include player A's honor; the Nash equilibrium is still for each player to play Steal, resulting in a payoff of $(2,2)^*$.

How do these two changes in A's preferences affect the strategy choices of the players in Game 3? First, because A was able to credibly promise to Steal, he could surmise that B would not choose Steal himself, which would leave both players with nothing. But if Steal is no longer viable for B in a game of incomplete information (it would be an undominated strategy in a game of complete information), A's optimal choice is to choose the best of his three strategies associated with B's choice of Split, which is his own strategy of Split that yields the outcome $(4,3)$. Indeed, these are exactly the choices the players made in the TV game.

Anomalously, A upholds his honor by breaking his promise to Steal and Splitting instead. This choice relieved him of the need to keep his pledge to Split later if B also Split, which would not have been immediately visible to the audience whose approbation he relished.

Our analysis shows that B never really understood A's desire to behave honorably; rather, he was duped by A's announcement into believing that A would Steal, giving him little choice but to Split to try to salvage something for himself. Although we implicitly assumed at the outset that *Golden Balls* was a game of complete information, it was not for B in the episode we have analyzed. But to reach the mutual choice of Split turned out to be tortuous for B, so it's little wonder that he asked A, plaintively, at the end of play: "Why did you do that to me?"

A's announcement succeeded not only in persuading B to Split but also, by choosing Split himself, providing a public display that his promise was not just cheap talk. Indeed, A's choice of Split spoke louder than his words, making him appear even more honorable—publicly rather than just privately (if he had Split later)—while achieving the same outcome.

Our game-theoretic explanation of what happened can be summarized as follows:

(1) B, believing that A will do what he says he will do—choose Steal, then Split (if possible)—prefers Split (3) to Steal (2) and so Splits.
(2) Anticipating B's choice of Split, A's best choice (4) in Game 3 is to Split himself, which he chooses.

Note that if A had believed the game to be Game 2, he would have been torn between choosing one or the other of his first two strategies, both of which give him 3. But Splitting breaks this tie in Game 3 and gives A a best outcome of 4 instead—demonstrating his honorability—so A chooses it.

5. Conclusions

Game 3 shows A's astuteness in using a lie to escape the Prisoners' Dilemma inherent in *Golden Balls* and turn it into a game that he could exploit. It helps to explain why he announced he would Steal in the original game: It enabled him credibly to cow B into "submission," after adding the sop that he would Split the jackpot he won after the show (this promise was not so credible).

But by Splitting at the outset, A could honor his promise to Split the jackpot publicly, without actually having to do so privately, so we never learn whether or not A was lying that he would have Split the jackpot in the end. Clearly, A's announcement was not only brilliantly devious, but it also worked.

Game 3 also explains how A overcame his credibility problem, given that his pledge to Split later was cheap talk. When the shell-shocked moderator did not object to A's announcement—presumably, he could have declared, "No private deals of any kind!"—B's perception of the game changed to one in which Steal was no longer viable: It was rational for him to Split in a game of incomplete information to try to win half the jackpot, as A had promised he would do if he Split (Talwalkar [**11**] reaches a similar conclusion, based on a different model).

It is worth noting that A's announcement has a precedent in the famous Bible story in which two women claimed maternity of a baby. When King Solomon announced that he would split the disputed baby in two if both women refused to give up their claims, he elicited from them responses that revealed who the true mother was, so there was no need to split the baby—just as A's announcement in *Golden Balls* enabled both contestants to escape a Prisoners' Dilemma and not leave empty-handed if they had both chosen Steal.

To be sure, Solomon's edict was not by one of the women but instead set the stage for one of the women (the mother) to speak up and offer the baby to the other woman. The other woman was somewhat akin to B, who was also deceived: She incorrectly thought that Solomon would stick to his edict and that she would win his favor. Little did she know that Solomon preferred a different outcome once he deduced who the true mother was, just as A preferred not to postpone his choice of Split but to do so in the game itself. (For a more robust action that Solomon might have taken to settle the women's dispute, which could be repeated in future disputes—unlike the threat he reneged on—see [**2**, ch.10].)

Other precedents include the naval blockade of Cuba by the United States during the October 1962 missile crisis with the Soviet Union. At the end of the crisis, which lasted 13 days, President Kennedy secretly pledged to eventually remove American missiles from Turkey, provided Premier Khrushchev immediately withdrew his missiles from Cuba. Khrushchev could not know for sure that Kennedy would honor his word, but a pledge by the American president was credible enough. The alternative—maintaining missiles on the island—was bound to lead to a confrontation in which both sides would lose. (For additional cases in international relations in which honor, as in *Golden Balls*, played a key role, see [**9**].)

Other choices by actors—sometimes threatened, other times carried out—have prevented conflicts, in a variety of situations, from escalating and being disastrous for both sides. We intend to look more closely at some of these in future work.

References

[1] Steven J. Brams, *Theory of moves*, Cambridge University Press, 1994.

[2] Steven J. Brams, *Divine games: Game theory and the undecidability of a superior being*, MIT Press, Cambridge, MA, 2018. MR3837506

[3] Steven J. Brams and Mehmet S. Ismail, *Every normal-form game has a Pareto-optimal nonmyopic equilibrium*, Theory and Decision **92** (2022), no. 2, 349–362, DOI 10.1007/s11238-021-09824-1. MR4381766

[4] Steven J. Brams and Alan D. Taylor, *Divide the dollar: Three solutions and extensions*, Theory and Decision **37** (1994), no. 2, 211–231.

[5] Maxwell N. Burton-Chellew and Stuart A. West, *Correlates of cooperation in a one-shot high-stakes televised Prisoners' Dilemma*, PloS one **7** (2012), no. 4, e33344.

[6] The Cornell University Course Blog, *Split or steal: An analysis using game theory*, Available at `https://blogs.cornell.edu/info2040/2012/09/21/split-or-steal-an-analysis-using-game-theory/` (2021/11/20), 2012.

[7] Einav Hart, *Steal the show: Payoff effect on accuracy of behavior-prediction in real high-stake dilemmas*, Hebrew University of Jerusalem, 2010.

[8] Boris Nikolaev et al., *Using experiments and media to introduce game theory into the principles classroom*, Journal of Private Enterprise **29** (2014), 149–160.

[9] Barry O'Neill, *Honor, symbols, and war*, University of Michigan Press, 2001.

[10] Radiolab, *The Golden Rule*, Available at `https://radiolab.org/episodes/golden-rule`, Feb. 25 2014.

[11] Presh Talwalkar, *How to beat the Prisoner's Dilemma in the TV game show Golden Balls*, Available at `https://mindyourdecisions.com/blog/2012/04/24/how-to-beat-the-prisoners-dilemma-in-the-tv-game-show-golden-balls/` Accessed: 2021/11/20, 2012.

[12] Uyanga Turmunkh, Martijn J. Van den Assem, and Dennie Van Dolder, *Malleable lies: Communication and cooperation in a high stakes TV game show*, Management Science **65** (2019), no. 10, 4795–4812.

[13] Martijn J. Van den Assem, Dennie Van Dolder, and Richard H. Thaler, *Split or steal? Cooperative behavior when the stakes are large*, Management Science **58** (2012), no. 1, 2–20.

Department of Politics, New York University, New York, New York 10012
Email address: `steven.brams@nyu.edu`

School of Political Sciences, Division of International Relations, University of Haifa, Haifa, Israel
Email address: `b.mor@poli.haifa.ac.il`

Contemporary Mathematics
Volume **795**, 2024
https://doi.org/10.1090/conm/795/15965

Conditions for fairness anomalies in instant-runoff voting

Adam Graham-Squire

ABSTRACT. In an instant-runoff election there are well-known necessary and sufficient conditions for a 3-candidate, fully-ranked upward monotonicity anomaly. We extend previous research to find necessary and/or sufficient conditions in a number of related areas: (1) monotonicity anomalies in 3-candidate elections when ballots are not necessarily fully-ranked, (2) monotonicity anomalies in 4-candidate and higher instant-runoff elections, and (3) participation (no-show) anomalies in 3-candidate elections, both fully-ranked and not. These conditions are useful for theoretical purposes as well as for computational methods used to find anomalies in election data.

1. Introduction

Instant Runoff Voting (IRV) has recently become a more common voting method in various municipalities across the United States and has been used for many years in other countries. IRV (also called 'ranked-choice', 'preferential' or 'transferable' voting, or 'the alternative vote') asks voters to rank candidates, as opposed to voting for a single candidate. Any candidate with a majority of first-place votes wins. If no candidate has a majority, the candidate with the fewest first-place votes is eliminated, and their votes are transferred to other candidates on the basis of second-place votes. If any candidate now has a majority of first-place votes, that candidate wins the election. If not, the process of elimination and transfer continues until one candidate has a majority, including first-place and transferred votes.

Despite having a number of theoretical and practical advantages over voting for a single candidate or party, IRV also has some drawbacks. Two prominent flaws are that IRV fails Monotonicity and Participation criteria. This means that there are situations in which, in an IRV election, a voter could hurt their preferred candidate by ranking that candidate higher on the ballot (a violation of the upward Monotonicity criterion), or a voter's preferred candidate would do better if the voter did not show up to vote (a violation of the Participation criterion). Following the convention of [**3**], we will refer to an instantiation of failure of the Monotonicity criterion as a *monotonicity anomaly* and failure of the Participation criterion as a *no-show anomaly*.

To simplify the analysis in this paper, we utilize the following conventions for the entirety of the paper:

2020 *Mathematics Subject Classification.* Primary 91B14; Secondary 91B12.

- We only investigate upward monotonicity anomalies, so any reference to monotonicity anomaly can be assumed to refer only to upward anomalies, not downward.
- Candidates are labeled in alphabetical order based on order of elimination. Thus A is the winner of an election, B comes in second-place, etc.
- Capital letters are both used to indicate candidates as well as their total number of first-place votes. In a given situation, it should be clear from context which meaning is intended. Votes for only a single candidate will be designated with underscores to differentiate them from a candidate or a candidate's first-place vote total. For example, if 7 people vote BAC, 5 vote BCA, and 3 vote only for B, we would say $B = 15$ and $B\_\_ = 3$.
- We indicate modified election results with $*$, thus A^* indicates the number of first-place votes A receives in a modified election.
- We also make the following assumptions to simplify the presentation of the material. Actual elections that would differ from these assumptions are very few in number:
  (a) Our elections generally have numbers that add up to 100 that can be thought of as either votes or percentages. For certain proofs, we treat votes as continuously variable percentages, including infinitesimally small votes of size ϵ. However, in empirical situations votes are actually discrete, not continuous. Thus it is possible, for example, that an election may meet conditions for an anomaly (according to the percentages) that is not realizable in a practical manner.
  (b) We ignore the possibility of ties. Thus, for example, $A \not> B$ means $A < B$.

We begin with precise definitions of each kind of anomaly, followed by an example to illustrate how IRV works, and how monotonicity and no-show anomalies can arise in IRV elections.

DEFINITION 1. A ballot profile for an election is the collection of all different ballots cast, along with the number (or percentage) of voters who cast that particular ballot.

An example of a ballot profile is given in Example 1 and Table 1.

DEFINITION 2. For a given election with ballot profile P and IRV winner A, a *monotonicity anomaly* occurs if

(1) P can be modified by raising only A's rank to be higher in some ballots to a ballot profile P', and
(2) A now loses the IRV election under the profile P'.

In looking for monotonicity violations, note that only the IRV winner may be raised higher in ranking; no other changes to candidate order are allowed. Thus a ballot of the form BAC can be modified to ABC, but not to ACB. Similarly, in a 4-candidate election, a ballot $BCAD$ could be changed to $ABCD$, but not to AB. We also note that for practical purposes, monotonicity anomalies only occur due to a change in the order of elimination of candidates, thus while it is theoretically possible to change a vote of $B\_\_$ to a vote of BA, in practical terms such a change would never cause an anomaly. Thus for this paper, the only modification to a vote of $B\_\_$ would be to change it to AB. We further note that in a 3-candidate IRV election, a vote of AB is equivalent to a vote of ABC, and we will refer to such ballots interchangeably.

DEFINITION 3. For a given election with ballot profile P and IRV winner A, a *no-show anomaly* occurs if

(1) P can be modified by removing certain ballots in which candidate X is ranked higher than A to a ballot profile P', and
(2) X now wins the IRV election under the profile P'.

Under the no-show anomaly, voters whose ballots were removed to create P' received a better result by *not* voting—the election results in their more-preferred candidate winning when those voters do not show up. In this paper we only look at no-show anomalies for 3-candidate elections, so for practical purposes our no-show anomalies are only present when BCA votes are removed from P, and that results in C winning the election.

We now look at an example to illustrate how IRV works, and how monotonicity and no-show anomalies can arise.

TABLE 1. Ballot profiles for Example 1; original election profile (Left), modified election profile 1 (Middle) and modified election profile 2 (Right).

	40	33	27
1st	A	B	C
2nd	C	C	A
3rd	B	A	B

	40	26	7	27
1st	A	B	A	C
2nd	C	C	B	A
3rd	B	A	C	B

	40	26	27
1st	A	B	C
2nd	C	C	A
3rd	B	A	B

EXAMPLE 1. Consider the three-candidate election with voter profile as given in Table 1 (Left); 40 voters have preference ACB, 33 voters have preference BCA and 27 have preference CAB. In round 1 of the IRV election, C is eliminated with only 27 first-place votes. All of C's votes go to A, who now has 67 first-place votes and wins the election in round 2. Thus, for this profile, A is the IRV election winner.

Suppose now that 7 of the BCA voters chose to favor A by raising A in their ballots to vote ABC, as illustrated in Table 1 (Middle). In this modified election, $A^* = 47$, $B^* = 26$ and $C^* = 27$. Candidate B is eliminated first and C receives 26 first-place votes. Candidate C now has 53 votes, a majority, and wins in round 2. Thus raising A's position in some ballots causes A to lose the election, which is an example of an (upward) monotonicity anomaly.

Alternatively, suppose 7 BCA voters chose to not vote at all, as illustrated in Table 1 (Right). In this second modification, the results are the same as in the first modification: B is eliminated first and C receives an additional 26 first-place votes. As before, C wins in round 2. We thus have the following situation: When the 7 BCA voters show up and vote, their last-place candidate (A) wins. When the 7 BCA voters *do not* show up, their second-place candidate (C) wins. We thus conclude those 7 voters receive a better result by *not* voting, which is an example of a no-show anomaly.

These examples demonstrate anomalies when all the ballots are fully ranked. The situation is more complicated when not all ballots are fully ranked. Ballots that are not fully-ranked can result in ballot exhaustion, where ballots do not continue on to the next round of voting because all candidates ranked on the ballot were eliminated. Ballot exhaustion can be caused by systemic truncation, where voters are not allowed to rank all candidates (for example, if there are 7 candidates running in an election, but voters are only allowed to rank their top 3 candidates) or by individual truncation, where individual voters choose not to rank all candidates. We will refer to an individually truncated ballot as an *incomplete* ballot, and we do not deal with systemic truncation in this paper. For further information related to the effect of truncated ballots, see [**1**] and [**4**].

Most research related to IRV and monotonicity anomalies has focused on the fully-ranked three-candidate case, because necessary and sufficient conditions for the existence of monotonicity anomalies have been established. Specifically, the third-place candidate must have greater than 25% of the first-place vote and beat the IRV winner in a head-to-head contest; cf. [**6**], [**7**], and [**8**]. Note that in [**6**], [**7**], and [**8**] the two conditions were primarily used, along with computer simulation, to calculate the frequency of monotonicity anomalies in 3-candidate IRV elections. We will identify the first condition as $L25$ and the second condition as $H2H$. Following the suggestions made in [**2**], we extend and generalize the necessary and sufficient IRV conditions by investigating: (1) necessary and sufficient conditions for monotonicity anomalies in 3-candidate elections in which some ballots are incomplete, (2) necessary conditions for monotonicity anomalies in elections with 4 or more candidates, and (3) necessary and sufficient conditions for fully-ranked, 3-candidate no-show anomalies.

Some preliminary research regarding extending and generalizing conditions for certain anomalies already exists. In [**3**], Graham-Squire and Zayatz: (1) prove that a necessary condition for a monotonicity anomaly at the 4-candidate level is that the fourth-place candidate must have 12.5% of the vote, and (2) state, without proof, that for a 3-candidate election, when allowing incomplete ballots, the 25% condition continues to be necessary but is no longer sufficient. In this paper we extend the proof of (1) to find a necessary condition for n candidates and prove (2), as well as find a new condition which is both necessary and sufficient at the 3-candidate level.

2. Three-candidate elections where some ballots are incomplete

We note that there is only one way for a 3-candidate monotonicity anomaly to occur: First, enough ballots with B listed in first place must be modified to move A up into first place so that B will be eliminated before C. Then, in the final round of the election, C must beat A in a head-to-head contest. This is true whether ballots are fully-ranked or not. Any other modification of ballots would not make a difference; moving A up in ballots that have C in first place (for example, a CAB ballot could be changed to ACB) still leads to C being eliminated first and to A winning the resulting head-to-head against B.

When ballots are not fully-ranked, the proofs found in [**6**], [**7**], and [**8**] still show that conditions $L25$ and $H2H$ are necessary. However, the following example, with the incomplete ballots $A\_\_$ and $B\_\_$, shows that the two conditions are no longer sufficient to cause a monotonicity anomaly.

EXAMPLE 2. Consider the election with voter profile given in Table 2 (Left). Candidate C is eliminated in the first round, and A is the IRV winner. Candidate C has greater than 25% of the first-place vote and C beats A in a head-to-head by 46 to 45 votes. To modify the election to cause a monotonicity anomaly, we would need to raise A above B in some ballots, changing either BCA votes to ABC or changing $B\_\_$ votes to AB. Changing BCA votes, however, is more damaging to C because it takes away head-to-head votes where C beats A, and converts them to votes where A beats C. Thus we choose to change $B\_\_$ votes to AB. We must modify more than 3 of the $B\_\_$ votes to AB in order for B to drop out before C. Changing 4 such votes results in the modified election shown in Table 2 (Right).

In this modified election, B is eliminated first and gives most of their votes to C, but C still loses in the head-to-head against A, 49 to 46. One can easily verify that modifying any other number of votes (from either BCA or $B\_\_$) will never result in an anomaly. Thus there is no monotonicity anomaly, even though the original election satisfies both conditions $L25$ and $H2H$.

TABLE 2. Ballot profile for Example 2; original election profile (Left) and modified election profile (Right).

	45	20	9	26
1st	A	B	B	C
2nd		C		A
3rd		A		B

	45	4	20	5	26
1st	A	A	B	B	C
2nd		B	C		A
3rd			A		B

Example 2 illustrates why the $H2H$ condition fails to be sufficient in this case: when the $B\_\_$ votes are modified to AB, new votes are created for A that did not previously exist, and those new votes are enough for A to now beat C in a head-to-head, and thus for A to win any modified election. Generally speaking, when trying to create a monotonicity anomaly, one must take care to choose the correct votes to change. Modifying BAC votes to ABC is best, because adjusting those votes helps eliminate B first, without changing the relative number of votes in the head-to-head matchup between A and C. Each $B\_\_$ vote changed to AB adds one more vote to A, and each BCA vote changed to ABC adds one vote to A and removes one vote from C in their head-to-head matchup. We will formalize this idea of the creation of new votes to find a condition which will be sufficient for elections with incomplete ballots.

To proceed, we need some notation. Let g be the gap in first-place votes between B and C (that is, $g = B - C$). We define B_A (B_C) to be the number of BAC (BCA) votes, and $v = \max(0, g - B_A)$. Thus v is the number of ballots that need to be changed, in addition to all BAC votes, in order to modify the election in such a way that B drops out before C. The variable v is suggestive, as these votes come from the "void" (either $B\_\_$ or BCA votes), that is, from creating "new" votes where A beats C head-to-head. Finally, let $h = C + B_C - (A + B_A)$. Note that $h > 0$ if and only if C beats A in a head-to-head matchup.

Colloquially speaking, v is the number of votes detrimental to C that must be changed so that B is eliminated first, and h is the maximum number of votes that can be changed before A beats C in a head-to-head matchup. Thus the condition

$v < h$ is a sweet spot for anomalies: enough votes can be modified to change the order of elimination, but not so many that it changes C beating A in a head-to-head.

THEOREM 1. *(joint work with Vicky Sanderson) For any 3-candidate IRV election, the following conditions are necessary and sufficient for a monotonicity anomaly to be present:*

(1) *The third-place candidate has greater than 25% of the first-place vote (L25) and*
(2) $v < h$ *(VH condition).*

Note that in Example 2, $h = 46 - 45 = 1$ and $v = 29 - 26 = 3$. Thus $v > h$ and condition (2) fails, so no monotonicity anomaly exists.

PROOF OF THEOREM 1. The necessity for the $L25$ condition is identical to that proved in [**6**], [**7**], and [**8**]. To prove the necessity of the VH condition, we assume $v > h$ and show the original IRV winner (A) will still win any modified election. We ignore the $v = h$ case as it could lead to a tie.

We first prove that if $v > h$, then there is no monotonicity anomaly. If $h < 0$, then C loses to A in a head-to-head matchup and A would win any modified election (this is identical to the necessity of the $H2H$ condition proved in [**6**], [**7**], and [**8**]). Thus we can assume $v > h > 0$. Let $\epsilon > 0$ such that $C > B - g - \epsilon$. In any election modified so that B is eliminated first, we need to move A up to first place in at least $g + \epsilon$ ballots. Thus $A^* = A + g + \epsilon$, $g = B_A + v$, $v > h$ and $h = C + B_C - A - B_A$ give

$$A+g+\epsilon = A+B_A+v+\epsilon > A+B_A+h = A+B_A+(C+B_C-A-B_A) = C+B_C.$$

So A would beat C in the final round of the modified election, creating no anomaly. This shows the VH condition is necessary.

To prove sufficiency, suppose both conditions hold and let $0 < \epsilon < h - v$. Since $C > 25$, we choose ϵ to be such that $C > B^* = B - g - \epsilon > 25$.

If $v = 0$, the result is straightforward. When $v = 0$, we have $g \leq B_A$ and $h > 0$. Modifying all BAC votes to ABC, as well as $h/3$ B_{--} votes (to AB) or BCA votes (to ABC), results in B dropping out first and C beating A in a head-to-head with a margin of at least $h/3$, causing a monotonicity anomaly.

If $v > 0$, then $B_A < g$. Modify all BAC votes to ABC as well as $(v+\epsilon)$ B_{--} or BCA votes to ABC. Thus $A^* = A+B_A+v+\epsilon$. As mentioned previously, to create anomalies B_{--} votes should be modified to AB before BCA votes are modified to ABC. We now look at two cases, based on which votes need to be modified.

CASE 1. If only B_{--} votes are modified, then no BCA votes are modified. Thus $B_C^* = B_C$ and using $\epsilon < h - v$ and $h = C + B_C - A - B_A$ gives

$$A^* = A+B_A+v+\epsilon < A+B_A+(h-\epsilon)+\epsilon = A+B_A+(C+B_C-A-B_A) = C+B_C^*$$

Thus C will beat A in the final round (because $A^* < C + B_C^*$), and there will be a monotonicity anomaly.

CASE 2. If BCA votes must be modified, then all BAC and B_{--} votes have already been modified, and all remaining B^* votes will be of the form BCA. Thus $C > B^* = B_C^* > 25$ and we conclude $C + B_C^* > 50$, so C will have a majority in the final round and win. We conclude there will be a monotonicity anomaly in this case as well. □

COROLLARY. *For any fully-ranked 3-candidate IRV election, the VH condition is equivalent to the $H2H$ condition.*

PROOF. The proof of the corollary is divided into two cases, based on the relationship between g and B_A.

CASE 1. $g < B_A$. Then $v = 0$ and $v < h$ if and only if $h > 0$, which is the $H2H$ condition.

CASE 2. $g \geq B_A$. Then BCA votes must be modified and $v > 0$. As Case 2 in Theorem 1 above indicates, when BCA votes must be modified and $h > v$, this will always result in a monotonicity anomaly. Then $v > 0$ and $h > v$, so $h > 0$ as desired. □

3. Conditions for monotonicity anomalies in 4-candidate IRV elections

In this section we prove some preliminary results for the 4-candidate case, as well as create a framework for approaching the variety of ways that a monotonicity anomaly can arise with 4 candidates. In order to prevent double-counting of anomalies, we follow the convention of [**3**] to define an anomaly at the *n-candidate level* to be an anomaly present when there are n candidates left, but not present for any number of candidates less than n. For example, suppose the election of Example 1 were modified to change one BCA vote to be DBC. Then candidate D would have 1% of the vote, and the same modifications (moving 7 BCA votes to ABC) would still cause a monotonicity anomaly. That anomaly, however, does not involve D in any way, as the same anomaly is also present in the round of 3 candidates. Thus we would say it is an anomaly at the 3-candidate level, and not an anomaly at the 4-candidate level. Note that we are considering both fully-ranked and incomplete ballots in this section.

THEOREM 2 (Graham-Squire and Zayatz [**3**]). *For a monotonicity anomaly to be present in an IRV election at the 4-candidate level it is necessary that the 4th-place candidate have greater than* 12.5% *of the first-place vote.*

Theorem 2 was proved in the appendix to [**3**]. The idea of the proof is that if D has 12.5% or fewer first-place votes, and another candidate (B or C) has votes modified such that they drop out before D, D would have less than 25% in the next round 3-candidate election, precluding a monotonicity anomaly. Using similar logic leads to a generalization of the $L25$ condition for n candidates.

THEOREM 3. *To have a monotonicity anomaly at the n-candidate level in an IRV election, it is necessary that the nth-place candidate have greater than $\frac{100}{2^{n-1}}\%$ of the first-place vote.*

PROOF. We prove this theorem in 4 steps: in step 1, we assume fully-ranked ballots and show that any candidate with less than or equal to $\frac{100}{2^{n-1}}\%$ of the first-place vote at the n-candidate level cannot win any modified election. In step 2, we prove the same when incomplete ballots are allowed. In step 3 we prove a corollary that candidate A, the IRV winner, can never have less than $\frac{100}{2^{n-1}}\%$ of the first-place vote at the n-candidate level. Finally, in step 4 we show in full generality that there cannot be an anomaly at the n-candidate level if X_n has less than or equal to $\frac{100}{2^{n-1}}\%$ of the first-place vote.

STEP 1. Suppose ballots are fully-ranked and let X be any candidate such that $X \leq \frac{100}{2^{n-1}}$ in the round of n candidates. Assuming the most favorable conditions for X (that all candidates who are eliminated will give all of their votes to X), in any regular or modified election in which another candidate drops out at the n-candidate level, X will receive less than $\frac{100}{2^{n-1}}$ from the eliminated candidate. Thus, in the round of $(n-1)$ candidates, $X < \frac{100}{2^{n-2}}$. Continuing this process and assuming X is not eliminated, in the round of 2 candidates $X < \frac{100}{2^{2-1}} = 50$. Since the ballots are fully-ranked, the other candidate will have a majority of first-place votes and X will lose the election.

STEP 2. Suppose some ballots are incomplete. We note that in Step 1, we assumed the most favorable conditions for candidate X—specifically that all votes from eliminated candidates would go to X, and none to other candidates. If some ballots are not fully-ranked, X will receive equal or fewer transfer votes than in Step 1, and all other candidates will have the same number of votes as in Step 1 (since they receive no votes from eliminated candidates by assumption). Thus X will still have less votes than the other candidate in the round of 2 candidates, were X to make it that far, and thus X would still lose.

STEP 3. As a corollary to Step 1, we note that candidate A can never play the role of X, as then A would lose any regular or modified election. This is a contradiction since we know A wins the regular election.

STEP 4. Let X_n be the candidate who is eliminated at the n-candidate level. Combining Steps 1 and 3 we see that, in an election where $X_n \leq \frac{100}{2^{n-1}}$ in the round of n candidates, A will always have a larger vote share than X_n in every round. It follows that A will never be eliminated before X_n given those conditions. Thus the only way to have a monotonicity anomaly is for A to be eliminated in some round *after* X_n has been eliminated. In that case, however, X_n is playing no role in the elimination of A, and thus the anomaly must be present at some level *less* than at the n-candidate level.

We conclude that to have a monotonicity anomaly at the n-candidate level in an IRV election, $X_n > \frac{100}{2^{n-1}}$. □

3.1. No global *H2H* condition for 4-candidate elections. Theorem 2 shows that there is a modification of the $L25$ condition present at the 4-candidate level. The following example illustrates that there can be a monotonicity anomaly present at the 4-candidate level even when the IRV winner is a Condorcet winner. (Recall that a Condorcet winner, if one exists, is a candidate that defeats every other candidate in a head-to-head election.) This demonstrates that there is no global *H2H* necessary condition at the 4-candidate level.

EXAMPLE 3. Consider the election with voter profile given in Table 3 (Left). Candidate A is the IRV winner and also the Condorcet winner (A beats B by 63 votes to 37 votes; A beats C by 53 to 47; and A beats D by 51 to 49).

Now suppose 4 BDA voters change their ballots to ABD as illustrated in Table 3 (Right). In the modified election, B is eliminated first and gives 12 votes to C

and 10 to D. In round 2, A is eliminated with only 31 votes (C has 36 and D has 33) and D then wins. This demonstrates a monotonicity anomaly in a 4-candidate election even though no candidate beats the IRV winner in a head-to-head contest. We conclude there is no global $H2H$ requirement for a monotonicity anomaly at the 4-candidate level.

TABLE 3. Ballot profile for Example 3; original election profile (Left) and modified election profile (Right).

	27	14	12	24	6	5	12
1st	A	B	B	C	D	D	D
2nd		D	C	A	B	C	A
3rd		A	D	D	C	B	B

	27	4	10	12	24	6	5	12
1st	A	A	B	B	C	D	D	D
2nd		B	D	C	A	B	C	A
3rd		D	A	D	D	C	B	B

3.2. Framework for 4-candidate monotonicity anomalies. In 3-candidate elections, there is only one way to cause an anomaly: votes are modified so that B drops out first, and then C beats A in a head-to-head contest. Thus if the regular dropout order is ABC (so that C drops out first, then B, and finally A), the anomaly dropout order must be CAB. In a 4-candidate election, however, there are six different potential anomaly dropout orders: $CDAB$, $DCAB$, $DBAC$, $BDAC$, $DABC$, and $DACB$. To justify that these are the only possible anomaly dropout orders, we first note there are $4! = 24$ different ways to arrange the letters A, B, C, and D. For a modified dropout order to lead to an anomaly, the winner (first letter) cannot be A, and the last letter (the candidate who, after ballots are modified, is eliminated first) must be B or C. There are only 8 such orderings, and two of them, $CADB$ and $BADC$, do not represent possible anomalies. $CADB$ is not a possible anomaly at the 4-candidate level because B has votes modified and then D drops out in the round of three candidates. Thus this type of anomaly does not actually involve D as an inherent part of the anomaly—it is actually a 3-candidate anomaly which is thus also present at the the 4-candidate level, but is *not* a 4-candidate level anomaly, according to the definition. $BADC$ is not a potential anomaly, because it indicates that B beats A in a head-to-head contest. However, we already know according to the original election (dropout order $ABCD$) that A beats B head-to-head, so $BADC$ is not a potential anomaly ordering.

The framework for analyzing 4-candidates anomalies, then, is to analyze each of the 6 categories of potential dropout orders to see what specific conditions apply to that category of election. For example, in the first four categories ($CDAB$, $DCAB$, $DBAC$, $BDAC$), A is eliminated in the round of three candidates, and thus there would be no necessity of an $H2H$ condition. The last two options ($DABC$ and $DACB$) would both require that D beat A in a head-to-head contest. We also know that $D > 12.5\%$ is a necessary condition for a 4-candidate anomaly, but for certain categories there is a higher threshold for an anomaly.

THEOREM 4. *In modified 4-candidate IRV elections with dropout order $CDAB$ or $DCAB$, it is necessary that D have greater than $16\frac{2}{3}\%$ of the first-place vote.*

PROOF. The original election order is $ABCD$, which implies that C must drop out in the round of three candidates. This means that $C + D_C < 33\frac{1}{3}$ and thus

$C < 33\frac{1}{3}$. In the modified election, A is eliminated in the round of three, thus $A^* < 33\frac{1}{3}$. It follows that $A^* + C < 66\frac{2}{3}$, so

$$B^* + D > 33\frac{1}{3}. \tag{3.1}$$

Since B is eliminated first in the modified election, we have $B^* < D$, thus

$$B^* + D < 2D \tag{3.2}$$

Combining equations (3.1) and (3.2), we have

$$2D > B^* + D > 33\frac{1}{3},$$

so $D > (33\frac{1}{3})/2 = 16\frac{2}{3}$, as desired. □

Note that the proof of Theorem 4 indicates that $16\frac{2}{3}$ is a lower bound for D, but not necessarily the maximum lower bound. We now show that $16\frac{2}{3}$ is the greatest lower bound by giving an example where $D = 16\frac{2}{3}+\epsilon$ is sufficient for a monotonicity anomaly to occur.

EXAMPLE 4. Let $\epsilon > 0$. Consider the election in Table 4 (Top). Candidate D drops out first with $16\frac{2}{3} + 2\epsilon$ votes, slightly fewer than B's total of $16\frac{2}{3} + 3\epsilon$ votes. Most of D's votes go to B. Candidate C then drops out in the next round with ϵ fewer votes than A. Candidate A then beats B by $50 + \epsilon$ to $50 - \epsilon$. We note that A is also the Condorcet winner, although this is not necessary.

Now suppose we modify 2ϵ votes from $BCAD$ to $ABCD$ as illustrated in Table 4 (Bottom). Candidate B is eliminated first with only $16\frac{2}{3} + \epsilon$ votes. Most of B's votes go to D and A is eliminated in round 2. The winner is C, but one could easily change the winner to D by modifying some A votes to AD. Such a change would not affect anything else, so this example could be adapted to demonstrate anomalies for both dropout orders $DCAB$ and $CDAB$.

TABLE 4. Ballot profile for Example 4; original election profile (Top) and modified election profile (Bottom).

	$33\frac{1}{3} - 3\epsilon$	4ϵ	$16\frac{2}{3} - \epsilon$	$16\frac{2}{3} + 2\epsilon$	$16\frac{2}{3} - 4\epsilon$	2ϵ	$16\frac{2}{3}$
1st	A	B	B	C	C	D	D
2nd		C	D	A	B	A	B
3rd		A	A				A
4th		D	C				C

	$33\frac{1}{3} - 3\epsilon$	2ϵ	2ϵ	$16\frac{2}{3} - \epsilon$	$16\frac{2}{3} + 2\epsilon$	$16\frac{2}{3} - 4\epsilon$	2ϵ	$16\frac{2}{3}$
1st	A	A	B	B	C	C	D	D
2nd		B	C	D	A	B	A	B
3rd		C	A	A				A
4th		D	D	C				C

4. Conditions for 3-candidate no-show anomalies

We now investigate necessary and sufficient conditions for no-show anomalies in 3-candidate IRV elections. We note that the $H2H$ condition will still be necessary, for the same reason it was necessary when analyzing monotonicity anomalies in 3-candidate elections. The main questions are what, if any, analogue exists for the $L25$ condition, and whether or not such conditions are both necessary and sufficient. Note that we are considering elections with both fully-ranked and incomplete ballots in this section. We first give an analogue for the $L25$ condition that we will refer to as the $L16.6$ condition.

THEOREM 5. *In a 3-candidate IRV election, it is necessary for the 3rd-place candidate to have greater than $16\frac{2}{3}\%$ of the first-place vote to have a no-show anomaly.*

PROOF. Suppose $C \leq 16\frac{2}{3}$. Then $A + B > 83\frac{1}{3}$. Since no candidate can have 50% of the vote at this point, we conclude that $B < 50$ and thus $A > 33\frac{1}{3}$. To have a no-show anomaly, in the modified election one needs $B^* < C \leq 16\frac{2}{3}$, and it follows that $B^* + C < 33\frac{1}{3} < A$. Thus A would win the modified election, so there could not be a no-show anomaly. We conclude that $C > 16\frac{2}{3}$ is necessary for a no-show anomaly. □

As we did in Example 4 for a monotonicity anomaly, we now show that $16\frac{2}{3}$ is the greatest lower bound for a 3-candidate no-show anomaly.

EXAMPLE 5. Let $\epsilon > 0$. Consider the election in Table 5 (Left). Candidate C is eliminated first and their votes go to A, who wins. Now suppose $33\frac{1}{3} - \epsilon$ of the BCA voters do not show up. We then have $B^* = 50 - \epsilon - (33\frac{1}{3} - \epsilon) = 16\frac{2}{3}$ and B is eliminated first. All BCA votes go to C and C beats A with $C = 33\frac{1}{3} + \epsilon$ votes, creating a no-show anomaly. We conclude C can be arbitrarily close to $16\frac{2}{3}$ and still cause a no-show anomaly.

For monotonicity, the $L25$ and $H2H$ conditions are necessary and sufficient for an anomaly to be present. We see below that the analogous conditions $L16.6$ and $H2H$ are *not* sufficient for a no-show anomaly to be present in a 3-candidate election.

EXAMPLE 6. Consider the election in Table 5 (Right). Candidate C has 32% of the first-place vote, satisfying the $L16.6$ condition, and C beats A in a head-to-head contest 51 to 49, satisfying $H2H$. Despite this, a no-show anomaly is not present in the election. To create such an anomaly, we can only remove votes for which C is favored over A, and we need to remove enough so that B drops out before C. Thus $3 + \epsilon$ of the BCA voters would need to not show up so that B is eliminated before C. Once such a modification occurs, though, C would now lose a head-to-head contest to A by 49 to $48-\epsilon$. Thus no anomaly is present, and we conclude that the $L16.6$ and $H2H$ conditions are *not* sufficient for a no-show anomaly. Note that although this example has $A\_\_$ ballots, adding rankings below the 33 A votes would not change any of the results. Thus this conclusion holds for elections with both fully-ranked and incomplete ballots.

We found a different condition, though, that when paired with the $H2H$ condition, will give necessary and sufficient conditions for a fully-ranked 3-candidate no-show anomaly.

TABLE 5. Ballot profiles for Examples 5 (Left) and 6 (Right).

	$50-\epsilon$	$33\frac{1}{3}$	$16\frac{2}{3}+\epsilon$
1st	B	A	C
2nd	C		A
3rd	A		B

	19	16	33	32
1st	B	B	A	C
2nd	C	A		A
3rd	A	C		B

THEOREM 6. *Let A be the winning candidate, C the candidate who is eliminated in the round of three, and B_A be the number of BAC votes. To have a no-show anomaly in a fully-ranked 3-candidate IRV election, the following conditions are necessary and sufficient:*

(1) *$C > B_A + \frac{A}{2}$ and*
(2) *C beats A in a head-to-head.*

PROOF. We first show that conditions 1 and 2 are sufficient. That is, if both conditions hold there exists a modified election where removing certain BCA ballots will result in B dropping out before C, and then C beating A in the final head-to-head round, creating a no-show anomaly. Recall that B_A (respectively B_C) is the number of BAC (respectively BCA) votes, and since we are fully ranked we have $B = B_A + B_C$. Let $\epsilon > 0$ and $g = B - C + \epsilon$ be the smallest number of BCA votes who need to not show up so B drops out before C. Removing g such votes from BCA gives $B_C^* = B_C - g$. Note that at least g such B_C votes must exist. This follows because, since B does not win the election, we know $B < 50$ implies $B - C < 50 - C$. Since the ballots are fully-ranked, condition 2 guarantees that $C + B_C > 50$ implies $B_C > 50 - C$. Thus $B_C > 50 - C > B - C$, so $B_C > g$ for an appropriate ϵ. Starting with condition 1 in Theorem 6, $C > B_A + \frac{A}{2}$ is equivalent to $2C > 2B_A + A$, which means that there exists an $\epsilon > 0$ such that

$$2C - \epsilon > 2B_A + A. \tag{4.1}$$

Adding $(B_C - B_C)$ to the left side of Equation 4.1 and subtracting B_A for both sides yields

$$2C - \epsilon + B_C - B_C - B_A > B_A + A. \tag{4.2}$$

However, Equation 4.2 is equivalent to

$$B_C - (B_C + B_A - C + \epsilon) + C = B_C - (B - C + \epsilon) + C = B_C - g + C > B_A + A.$$

Using the definition of g, this last inequality reduces to

$$B_C^* + C > B_A + A.$$

The inequality $B_C^* + C > B_A + A$ indicates that C beats A in the head-to-head round once g of the BCA voters do not show up and B is eliminated. This verifies that conditions 1 and 2 are sufficient for a no-show anomaly to be present. We note that if $C < B_A + \frac{A}{2}$, the same series of implications as above will result in $B_C^* + C < B_A + A$. Thus C would lose the resulting head-to-head matchup against A, and no anomaly would occur. This verifies the necessity for condition 1. As stated previously, the necessity for condition 2 is identical to the monotonicity anomaly case. □

REMARK. Condition 1 implies that $C > 16\frac{2}{3}$ is necessary for a no-show anomaly when ballots are fully ranked. If $C \leq 16\frac{2}{3}$, then condition 1 implies that $A < 33\frac{1}{3}$. It follows that $A + C < 50$, and thus $B > 50$, which contradicts A being the IRV winner. Theorem 5 is a stronger result, however, as it holds for ballots that are not necessarily fully-ranked.

The necessity of the conditions stated in Theorem 6 are illustrated by our previous examples. In Example 1, we had $C = 27$, $B_A = 0$ and $A = 40$. Thus $C > B_A + A/2$, and as the $H2H$ condition also held, a no-show anomaly was present. In Example 6, we had $C = 32$, $B_A = 16$, and $A = 33$, so $C = 32 < 32.5 = B_A + A/2$ and thus no anomaly was present. These conditions, however, are sufficient only for fully-ranked ballots, as demonstrated by the following well-known example.

EXAMPLE 7. The 2009 Burlington, Vermont, mayoral election famously provides an example of a real-world monotonicity anomaly [**5**]. However, a no-show anomaly was not present in that election even though it satisfied conditions 1 and 2 of Theorem 6, because some ballots were incomplete. Table 6 shows the voter profile after all but three candidates were eliminated.

Candidate C beats A in a head-to-head 4067 to 3477, and $C = 2554 > 1986 = B_A + A/2$. To create a no-show anomaly, 744 of the BCA voters would need to be removed so that B would be eliminated before C. Doing so, however, would reduce C's head-to-head vote count and $C^* = C + B_C^* = 3323 < 3477 = A + B_A$, so A would remain the winner, and there would not be a no-show anomaly.

TABLE 6. Final stages of the ballot profile for the 2009 mayoral race in Burlington, VT.

	1332	767	455	2043	371	568	1513	495	1289
1st	C	C	C	A	A	A	B	B	B
2nd	A	B		C	B		C	A	
3rd	B	A		B	C		A	C	

5. Conclusion

Our work expands necessary and sufficient conditions for monotonicity and no-show anomalies in a number of situations, specifically finding conditions for monotonicity anomalies when there are incomplete ballots and 3-4 candidates and for no-show anomalies for 3 candidates. These results will be useful for theoretical and empirical investigations regarding the frequency of such anomalies. For example, in the same way that the well-known conditions for fully-ranked 3-candidate elections have been used in programs to find monotonicity anomalies, similar investigations could be done to search for no-show anomalies. We will continue to investigate conditions for 4-candidate monotonicity anomalies using the framework discussed in this article, as well as look into conditions for 4-candidate no-show anomalies. Further research can be done to expand necessary and sufficient conditions to other voting methods and types of anomalies, particularly downward monotonicity anomalies. While some anomalies are easy to find (for example when Borda Count fails the majority criterion), other anomalies may have less obvious, but still useful, conditions similar to what we have found in IRV elections.

References

[1] Craig M. Burnett and Vladimir Kogan, *Ballot (and voter) "exhaustion" under instant runoff voting: An examination of four ranked-choice elections*, Electoral Studies **37** (2015), 41–49.

[2] Dan S. Felsenthal, *On paradoxes afflicting voting procedures: Needed knowledge regarding necessary and/or sufficient condition (s) for their occurrence*, The Future of Economic Design, Springer, 2019, pp. 83–87.

[3] Adam Graham-Squire and N. Zayatz, *Lack of monotonicity anomalies in empirical data of instant-runoff elections*, Representation (2020), 1–9.

[4] Eric Kamwa and Issofa Moyouwou, *Susceptibility to manipulation by sincere truncation: the case of scoring rules and scoring runoff systems*, Evaluating voting systems with probability models—essays by and in honor of William Gehrlein and Dominique Lepelley, Stud. Choice Welf., Springer, Cham, [2021] ©2021, pp. 275–295, DOI 10.1007/978-3-030-48598-6_12. MR4305684

[5] Juho Laatu and Warren D. Smith, *The rank-order votes in the 2009 burlington mayoral election*, 2009, [Online; accessed 10-May-2015], `http://www.rangevoting.org/JLburl09.txt`.

[6] Dominique Lepelley, Frédéric Chantreuil, and Sven Berg, *The likelihood of monotonicity paradoxes in run-off elections*, Math. Social Sci. **31** (1996), no. 3, 133–146, DOI 10.1016/0165-4896(95)00804-7. MR1394645

[7] Nicholas R. Miller, *Closeness matters: monotonicity failure in IRV elections with three candidates*, Public Choice **173** (2017), no. 1-2, 91–108.

[8] Joseph T. Ornstein and Robert Z. Norman, *Frequency of monotonicity failure under Instant Runoff Voting: estimates based on a spatial model of elections*, Public Choice **161** (2014), no. 1-2, 1–9.

High Point University, 1 University Parkway, High Point North Carolina 27268
Email address: `agrahams@highpoint.edu`

Contemporary Mathematics
Volume **795**, 2024
https://doi.org/10.1090/conm/795/15966

Piercing numbers in circular societies

Kristen Mazur, Mutiara Sondjaja, Matthew Wright, and Carolyn Yarnall

ABSTRACT. In the system of approval voting, individuals vote for all candidates they find acceptable. Many approval voting situations can be modeled geometrically, and thus geometric concepts such as the piercing number have a natural interpretation. In this paper, we explore piercing numbers in the setting where voter preferences can be modeled by congruent arcs on a circle, i.e., in fixed-length circular societies. Given a number of voters and the length of the voter preference arcs, we give bounds on the possible piercing number of the society. Further, we explore which piercing numbers are more likely. Specifically, under the assumption of uniformly distributed voter preference arcs, we determine the probability distribution of the piercing number of societies in which the length of the arcs is sufficiently small. We end with simulations that give estimated probabilities of piercing number for societies with larger voter preference arcs.

1. Introduction

Approval voting is a system in which each individual casts a vote for all candidates of which they approve, and the winning candidate is the one that receives the most votes. This system is of interest in that it has many advantages over the more common plurality voting systems and it can be applied outside of the political arena [**2**]. For instance, approval voting is often the way that committees approach scheduling problems: each committee member indicates the times they are available, and the meeting is held at the time most members are able to meet. This situation can be modeled mathematically by considering each person's availability as an interval on a line or an arc on a circle, the latter corresponding to a 24-hour clock which may be especially useful for scheduling virtual meetings occurring across several time zones.

Previous research ([**1**], [**5**], [**8**]) developed bounds on the maximum number of voters who approve of a single candidate in various approval voting scenarios. For example, when we model the scheduling problem above using intervals on a line, it is not hard to show that if two out of every three committee members agree on a meeting time then there is a time that works for at least half of the committee members [**1**]. Alternatively, when we model the problem using arcs on a circle, if

2020 *Mathematics Subject Classification.* Primary 52A35, 60D05, 91B12.
Key words and phrases. Approval voting, piercing number, circular societies, discrete geometry, applied topology, probability.

two out every three members agree on a meeting time, then there is a time that works for at least one third of the committee members [**5**].

This paper explores a different question. Instead of the maximum number of voters who agree on a single candidate, we develop a bound on the number of candidates needed to satisfy all voters. Specifically, expanding upon the scheduling problem above, suppose the committee requires all of its members to attend a training session. They plan to hold the same training session at multiple times so that all members will be able to attend one of the sessions. After viewing each member's availability the committee then asks, "what is the minimum number of training sessions needed so that each member can attend at least one of the sessions?" In [**10**], Su and Zerbib call this collection of training sessions a *representative candidate set* or, using language from convex geometry, a *piercing set*. In this paper we use the latter terminology. More formally, given a collection of sets, a *piercing set* is a collection of points such that each set contains at least one of the points. The size of the smallest possible piercing set is the *piercing number* of the collection of sets. Thus, if we model the above scenario geometrically, then the collection of time intervals at which the members are available is a collection of sets, and the minimum number of training sessions needed is the piercing number.

Su and Zerbib recently contextualized piercing numbers in terms of approval voting [**10**]. They present a collection of piercing results that best apply to approval voting scenarios that are modeled using intervals on a line. Most notably, they cite Hadwiger and Debrunner's [**4**] classic 1957 result that provides an upper bound on the piercing number of convex sets in $\mathbb{R}^d$ based on a local intersection property and state its implications in approval voting. Su and Zerbib then leverage this result to make statements regarding piercing sets and piercing numbers in various approval voting scenarios. For example, in a scenario modeled with a collection of intervals on the real number line, if two out of every four voters agree on a candidate, then the piercing number is at most three. Thus, there is a piercing set that contains three candidates.

Such a result does not apply to the above training session scenario because this scenario is best modeled using arcs on a circle (instead of intervals on the real line), to account for the periodic nature of the calendar year (e.g., members are able to select an availability arc of December–January) or a clock. In approval voting we call this model a *circular society*. The piercing number for circular societies has scarcely been explored (see Final Remark 1 in [**10**]).

This paper explores piercing numbers of circular societies. After reviewing the mathematical background on approval voting in Section 2, we provide a complete characterization of possible piercing numbers in circular societies in which all arcs have the same fixed length in Section 3. Then, in Section 4, we explore the probability of certain piercing numbers in circular societies when arcs of a fixed length are randomly distributed around the circle.

2. Background

Any approval voting scenario consists of a collection of voters and a collection of candidates that voters may approve of. We refer to the set of candidates as the *spectrum* X. The set of candidates acceptable to an individual voter is called an *approval set*. Thus, each voter has an approval set that is a subset of the spectrum.

DEFINITION 2.1. A *society* $\mathcal{S}$ of n voters is a pair $(X, \mathcal{A})$ in which X is a spectrum and $\mathcal{A}$ is a collection of n approval sets.

Despite the term "candidate" seemingly implying discreteness, we study societies in which the spectra are continuous. In [**1**], Berg et al. introduced the language of a society, focusing on societies in which the spectrum is the real number line and approval sets are intervals. These societies are called *linear societies.* Alternatively, in a *circular society,* the spectrum is a circle and approval sets are arcs on the circle. Circular societies were first studied by Hardin [**5**]. The authors of this paper previously studied products of linear and circular societies [**8**].

Given a society $\mathcal{S} = (X, \mathcal{A})$, since $\mathcal{A}$ is a collection of sets, we can consider a piercing set of $\mathcal{A}$ or the piercing number of $\mathcal{A}$. In the context of an approval voting society $\mathcal{S}$, a piercing set of $\mathcal{A}$ is a collection of candidates in X such that each voter approves of at least one candidate in the collection. The piercing number is the smallest number of candidates needed so that every voter is happy. Thus, we refer to the piercing number of $\mathcal{A}$ as the piercing number of the society $\mathcal{S}$ and denote it $\tau(\mathcal{S})$. Moreover, if an approval set A contains a candidate x we say that x *pierces* A.

Su and Zerbib [**10**] use the notion of (k, m)-agreeability to give upper bounds on the piercing numbers of linear and circular societies. A society is (k, m)-agreeable [**1**] if k out of every m approval sets intersect. For example, in a $(2, 3)$-agreeable society, two out of every three voters agree on a candidate. Su and Zerbib's result, given in Theorem 2.2 below, follows directly from a theorem of Hadwiger and Debrunner [**4**] that gives conditions of overlap for families of convex sets. (See Theorem 5 in [**10**].)

THEOREM 2.2 (Piercing Numbers in Linear Societies [**10**]). *If $\mathcal{S}$ is a (k, m)-agreeable linear society then we can find a piercing set that contains $m - k + 1$ candidates. Hence, $\tau(\mathcal{S}) \leq m - k + 1$.*

To obtain an analogous result for a circular society $\mathcal{S}$, we can arbitrarily remove a candidate on the spectrum, cut the circle at that point and "unroll" it to produce a linear society. Applying Theorem 2.2 and including the removed candidate as a piercing point, we obtain an upper bound for $\tau(\mathcal{S})$.

COROLLARY 2.3 (Piercing Numbers in Circular Societies [**10**]). *Suppose $\mathcal{S}$ is a (k, m)-agreeable circular society. Then $\mathcal{S}$ has a piercing set that contains $m - k + 2$ candidates. Hence, $\tau(\mathcal{S}) \leq m - k + 2$.*

While Theorem 2.2 and Corollary 2.3 are mathematically interesting, results that give information about piercing based on (k, m)-agreeability are difficult to apply to approval voting in practice. In an approval voting scenario, in order to determine if the society is (k, m)-agreeable we must know all voter preferences and how they interact with one another. Therefore, we would likely also be able to determine the piercing number.

In this paper, we study piercing numbers through a different lens. Rather than working with (k, m)-agreeable circular societies, we shift our focus to circular societies in which all approval sets have the same length. For example, in a committee scheduling scenario, we could require that all members choose a 3-hour time block during which they are available.

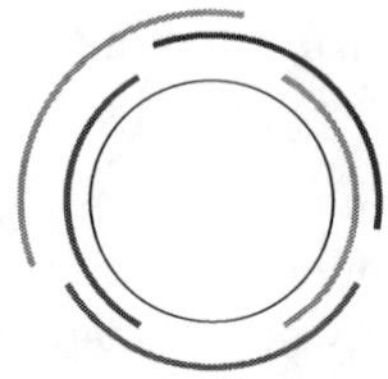

FIGURE 1. A fixed-length circular society with five voters. The black circle is the spectrum, and each colored arc represents the approval set of one voter.

DEFINITION 2.4. Let $\mathcal{S} = (X, \mathcal{A})$ be a circular society. We say that $\mathcal{S}$ is a *fixed-length* circular society if all approval sets in $\mathcal{A}$ have the same length. We call this fixed length the *approval set length* of $\mathcal{S}$ and denote it p.

In order to work with such societies more clearly and concisely, we lay out some useful conventions for circular societies that we use throughout this paper.

REMARK 2.5 (Circular Societies as $\mathbb{R}/\mathbb{Z}$). We use the classic topological property that a circle is equivalent to $\mathbb{R}/\mathbb{Z}$, identifying the points x and $x+n$ for any real number x and integer n. In other words, given a society $\mathcal{S} = (X, \mathcal{A})$, we represent the circular spectrum X as the interval $[0, 1)$, identifying 0 and 1. In doing so, we force the circumference of X to be 1, rescaling all approval sets as needed. Thus, in a fixed-length circular society, the approval set length p is less than 1 and can be thought of as a proportion of the circumference of the spectrum. For example, suppose we are modeling time preferences on a 24-hour clock. If we require individual approval sets to be 3-hour time windows, then since the spectrum circumference is 1, the approval set length p is $\frac{1}{8}$.

REMARK 2.6 (Interval Notation of Approval Sets). Following Remark 2.5, identifying the spectrum with $\mathbb{R}/\mathbb{Z}$ allows us to depict fixed-length circular societies "linearly" with approval sets as intervals (or a union of two intervals if the corresponding arc crosses the point corresponding to 0) of length p, as in Figure 2. We also assume that approval sets are closed unless otherwise noted.

Further, we often label an approval set as $A = [\ell, r]$, referring to ℓ as the "left" endpoint and r as the "right" endpoint. This is natural when $\ell < r$ but might seem a bit backwards when considering intervals that contain the point $0 = 1$. For example, in Figure 2, the left endpoint of the brown interval labeled A is to the right of the right endpoint in the linear depiction. Alternatively, we can think of ℓ as the endpoint that is counterclockwise from the center of the interval A and r as the endpoint that is clockwise from the center of A.

3. Possible piercing numbers

If a fixed-length circular society $\mathcal{S}$ has approval length $p \geq \frac{1}{k}$ then we can form a piercing set by taking k candidates that are evenly spaced around the spectrum. Hence, $\tau(\mathcal{S}) \leq k$. However, in order to completely characterize all possible piercing numbers in circular societies we need to consider *both* the size of the approval sets *and* the number of voters.

Intuitively societies with shorter approval sets may require more candidates to pierce all sets. For example, consider societies with five voters. If the approval set

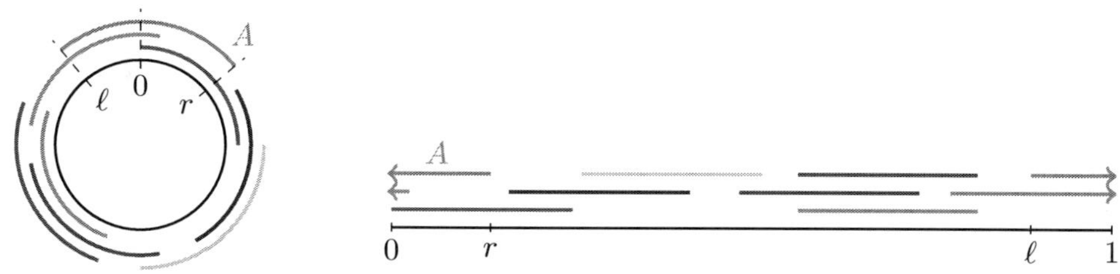

FIGURE 2. Two depictions of the same fixed-length circular society with approval set length $p = \frac{1}{4}$. Endpoints of approval set A are labeled. We cut the figure on the left at the dashed line and "unroll" to obtain the figure on the right. The arrows on two approval sets on the right represent that these approval sets "wrap around."

length is very small (more precisely, less than $\frac{1}{5}$), then the piercing number can range from 1 to 5, depending on how the approval sets overlap. However, if the approval set length is very large (greater than $\frac{4}{5}$), then the piercing number must be 1; this is because the complements of all approval sets cannot cover the entire circle, so there is a point on the circle that is not in the complement of any approval set, and is thus contained in every approval set. The story becomes murkier when approval set lengths are between $\frac{1}{5}$ and $\frac{4}{5}$. For example, in Theorem 3.1 we show that when the approval set length is between $\frac{2}{5}$ and $\frac{4}{5}$, the piercing number is less than or equal to two. However, in Theorem 3.2 we construct a society with approval set length $\frac{2}{5} - \epsilon$ in which the piercing number equals three.

Changing perspective, we can consider what happens when we fix the approval set length and vary the number of voters. In general, a society with more voters may have a larger piercing number. For example, a society with approval set length of $\frac{1}{3}$ and four voters will have a piercing number of at most two, while a society with five voters and approval set length of $\frac{1}{3}$ can have a piercing number of three, as demonstrated in Figure 1. The reasoning behind the first result is nontrivial; the justification relies on the proof of Theorem 3.1.

Figure 3 illustrates our main results in this section, stated formally in Theorem 3.1 and Theorem 3.2. In the figure we see how the maximum possible piercing number for any fixed-length circular society depends on the number of voters n and the approval set length p. Further, the figure is best understood by observing the points where the color changes, drawn as dots in the figure. These dots give thresholds in the (n, p)-space at which the maximum piercing number changes by one. We make this idea precise as follows. Given any $n \in \mathbb{N}$ and $k \in \{1, 2, \dots, n\}$, there exists a *critical length*

$$Q(n,k) = \frac{\left\lceil \frac{n}{k} \right\rceil - 1}{k \left\lceil \frac{n}{k} \right\rceil - k + 1} \tag{3.1}$$

that gives the threshold p below which piercing number $\tau = k + 1$ is possible for a fixed-length circular society of n voters. Note that when $k = n$, the critical length is zero, which aligns with the fact that the piercing number k will not be larger than the number of voters n. The following theorems state the relationship between $Q(n, k)$ and possible piercing numbers.

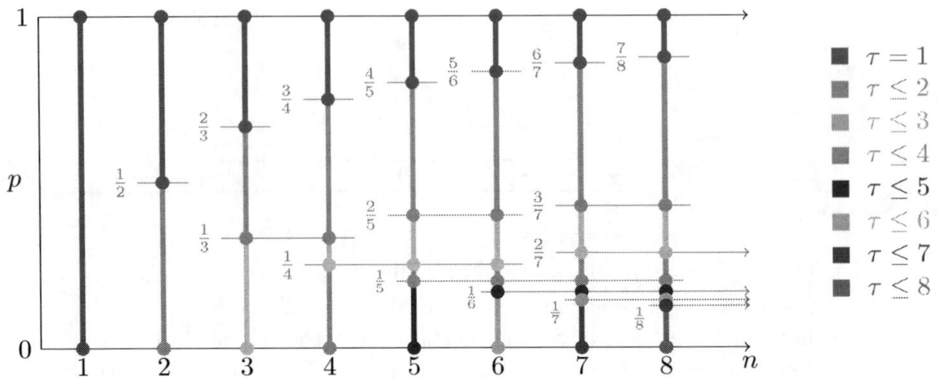

FIGURE 3. Maximum possible piercing numbers for fixed-length circular societies with n voters and approval set length p.

THEOREM 3.1. *Let $n \in \mathbb{N}$ and $k \in \{1, 2, \dots, n\}$. Then a fixed-length circular society $\mathcal{S}$ with n voters and approval length $p \geq Q(n, k)$ has piercing number $\tau(\mathcal{S}) \leq k$.*

THEOREM 3.2. *Let $n \in \mathbb{N}$, $k \in \{1, 2, \dots, n-1\}$, and $0 < p < Q(n, k)$. There exists a fixed-length circular society $\mathcal{S}$ with n voters, approval set length p, and piercing number $\tau(\mathcal{S}) = k + 1$.*

Before proving these theorems, we offer some comments about $Q(n, k)$. We find the expression for $Q(n, k)$ in Equation (3.1) to be non-obvious but descriptive of certain patterns in possible piercing numbers, most notably the stepped asymptotic behavior of the threshold values that appear in Figure 3.

Intuitively, with more voters, longer approval sets are needed to guarantee a specific piercing number. Thus, as the number of voters n increases, the threshold for p below which piercing number k is possible increases as well. That is, for a fixed piercing number k, $Q(n, k)$ increases with n, but this increase is strictly monotonic only for $k = 1$. In general, we find that $Q(n, k)$ increases in steps of size k. Figure 3 highlights this by horizontal line segments connecting equal values of $Q(n, k)$. For example, in a society of 4 voters, the approval set length must be $p \geq \frac{1}{4}$ in order to guarantee a piercing number of 3 or less, and this minimum length holds if a fifth or sixth voter joins the society. However, if a seventh voter joins the society, then the approval set length must be at least $\frac{2}{7}$ to guarantee a piercing number of 3 or less.

Furthermore, for fixed k, $Q(n, k)$ is bounded above by an asymptote at $\frac{1}{k}$. Specifically,

$$\lim_{n \to \infty} Q(n, k) = \frac{1}{k}.$$

For example, a piercing number of 1 is only guaranteed when $p \geq \frac{n-1}{n}$; below this, piercing number 2 is possible. Hence, since $\lim\limits_{n \to \infty} \frac{n-1}{n} = 1$, if a society has a large number of voters then a piercing number of 2 is possible even when the approval set length is very close to 1. Likewise, as $n \to \infty$ a piercing number of 3 becomes possible for approval set lengths p approaching $\frac{1}{2}$ from below, yet a piercing number of 3 is never possible for $p \geq \frac{1}{2}$. Similar results hold for greater piercing numbers:

as $n \to \infty$, a piercing number of $k+1$ is only possible for p approaching $\frac{1}{k}$ from below.

We now prove Theorem 3.1.

PROOF OF THEOREM 3.1. Let $\mathcal{S} = (X, \mathcal{A})$ be a fixed-length circular society with n voters and approval set length $p \geq Q(n,k)$. We will show that $\tau(\mathcal{S}) \leq k$. To simplify notation, let $q = \lceil \frac{n}{k} \rceil$, so $Q(n,k) = \frac{q-1}{kq-(k-1)}$. We proceed in two cases, depending on whether there exists an open arc $C \subseteq X$ of length $\frac{1}{kq-(k-1)}$ that contains no *left* endpoints of approval sets. (Recall from Remark 2.6 that we call the counterclockwise endpoint of an approval set the left endpoint.)

For our first case, suppose there exists an open arc C of length $\frac{1}{kq-(k-1)}$ that contains no left endpoints of approval sets. If C contains the right endpoints of any approval sets, let $x_1 \in C$ be the leftmost of these right endpoints, as illustrated on the left in Figure 4. Thus, x_1 pierces all approval sets that intersect C. If C contains no right endpoints of any approval sets, then either C is disjoint from all approval sets or is contained within all approval sets that it intersects; in either case, choose $x_1 \in C$ arbitrarily.

Any approval set not pierced by x_1 is contained in the complement $\overline{C}$ of C. The length of $\overline{C}$ is $1 - \frac{1}{kq-(k-1)} = \frac{kq-k}{kq-(k-1)}$. There are $k-1$ equally-spaced points $x_2, x_3, \ldots, x_k$ in $\overline{C}$ that divide $\overline{C}$ into k intervals of equal length $\frac{q-1}{kq-k+1}$. Any approval set of length $p \geq \frac{q-1}{kq-(k-1)}$ must be pierced by at least one of these points. Thus, $\{x_1, x_2, x_3, \ldots, x_k\}$ is a piercing set, so $\tau(\mathcal{S}) \leq k$.

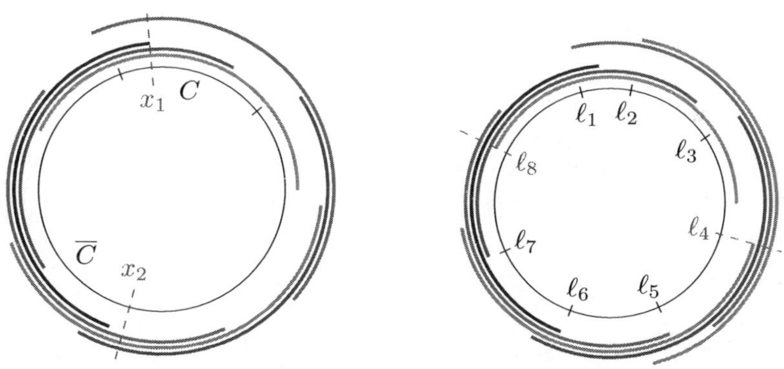

FIGURE 4. In the proof of Theorem 3.1, either there exists an open arc C of length $\frac{1}{kq-(k-1)}$ containing no left endpoints of approval sets (illustrated at left, where $n = 7$, $p = \frac{3}{7}$, and $k = 2$), or there does not exist such an arc C (illustrated at right, where $n = 8$, $p = \frac{3}{7}$, and $k = 2$).

For the second case, suppose there does not exist an (open) arc $C \subseteq X$ of length $\frac{1}{kq-(k-1)}$ that contains no *left* endpoints of approval sets. This implies $n \geq kq - (k-2)$, for if $n < kq - (k-2)$ then there are too few left endpoints of approval sets in X to avoid a gap of size $\frac{1}{kq-(k-1)}$. Label the left endpoints of approval sets consecutively $\ell_1, \ell_2, \ldots, \ell_n$, as illustrated on the right in Figure 4.

Since every two consecutive left endpoints of approval sets are separated by an arc of length less than $\frac{1}{kq-(k-1)}$, the arc from ℓ_1 to ℓ_q has length less than $\frac{q-1}{kq-(k-1)}$.

Since approval sets have length $p \geq \frac{q-1}{kq-(k-1)}$, this implies that the point ℓ_q pierces the approval sets whose left endpoints are $\ell_1, \ell_2, \ldots, \ell_q$. Similarly, consider the left endpoints $\ell_{2q}, \ell_{3q}, \ldots, \ell_{mq}$, where $m = \lfloor \frac{n}{q} \rfloor$; each of these pierces q approval sets. If k divides n, then $n = kq = mq$, and thus $\{\ell_q, \ell_{2q}, \ell_{3q}, \ldots, \ell_{mq}\}$ forms a piercing set of size k. If k does not divide n, then $mq < n$, so approval sets whose left endpoints are ℓ_i for $mq < i \leq n$ might remain unpierced. However, ℓ_n pierces all of these remaining sets since the arc length from ℓ_{mq+1} to ℓ_n is less than $\frac{q-1}{kq-(k-1)}$. Thus, $\{\ell_q, \ell_{2q}, \ell_{3q}, \ldots, \ell_{mq}, \ell_n\}$ forms a piercing set. Since k does not divide n,

$$\frac{n}{q} = \frac{n}{\lceil n/k \rceil} < \frac{n}{n/k} = k,$$

which implies $m = \lfloor \frac{n}{q} \rfloor \leq k - 1$, and we have a piercing set of size at most k. Therefore, in either case, $\tau(S) \leq k$. □

Our proof Theorem 3.2 relies on the construction of fixed-length circular societies whose approval sets are spaced around the circle as evenly as possible. Our construction is similar to the uniform societies defined by Hardin [**5**], but in contrast to Hardin, our approval sets are closed rather than half-open. We give the formal definition below.

DEFINITION 3.3. Let N and T be positive integers such that $N > T$ and let ϵ be a parameter such that $0 < \epsilon < \frac{1}{N}$. Define the *uniform society* $U_\epsilon(N, T)$ to be the fixed-length circular society with N voters, approval set length $\frac{T}{N} - \epsilon$, and approval sets defined as follows. Identifying the spectrum with the real numbers modulo 1 as in Remark 2.5, the approval set for voter $i \in \{0, 1, \ldots, N-1\}$ is the interval $A_i = \left[\frac{i}{N}, \frac{i+T}{N} - \epsilon\right]$.

We give an example of a uniform society in Figure 5. Note that the combined total length of all approval sets in a uniform society $U_\epsilon(N, T)$ is $N \cdot \left(\frac{T}{N} - \epsilon\right) = T - N\epsilon$. When we assume ϵ to be small, the total length is approximately T. Furthermore, every point on the spectrum is contained in either T or $T-1$ approval sets.

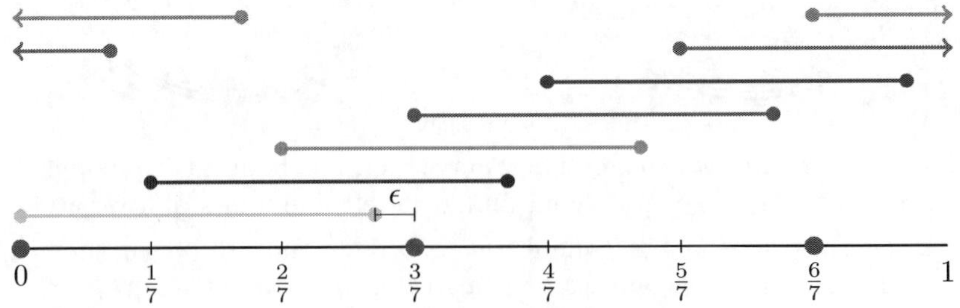

FIGURE 5. The uniform society $U_\epsilon(7, 3)$. The highlighted points on the spectrum form the piercing set constructed in Lemma 3.4. Hence, the piercing number is three.

LEMMA 3.4. *The uniform society* $U_\epsilon(N, T)$ *has piercing number* $\lceil \frac{N}{T} \rceil$.

PROOF. Since every point in the spectrum is contained in at most T approval sets, at least $\lceil \frac{N}{T} \rceil$ points are necessary to pierce all sets.

We claim that the collection

$$W = \left\{0, \frac{T}{N}, 2\frac{T}{N}, 3\frac{T}{N}, \dots, \left(\left\lceil \frac{N}{T} \right\rceil - 1\right)\frac{T}{N}\right\}$$

is a piercing set. To justify this claim, observe that W consists of $\lceil \frac{N}{T} \rceil$ points in $[0, 1)$, and the distance between neighboring points in W is $\frac{T}{N}$. Since $[0, 1) = \mathbb{R}/\mathbb{Z}$, the distance between the points 0 and $(\lceil \frac{N}{T} \rceil - 1)\frac{T}{N}$ on the circular spectrum is less than or equal to $\frac{T}{N}$ (for an example, see Figure 5). Since each approval set is of length $\frac{T}{N} - \epsilon$ with its left endpoint at a multiple of $\frac{1}{N}$, each approval set must contain a point in W. Specifically, piercing point $j\frac{T}{N}$ is contained in A_i for each $i \in \{jT - (T - 1), \dots, jT - 1, jT\}$. Thus, W is a piercing set, and the piercing number is $\lceil \frac{N}{T} \rceil$. □

We now prove Theorem 3.2.

Proof of Theorem 3.2. Let $n \in \mathbb{N}$, $k \in \{1, 2, \dots, n - 1\}$, and $0 < p < Q(n, k)$. We will construct a fixed-length circular society $\mathcal{S}$ with n voters, approval set length p, and piercing number $\tau(\mathcal{S}) = k + 1$.

First, we construct a uniform society with piercing number $\tau = k + 1$. Let $N = k\lceil \frac{n}{k} \rceil - k + 1$, the denominator of $Q(n, k)$; $T = \lceil \frac{n}{k} \rceil - 1$, the numerator of $Q(n, k)$; and $\epsilon = \min\left(\frac{T}{N} - p, \frac{1}{2N}\right)$. Then define the uniform society $\mathcal{U} = U_\epsilon(N, T)$. Lemma 3.4 implies that $\mathcal{U}$ has piercing number

$$\tau = \left\lceil \frac{N}{T} \right\rceil = \left\lceil \frac{k\left(\lceil \frac{n}{k} \rceil - 1\right) + 1}{\lceil \frac{n}{k} \rceil - 1} \right\rceil = \left\lceil k + \frac{1}{\lceil \frac{n}{k} \rceil - 1} \right\rceil = k + 1.$$

Moreover, $\mathcal{U}$ has $N \leq n$ voters. To see this, note that $k\lceil \frac{n}{k} \rceil$ is the smallest multiple of k that is greater than or equal to n, which implies $k\lceil \frac{n}{k} \rceil - n \leq k - 1$. Thus $N = k\lceil \frac{n}{k} \rceil - k + 1 \leq n$. Further, since $\epsilon \leq \frac{T}{N} - p$, the approval sets in $\mathcal{U}$ have length $\frac{T}{N} - \epsilon \geq p$.

While $\mathcal{U}$ has the desired piercing number, it may not be the society $\mathcal{S}$ that we set out to create because it may not have enough voters or the approval set length may be too big. However, both of these issues are easy to rectify. If $N < n$, then augment $\mathcal{U}$ with $n - N$ additional approval sets, each given by the interval $[0, p]$ to create a society with n voters and piercing number $k + 1$. If $\frac{T}{N} - \epsilon > p$, shrink each approval set around its piercing point to obtain a society with approval sets of length p and piercing number $k + 1$. After performing one or both of these operations we have created a society $\mathcal{S}$ as desired. □

For example, given $n = 8$ and $k = 2$, following the proof of Theorem 3.2, we can create a society with 8 voters, an approval set length less than $Q(n, k) = \frac{3}{7}$, and piercing number $\tau = k + 1 = 3$. Computing $N = 7$ and $T = 3$ as in the proof, we first construct the uniform society $U_\epsilon(7, 3)$ that has 7 voters and piercing number $\tau = 3$. Figure 5 illustrates $U_\epsilon(7, 3)$ with p slightly less than $\frac{3}{7}$. Adding the interval $[0, p]$ as an eighth approval set does not change the piercing number. Furthermore, we can modify $U_\epsilon(7, 3)$ to make p arbitrarily small while preserving the piercing number $\tau = 3$ by contracting each approval set around the piercing point it contains.

4. Probabilities of piercing numbers

In Section 3, we explored which piercing numbers are possible for fixed-length circular societies with a given number of voters n and a given approval set length p. For example, because $Q(5,2) = \frac{2}{5}$, any circular society with $n = 5$ voters and an approval set length $p = \frac{2}{5}$ has a piercing number of at most 2, and we can construct a circular society with $n = 5$ voters and approval set length $p = \frac{2}{5} - \epsilon$ (say $p = 0.3999$) that has piercing number 3. However, the method for constructing such a circular society in the proof of Theorem 3.2 is quite specific. Hence, intuitively, we might believe that most circular societies with $n = 5$ and $p = 0.3999$ would have a piercing number that is less than three.

In this section, we develop a probabilistic model for understanding the overall distribution of piercing numbers of fixed-length circular societies. In particular, we explore the space of fixed-length circular societies whose approval sets are randomly-generated with a uniform distribution on the spectrum X. We define these societies below.

DEFINITION 4.1. A *random fixed-length circular society* with n voters and approval set length $p \in (0,1)$ is a circular society $\mathcal{S} = (X, \mathcal{A})$ in which $\mathcal{A}$ consists of n approval sets of the form $A_i = [\ell_i, r_i]$ for $i = 1, \ldots, n$ where

- $\ell_1, \ldots, \ell_n$ are independent random variables, uniformly distributed on the spectrum X, and
- $r_i = \ell_i + p$.

Recall from Remark 2.5 that we identify X with $\mathbb{R}/\mathbb{Z}$. So, $r_i = \ell_i + p \mod 1$.

Our ultimate goal is to provide a complete probability distribution for the piercing number as a function of the number of voters, n, and the approval set length, p. In other words, we hope to answer the question below.

QUESTION 4.2. Given a random fixed-length circular society $\mathcal{S}$ with n voters and approval sets of length $p \in (0,1)$, and a positive integer k, what is the probability that $\tau(\mathcal{S}) = k$?

For an arbitrary n, we are able to answer this question only if the approval set length p is sufficiently large or sufficiently small. Theorem 3.1 tells us that if $p \geq Q(n,1) = \frac{n-1}{n}$, then the piercing number must be 1. If p is sufficiently small, then the approval sets may be pairwise disjoint, making it possible for the piercing number to equal the number of voters (i.e., $k = n$). The following theorem from Solomon [**9**, Equation (4.5)] gives us the probability that the piercing number equals the number of voters when $p < \frac{1}{n}$.

THEOREM 4.3 (Solomon). *The probability that n arcs of length $p \in [0, 1/n)$, randomly placed on the circle, are pairwise disjoint is $(1 - np)^{n-1}$.*

Determining the probability that the piercing number equals k, for k between 1 and the number of voters, is a difficult task. Theorem 4.4 below, which is the main result of this section, provides the probability that the piercing number is k assuming the approval set length is sufficiently small.

THEOREM 4.4. *Let k be a positive integer and $\mathcal{S}$ be a random fixed-length circular society with n voters and approval set length $p < \frac{1}{2k}$. The probability that*

$\tau(\mathcal{S}) = k$ is

$$\binom{n}{k}(1-kp)^{k-1}(kp)^{n-k}. \tag{4.1}$$

We prove Theorem 4.4 in Section 4.1. While Question 4.2 largely remains open for general values of n, k, and p, we are able to give a complete answer for some special cases, when n is small. We present results for these special cases in Section 4.2. In Section 4.3, we discuss simulation results, including those that suggest that Theorem 4.4 might apply for a greater range of values of p. Throughout, we use the observations summarized in the remark below.

REMARK 4.5. Let $\mathcal{S} = (X, \mathcal{A})$ be a random fixed-length circular society with n voters, approval set length p, and approval sets $A_i = [\ell_i, r_i]$ as defined in Definition 4.1. Since $\ell_1, \ldots, \ell_n$ are independent random variables that are uniform on $\mathbb{R}/\mathbb{Z} = [0, 1)$, each ℓ_i has probability density function $f_{\ell_i}(x) = \mathbf{1}_{[0,1)}(x)$. This means that

(1) given any interval $I \subseteq \mathbb{R}/\mathbb{Z}$, the probability that $\ell_i \in I$ is equal to the length of I, and
(2) the approval sets $A_1, \ldots, A_n$ are distinct with probability 1.

4.1. Proofs of piercing probabilities. Our main goal in this section is to prove Theorem 4.4, which gives the probability that $\tau(\mathcal{S}) = k$ when the length of approval sets satisfies $p < \frac{1}{2k}$. The proof relies on Theorem 4.3 and also on Lemma 4.6, below.

LEMMA 4.6. *Let $\mathcal{S} = (X, \mathcal{A})$ be a fixed-length circular society with n distinct approval sets of length $p < \frac{1}{2k}$. Then S has a piercing number $\tau(\mathcal{S}) = k$ if and only if there exists a unique collection of k pairwise-disjoint approval sets in $\mathcal{A}$ whose union contains the left endpoints of all other approval sets.*

PROOF OF LEMMA 4.6. Consider a fixed-length circular society $\mathcal{S} = (X, \mathcal{A})$ with distinct approval sets $A_1, \ldots, A_n$ of length $p < \frac{1}{2k}$. We follow Definition 4.1, denoting $A_i = [\ell_i, r_i]$, where $r_i = \ell_i + p$.

First, suppose there exists a unique collection of k pairwise-disjoint approval sets $A_1, \ldots, A_k$ whose union contains the left endpoints of all other approval sets. This means that each of the remaining $n-k$ approval sets contains one of the right endpoints of $A_1, \ldots, A_k$. Thus, the set of the right endpoints of $A_1, \ldots, A_k$ is a piercing set of size k. Since $A_1, \ldots, A_k$ are pairwise disjoint, a minimum piercing set of $\mathcal{S}$ must contain at least k points. Therefore, $\tau(\mathcal{S}) = k$.

Conversely, suppose $\tau(\mathcal{S}) = k$. Then, there is a minimum piercing set $W = \{x_1, \ldots, x_k\}$ of $\mathcal{S}$. For each $i = 1, \ldots, k$, let $\mathcal{S}_i$ denote the collection of all approval sets pierced by x_i. Let L_i denote the leftmost left endpoint among sets in $\mathcal{S}_i$ ($i = 1, \ldots, k$) and let R_i denote the rightmost right endpoint among sets in $\mathcal{S}_i$. Since each set in $\mathcal{S}_i$ is pierced by x_i, then L_i and R_i are each within distance p from x_i. Therefore, $R_i - L_i \leq 2p < \frac{1}{k}$. So, the union of all approval sets in each $\mathcal{S}_i$ is a closed interval of length strictly less than $\frac{1}{k}$.

This means that the union of all sets in $\mathcal{S}$ is a union of closed intervals whose total length is strictly less than $k\frac{1}{k} = 1$, which implies that not all points on the circle are covered by the approval sets of $\mathcal{S}$. Therefore, $\mathcal{S}$ is essentially a linear society: We can choose a point x not contained in any approval set; then, we can "cut" the spectrum at x and "unroll" it to produce a linear society. Note that choosing a different point x may have the effect of a cyclic permutation of the approval sets in the S, but does not affect the results of this theorem. We treat $\mathcal{S}$ as a linear society in the remainder of the proof.

For general collections of arbitrary sets, the task of finding minimum piercing sets is computationally hard. However, for linear societies, there is a simple linear-time algorithm that finds a minimum piercing set [**7**, Section 1.4]: Start by ordering the approval sets by their right endpoints. Find the approval set with the leftmost right endpoint, r, and add r into the piercing set; then, remove all approval sets pierced by r. Repeat with the remaining sets until all sets are pierced. The collection of all right endpoints chosen in this way is a minimum piercing set. In addition, the approval sets that correspond to these right endpoints are pairwise disjoint.

Since we can treat $\mathcal{S}$ as a linear society, we can apply this algorithm to $\mathcal{S}$, to produce a minimum piercing set, call it W'. The set W' might be different than the minimum piercing set W we start with, but it must also contain k points. Further, W' consists of k *right endpoints*, call them $r_1, \ldots, r_k$ (with $r_1 < \ldots < r_k$), the right endpoints of approval sets $A_1 = [\ell_1, r_1], \ldots, A_k = [\ell_k, r_k]$. The collection of approval sets $\{A_1, \ldots, A_k\}$ consists of k pairwise disjoint sets.

Next, we show that the union, $\cup_{i=1}^k A_i$, contains all other left endpoints: For each $i = 1, \ldots, k$, let $\mathcal{T}_i$ be the collection of approval sets that contain r_i but not r_j for any $j < i$. Since $A_1, \ldots, A_k$ are pairwise disjoint, then $A_i \in \mathcal{T}_i$. Since $\mathcal{S}$ is a fixed-length society with distinct approval sets, no approval set is contained in another; therefore, any other approval set in $\mathcal{T}_i$ must have its left endpoint in A_i. Since $r_1, \ldots, r_k$ form a piercing set, every approval set is in one of the $\mathcal{T}_i$. Thus, $A_1, \ldots, A_k$ collectively contains the left endpoints of all approval sets in the society.

Finally, it remains to show that $\{A_1, \ldots, A_k\}$ is the *unique* subcollection of k pairwise disjoint approval sets that contain the left endpoints of the remaining $n-k$ approval sets. First, A_1 must be the leftmost approval set in S, since no other approval set contains its left endpoint. Likewise, A_2 must be the leftmost approval set that is disjoint from A_1. Similarly, A_i must be the leftmost approval set that is disjoint from $A_1 \cup A_2 \cup \cdots \cup A_{i-1}$. This results in a unique collection of approval sets $A_1, \ldots, A_k$ that contain the left endpoints of all remaining approval sets. $\square$

We are now ready to prove Theorem 4.4.

PROOF OF THEOREM 4.4. Let $\mathcal{S} = (X, \mathcal{A})$ be a random fixed-length circular society with approval sets $A_1, \ldots, A_n$ of length $p < \frac{1}{2k}$. Denote $A_i = [\ell_i, r_i] = [\ell_i, \ell_i + p]$. We can assume that these approval sets are distinct, which occurs with probability 1 (Remark 4.5).

Let E denote the event that $\tau(\mathcal{S}) = k$. By Lemma 4.6, E is the event that there exists a unique collection of k pairwise disjoint approval sets whose union contains the left endpoints of the remaining $n-k$ sets. Our goal is to compute $P(E)$, by expressing E as a union of disjoint events.

Suppose T is a subset of $\{1, \ldots, n\}$ of cardinality k. Let D_T denote the event that the approval sets $\{A_i\}_{i \in T}$ are pairwise disjoint. Let C_T be the event that $\bigcup_{i \in T} A_i$ contains the left endpoints of the remaining $n-k$ approval sets. Then, $C_T \cap D_T$ denotes the event that $\{A_i\}_{i \in T}$ is pairwise disjoint and the union $\bigcup_{i \in T} A_i$

contains the left endpoints of the remaining sets. The uniqueness of the collection of k pairwise disjoint sets specified in Lemma 4.6 implies that the events $\{C_T \cap D_T\}_{T\subseteq\{1,\ldots,n\},\ |T|=k}$ are pairwise disjoint.

Thus, E is the union of the disjoint events $\{C_T \cap D_T\}_{T\subseteq\{1,\ldots,n\},\ |T|=k}$, which implies that

$$P(E) = P\left(\bigcup_{T\subseteq\{1,\ldots,n\},\ |T|=k} (C_T \cap D_T)\right) = \sum_{T\subseteq\{1,\ldots,n\},\ |T|=k} P(C_T \cap D_T).$$

It remains to compute $P(C_T \cap D_T) = P(D_T)P(C_T|D_T)$. Theorem 4.3 shows that the probability that k arcs of length $p < \frac{1}{k}$, randomly placed on the circle, are pairwise disjoint is $(1-kp)^{k-1}$. Thus, $P(D_T) = (1-kp)^{k-1}$. Next, we compute $P(C_T|D_T)$. Given that k approval sets $\{A_i\}_{i\in T}$ are disjoint, the length of $\bigcup_{i\in T} A_i$ is kp. Hence, the probability that the left endpoints of the remaining $n-k$ approval sets are in $\bigcup_{i\in T} A_i$ is $(kp)^{n-k}$. Therefore, $P(C_T \cap D_T) = P(D_T)P(C_T|D_T) = (1-kp)^{k-1}(kp)^{n-k}$.

Since the number of possible subsets $T \subseteq \{1,\ldots,n\}$ of cardinality k is $\binom{n}{k}$, the probability that $\tau(\mathcal{S}) = k$ is

$$P(E) = \sum_{T\subseteq\{1,\ldots,n\},|T|=k} P(C_T \cap D_T) = \binom{n}{k}(1-kp)^{k-1}(kp)^{n-k}.$$

□

4.2. Proofs of special cases. The proof of Lemma 4.6 relies on the fact that a fixed-length circular society $\mathcal{S}$ with $p < \frac{1}{2k}$ and $\tau(\mathcal{S}) = k$ is equivalent to a linear society. If this is not the case, then $\mathcal{S}$ might not have a unique collection of k disjoint approval sets that contain the left endpoints of all other approval sets. This complicates the task of finding the probability distribution of $\tau(\mathcal{S})$ when $p \geq \frac{1}{2k}$. However, when a society S has a small number of voters we can determine the probability of each possible piercing number regardless of approval set length.

For example, when a society has $n = 2$ voters, Theorem 3.1 and Theorem 4.4 allow us to completely describe the distribution of piercing numbers for all possible approval set lengths p. We state this in the following corollary.

COROLLARY 4.7. *For a random fixed-length circular society $\mathcal{S}$ with $n = 2$ voters and approval set length p,*

$$P(\tau(\mathcal{S}) = 1) = \begin{cases} 2(1-p)p & \text{if } p \in [0, 1/2), \\ 1 & \text{if } p \in [1/2, 1], \end{cases}$$

$$P(\tau(\mathcal{S}) = 2) = \begin{cases} 1-2p(1-p) & \text{if } p \in [0, 1/2), \\ 0 & \text{if } p \in [1/2, 1]. \end{cases}$$

PROOF. We start by computing the probability that $\tau(\mathcal{S}) = 1$. By Theorem 3.1, when $n = 2$ and $p \geq 1/2$, the piercing number must be 1. So, $P(\tau(\mathcal{S}) = 1) = 1$. Next consider the case when $p < 1/2$. We can apply Theorem 4.4 to compute the probability that the piercing number is $k = 1$:

$$P(\tau(\mathcal{S}) = 1) = \binom{2}{1}(1-p)p = 2p(1-p).$$

Finally, if the piercing number is not 1, it must be 2. So, $P(\tau(\mathcal{S}) = 2) = 1 - P(\tau(\mathcal{S}) = 1)$. □

As n increases, however, Theorem 3.1 and Theorem 4.4 only give us piercing number probabilities for either small or large approval set lengths. For example, restricting our attention to the probability that $\tau(\mathcal{S}) = 1$, Theorem 4.4 states that if $p < \frac{1}{2}$, then $P(\tau(\mathcal{S}) = 1) = np^{n-1}$. Theorem 3.1 states that $\tau(\mathcal{S}) = 1$ for any fixed-length circular society S with approval set length $p \geq \frac{n-1}{n}$. Thus, when a society consists of more than $n = 2$ voters, it remains to find $P(\tau(\mathcal{S}) = 1)$ when the approval set length is $\frac{1}{2} \leq p < \frac{n-1}{n}$.

Lemma 4.8 gives $P(\tau(\mathcal{S}) = 1)$ for societies with 3 voters and approval set length $\frac{1}{2} \leq p < \frac{2}{3}$. Figure 7 then shows this probability as a function of the approval set length p. This is the last piece of the puzzle needed for a complete picture of the piercing number probabilities for a society with 3 voters, which we state in Corollary 4.9.

LEMMA 4.8. *For a random fixed-length circular society S with $n = 3$ voters and approval set length $\frac{1}{2} \leq p < \frac{2}{3}$, $P(\tau(\mathcal{S}) = 1) = -9p^2 + 12p - 3$.*

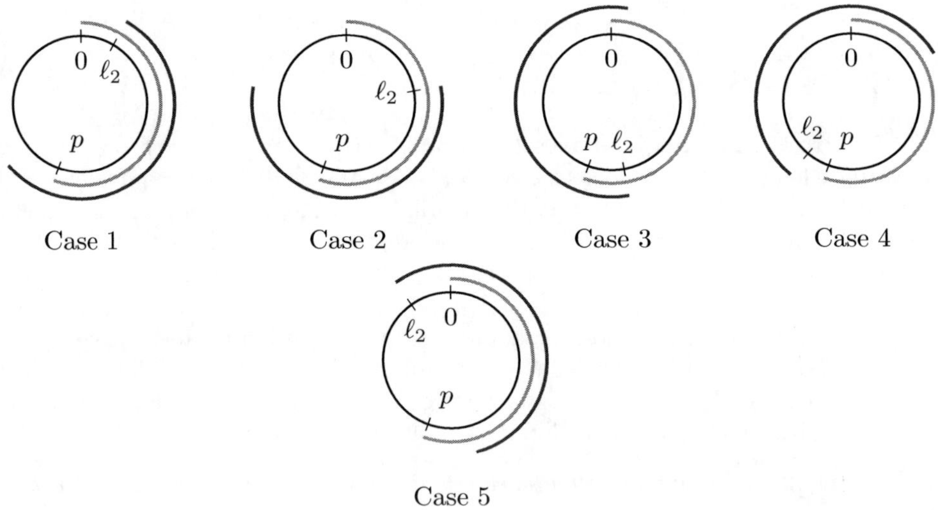

FIGURE 6. The five cases in the proof of Lemma 4.8. $A_1 = [0, p]$ is drawn in orange, and $A_2 = [\ell_2, \ell_2+p]$ in blue. A_3 is not shown, but A_3 *must* intersect $A_1 \cap A_2$ in Cases 1, 3, and 5; A_3 *might* intersect $A_1 \cap A_2$ in Cases 2 and 4.

PROOF. Without loss of generality, let $A_1 = [0, p]$. We consider five disjoint cases for the position of $A_2 = [\ell_2, \ell_2 + p]$, as illustrated in Figure 6.

CASE 1. Let E_1 be the event that $0 < \ell_2 < 2p - 1$. In this case, $A_1 \cap A_2$ is an interval of length greater than $1 - p$, so A_3 must intersect $A_1 \cap A_2$. Thus, $P((\tau = 1) \cap E_1) = P(E_1) = 2p - 1$.

CASE 2. Let E_2 be the event that $2p-1 < \ell_2 < 1-p$. In this case, $A_1 \cap A_2$ is an interval of length $p - \ell_2$, which is less than $1-p$. The probability that E_2 occurs and A_3 intersects $A_1 \cap A_2$ is

$$P((\tau = 1) \cap E_2) = \int_{2p-1}^{1-p} (2p - \ell_2)\, d\ell_2 = \frac{3}{2}\left(2p - 3p^2\right).$$

CASE 3. Let E_3 be the event that $1-p < \ell_2 < p$. In this case, $A_1 \cap A_2$ is a union of two disconnected intervals: $A_1 \cap A_2 = [0, \ell_2 + p - 1] \cup [\ell_2, p]$. Since one of these intervals contains 0 and the other contains $\frac{1}{2}$, A_3 *must* intersect $A_1 \cap A_2$. Thus, $P((\tau = 1) \cap E_3) = P(E_3) = 2p - 1$.

CASE 4. Let E_4 be the event that $p < \ell_2 < 2 - 2p$. In this case, $A_1 \cap A_2$ is an interval of length $\ell_2 + p - 1$, which is less than $1-p$. The probability that E_4 occurs and A_3 intersects $A_1 \cap A_2$ is

$$P((\tau = 1) \cap E_4) = \int_{p}^{2-2p} (2p - 1 + \ell_2)\, d\ell_2 = \frac{3}{2}\left(2p - 3p^2\right).$$

CASE 5. Let E_5 be the event that $2 - 2p < x < 1$. In this case, $A_1 \cap A_2$ is an interval of length greater than $1-p$. As in Case 1, $(A_1 \cap A_2) \cap A_3 \neq \emptyset$, so $P((\tau = 1) \cap E_5) = P(E_5) = 2p - 1$.

By the law of total probability,

$$P(\tau = 1) = \sum_{i=1}^{5} P((\tau = 1) \cap E_i) = 3(2p-1) + 2 \cdot \frac{3}{2}\left(2p - 3p^2\right) = -9p^2 + 12p - 3.$$

$\square$

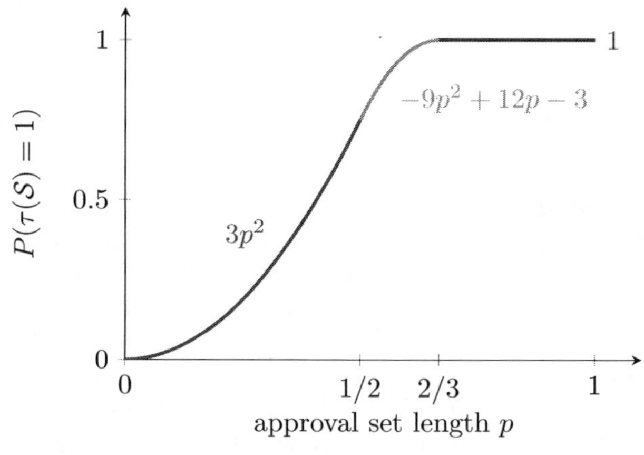

FIGURE 7. For fixed-length circular societies with $n = 3$ voters, this plot shows $P(\tau(\mathcal{S}) = 1)$ as a function of the approval set length p. The purple curve for $0 < p < \frac{1}{2}$ is given by Theorem 4.4, the orange curve for $\frac{1}{2} \leq p < \frac{2}{3}$ is given by Lemma 4.8, and the blue line for $\frac{2}{3} \leq p < 1$ by Theorem 3.1.

COROLLARY 4.9. *For a random fixed-length circular society $\mathcal{S}$ with $n = 3$,*

$$P(\tau(\mathcal{S}) = 1) = \begin{cases} 3p^2 & \text{if } p \in [0, 1/2), \\ -9p^2 + 12p - 3 & \text{if } p \in [1/2, 2/3), \\ 1 & \text{if } p \in [2/3, 1], \end{cases}$$

$$P(\tau(\mathcal{S}) = 2) = \begin{cases} 6p - 12p^2 & \text{if } p \in [0, 1/3), \\ 1 - 3p^2 & \text{if } p \in [1/3, 1/2), \\ 9p^2 - 12p + 4 & \text{if } p \in [1/2, 2/3), \\ 0 & \text{if } p \in [2/3, 1], \end{cases}$$

$$P(\tau(\mathcal{S}) = 3) = \begin{cases} (1 - 3p)^2 & \text{if } p \in [0, 1/3), \\ 0 & \text{if } p \in [1/3, 1]. \end{cases}$$

PROOF. We start by computing the probability that $\tau(\mathcal{S}) = 1$. Theorem 3.1 tells us that if $p \geq 2/3$, then $P(\tau(\mathcal{S}) = 1) = 1$. Theorem 4.4 tells us that if $p < 1/2$, then $P(\tau(\mathcal{S}) = 1) = 3p^2$. Therefore, along with Lemma 4.8,

$$P(\tau(\mathcal{S}) = 1) = \begin{cases} 3p^2 & \text{if } p \in [0, 1/2), \\ -9p^2 + 12p - 3 & \text{if } p \in [1/2, 2/3), \\ 1 & \text{if } p \in [2/3, 1]. \end{cases}$$

Next, we compute the probability that $\tau(\mathcal{S}) = 3$. Theorem 3.1 tells us that if $p \geq 1/3$, then $\tau(\mathcal{S}) \leq 2$; this means that $P(\tau(\mathcal{S}) = 3) = 0$ when $p \geq 1/3$. If $p < 1/3$, then Theorem 4.3 tells us that the probability that three arcs of length p are pairwise disjoint (that is, that the piercing number is 3) is $(1 - 3p)^2$. So,

$$P(\tau(\mathcal{S}) = 3) = \begin{cases} (1 - 3p)^2 & \text{if } p \in [0, 1/3), \\ 0 & \text{if } p \in [1/3, 1]. \end{cases}$$

Finally, $P(\tau(\mathcal{S}) = 2) = 1-(P(\tau(\mathcal{S}) = 1) + P(\tau(\mathcal{S}) = 3))$, and the result follows. □

Generalizing Lemma 4.8 to larger values of n is tedious due to the large number of cases, as the approval sets can intersect in many different ways. We leave this as a direction for future study.

4.3. Simulating societies. We can only compute the exact piercing number probability for fixed-length circular societies with small approval set length p or small number of voters n, however, we can use simulation to explore the probability distribution of piercing numbers of random fixed-length circular societies with larger p and/or larger n. This provides an empirical answer to Question 4.2. Given a fixed-length circular society $\mathcal{S}$ with n voters and approval set length p, we can estimate the probability that the piercing number is k by randomly generating N such societies and computing the piercing number of each. The proportion of the N circular societies whose piercing number is equal to k is our estimate of $P(\tau(\mathcal{S}) = k)$.

Theorem 4.4 states that the probability $P(\tau(\mathcal{S}) = k)$ for a random fixed-length circular society with approval set length $p < \frac{1}{2k}$ is given by Equation (4.1). Our simulations suggest that Equation (4.1) may give the correct probability even when p is somewhat larger than $\frac{1}{2k}$, at least for certain values of k. Figure 8 compares the simulated probability with the value given by Equation (4.1) for random fixed-length circular societies with $n = 4$ voters, each $k \in \{1, 2, 3, 4\}$, and a range of approval set lengths $p \in (0, 0.8)$. We see that for each $\tau = k$ the estimated probability agrees with Equation (4.1) for $p < \frac{1}{2k}$, as it should. However, for $\tau > 2$, these values agree for p beyond the bound stated in Theorem 4.4. For $\tau = 3$, the values seem to match well up to $p = \frac{1}{3}$, and for $\tau = 4$, they seem to match well up to $p = \frac{1}{4}$. This leads us to the following conjecture, which is further supported by additional simulations for larger n and k.

CONJECTURE 4.10. For a random fixed-length circular society S,

$$P(\tau(\mathcal{S}) = k) = \binom{n}{k}(1 - kp)^{k-1}(kp)^{n-k}$$

for p in an interval $(0, \frac{1}{2k} + \delta)$ for some $\delta > 0$.

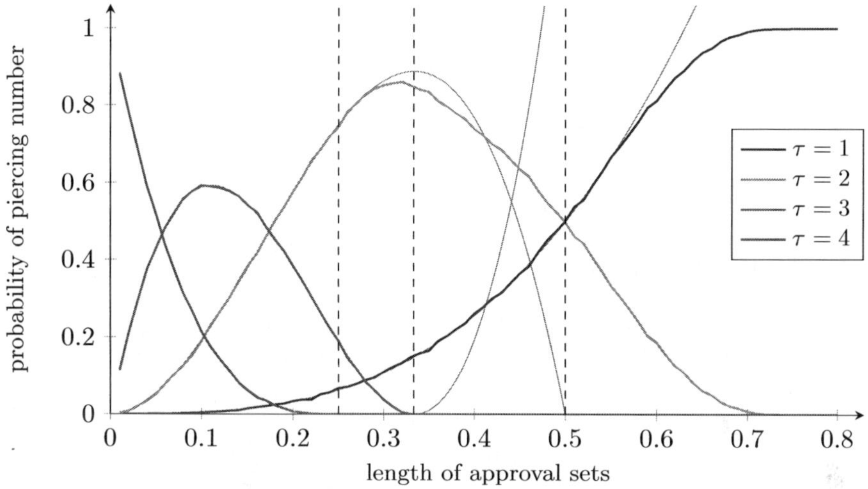

FIGURE 8. Comparison of simulated probabilities $P(\tau(\mathcal{S}) = k)$ with those given by Theorem 4.4 and Conjecture 4.10 for each $k = 1, \ldots, 4$ in fixed-length circular societies with $n = 4$ voters and approval set lengths $0 \leq p \leq 0.8$. Simulated probabilities are plotted in color; polynomial curves given by Equation (4.1) are plotted in gray. The simulated probabilities suggest that Equation (4.1) gives the correct probability for p somewhat larger than $\frac{1}{2k}$. The dashed vertical lines highlight $p = \frac{1}{4}$, $p = \frac{1}{3}$, and $p = \frac{1}{2}$. Note that when $\tau = 3$ the formula seems to hold for $p \leq \frac{1}{3}$, and when $\tau = 4$ it seems to hold for $p \leq \frac{1}{4}$.

Furthermore, Table 1 displays estimated probabilities $P(\tau(\mathcal{S}) = k)$ for certain values of p greater than $\frac{1}{2k}$. For each combination of n, k, and p listed in the table, we randomly generated $N = 100{,}000$ fixed-length circular societies and computed the piercing number for each. The table shows that the simulated probabilities $P(\tau(\mathcal{S}) = k)$ are close to the values given by Equation (4.1) and Conjecture 4.10. Hence, there is evidence that the upper bound on p should be greater than $\frac{1}{2k}$. However, we do not yet know the exact upper bound on p at which Conjecture 4.10 fails to hold. We state this as a open question.

TABLE 1. Estimated piercing number probabilities from simulation compared with probabilities given by Conjecture 4.10. Results are rounded to four decimal places; simulated results of zero are not shown.

n	p	k	estimated $P(\tau(\mathcal{S}) = k)$	probability given by Conjecture 4.10
5	0.15	4	0.1903	0.1920
5	0.15	5	0.0038	0.0039
8	0.12	5	0.3102	0.3040
8	0.12	6	0.0254	0.0250
8	0.12	7	0.0001	0.0001
10	0.10	5	0.4896	0.4922
10	0.10	6	0.2808	0.2787
10	0.10	7	0.0309	0.0300
10	0.10	8	0.0004	0.0004

QUESTION 4.11. What is the largest value of δ for which Conjecture 4.10 gives the probability that $\tau(\mathcal{S}) = k$? What is the relationship between δ, the number of voters n, and the piercing number k?

Our probabilistic approach also helps us understand the "average-case scenario" for piercing numbers of fixed-length circular societies. If $\mathcal{S}$ is a fixed-length circular society with n voters and approval set length $p < \frac{1}{2n}$, then Theorem 4.4 holds for all $k \in \{1, \ldots, n\}$, and the expected value of $\tau(\mathcal{S})$ is:

$$E[\tau(\mathcal{S})] = \sum_{k=1}^{n} k \binom{n}{k} (1 - kp)^{k-1} (kp)^{n-k}. \tag{4.2}$$

However, given Conjecture 4.10, the formula above may hold for p somewhat larger than $\frac{1}{2n}$. We simulated the average piercing number in random fixed-length circular societies with $p \geq \frac{1}{2n}$ and provide selected results in Table 2. We see that that the estimated average piercing number is close to the value given by Equation (4.2).

Returning to the scenario given in the introduction, suppose that a 25-member committee requires all members to attend a training session. They ask each member to select a three-hour interval during which they are available. (Given a 24-hour clock, this yields $p = \frac{1}{8}$). Theorem 3.1 tells us that as a worst case, the committee would need to schedule eight training sessions at different times to ensure that all members can attend one of the sessions. Assuming availability intervals are uniformly distributed around the circle, the average piercing number is about 6, so in an average case, the committee would only need to schedule six sessions in order to satisfy each member's preference. One the other hand, the uniform distribution is itself a sort of worst-case, as it spreads intervals uniformly around the clock. Realistically, committee members' availabilities are likely to be clustered around certain times of the day; hence, a uniform distribution may not be the best model for this situation. Consideration of other distributions is an interesting area for future work.

TABLE 2. For each combination of values of n and p, we simulated the average piercing number by randomly generating $N = 10{,}000$ fixed-length circular societies. We compare the simulation averages with the conjectured expected values, computed from Equation (4.2).

n	p	estimated average piercing number	expected value given by Equation (4.2)
5	0.100	3.494	3.492
5	0.200	2.633	2.632
5	0.250	2.340	2.324
25	0.050	11.207	11.222
25	0.100	7.198	7.201
25	0.125	6.099	5.986
45	0.020	23.823	23.816
45	0.050	13.916	13.912
45	0.080	9.822	9.816

We invite readers to explore our simulations of piercing number probabilities and average values using our circular society visualization and simulation tool, which is available at `https://github.com/tiasondjaja/circular_societies`.

References

[1] Deborah E. Berg, Serguei Norine, Francis Edward Su, Robin Thomas, and Paul Wollan, *Voting in agreeable societies*, Amer. Math. Monthly **117** (2010), no. 1, 27–39, DOI 10.4169/000298910X474961. MR2599465

[2] Steven Brams and Peter Fishburn, *Going from Theory to Practice: The Mixed Success of Approval Voting*, Social Choice and Welfare, vol. 25, no. 2 (2005), 457–474, DOI 10.1007/s00355-005-0013-y.

[3] Timothy M. Chan, *Polynomial-time approximation schemes for packing and piercing fat objects*, J. Algorithms **46** (2003), no. 2, 178–189, DOI 10.1016/S0196-6774(02)00294-8. MR1969718

[4] H. Hadwiger and H. Debrunner, *Über eine Variante zum Hellyschen Satz* (German), Arch. Math. (Basel) **8** (1957), 309–313, DOI 10.1007/BF01898794. MR92987

[5] Christopher S. Hardin, *Agreement in circular societies*, Amer. Math. Monthly **117** (2010), no. 1, 40–49, DOI 10.4169/000298910X474970. MR2599466

[6] Matthew J. Katz, Frank Nielsen, and Michael Segal, *Maintenance of a piercing set for intervals with applications*, Algorithmica **36** (2003), no. 1, 59–73, DOI 10.1007/s00453-002-1006-1. MR1957862

[7] Abdullah-Al Mahmood, *Approximation algorithms for rectangle piercing problems*, `https://uwspace.uwaterloo.ca/bitstream/handle/10012/1025/amahmood2005.pdf`.

[8] Kristen Mazur, Mutiara Sondjaja, Matthew Wright, and Carolyn Yarnall, *Approval voting in product societies*, Amer. Math. Monthly **125** (2018), no. 1, 29–43, DOI 10.1080/00029890.2018.1390370. MR3756327

[9] Herbert Solomon, *Geometric probability*, Conference Board of the Mathematical Sciences Regional Conference Series in Applied Mathematics, No. 28, Society for Industrial and Applied Mathematics, Philadelphia, Pa., 1978. Ten lectures given at the University of Nevada, Las Vegas, Nev., June 9–13, 1975. MR0488215

[10] Francis Edward Su and Shira Zerbib, *Piercing numbers in approval voting*, Math. Social Sci. **101** (2019), 65–71, DOI 10.1016/j.mathsocsci.2019.06.007. MR3982633

Department of Mathematics and Statistics, Elon University, Elon, North Carolina 27244
Email address: kmazur@elon.edu

OpenX Software, New York City, New York 10012
Email address: msondjaja@gmail.com

Department of Mathematics, Statistics, and Computer Science, St. Olaf College, Northfield, Minnesota 55057
Email address: wright5@stolaf.edu

Department of Mathematics, California State University, Dominguez Hills, Carson, California 90747
Email address: cyarnall@csudh.edu

Contemporary Mathematics
Volume **795**, 2024
https://doi.org/10.1090/conm/795/15967

Voting on cyclic orders, group theory, and ballots

Karl-Dieter Crisman, Abraham Holleran, Micah Martin, and Josephine Noonan

Abstract. A *cyclic order* may be thought of informally as a way to seat people around a table, perhaps for a game of chance or for dinner. Given a set of agents such as $\{A, B, C\}$, we can formalize this by defining a cyclic order as a permutation or linear order on this finite set, under the equivalence relation where $A \succ B \succ C$ is identified with both $B \succ C \succ A$ and $C \succ A \succ B$. As with other collections of sets with some structure, we might want to aggregate preferences of a (possibly different) set of voters on the set of possible ways to choose a cyclic order.

However, given the combinatorial explosion of the number of full rankings of cyclic orders, one may not wish to use the usual voting machinery. This raises the question of what sort of ballots may be appropriate; a single cyclic order, a set of them, or some other ballot type? Further, there is a natural action of the group of permutations on the set of agents. A reasonable requirement for a choice procedure would be to respect this symmetry (the equivalent of neutrality in normal voting theory).

In this paper we will exploit the representation theory of the symmetric group to analyze several natural types of ballots for voting on cyclic orders, and points-based procedures using such ballots. We provide a full characterization of such procedures for two quite different ballot types for $n = 4$, along with the most important observations for $n = 5$.

1. Introduction

1.1. New outputs and new ballots. Two dominant paradigms for voting theory (going back before [**3**] to names like Borda, Condorcet, Cusanus, and Llull) have long been the search for either a single winner or a full ranking of candidates. Although these are evidently very useful models, even the fact that one needs definitions of terms like 'resolute' and 'social choice function' (both of which have to do with the inevitable occurrence of multiple winning outcomes in a choice procedure) indicates that other paradigms might be useful.

One very relevant area that has attracted a fair amount of recent attention is voting for committees – to wit, fixed-size multiwinner elections. Many people live in polities which allow for outcomes (and/or ballots) which are fixed-size subsets of the set of candidates, from school boards to national representatives. Yet, as the

---

2020 *Mathematics Subject Classification.* Primary 91B12; Secondary 91B14,20C05.

Key words and phrases. Voting theory, cyclic orders, representation theory, social choice.

This work was partly supported by a Gordon College Provost's Summer Undergraduate Research Fellowship.

recent survey [**16**] of the issue in computational social choice declares, despite some initial work, this paradigm is still a relatively 'new challenge' for the community.

Similarly, there has been an increasing amount of discussion about different types of ballots. The most well-known alternate ballot is the simplest, coming from approval voting (see works from [**6**] to [**22**]). An approval ballot is one where an agent simply selects a subset of all possible candidates as 'approved'. This has been recognized as a ballot and not just a voting system, as indicated by its use in selecting committees in [**7**]. Other variant (but intuitive) ballot types include having three options ([**2**]) as well as ballots from systems such as range voting or cumulative voting.

Returning to committees, in [**30**] the case is largely made *against* submitting a full ranking of all outcomes, whereupon [**31**] instead restricts the ballot to picking one of a restricted subset of allowable committees. In a somewhat different context, [**13**] asks for ballots which are *subsets* of a committee. Which ballot is best? Suffice to say[1] that the door is just opening when it comes to what to do in voting *and* balloting on less typical combinatorial objects appearing in voting theory.

1.2. Some new scenarios. This paper addresses both questions – that of a new output type and that of which ballots to use – for *cyclic orders.* Consider the following possible scenarios.

SCENARIO 1.1. Your large family is gathering for some important holiday (such as Thanksgiving, in the United States). Everyone needs to be seated around a gigantic round table, but there are many things to keep in mind! Here is a list of just some of *your* priorities.

- Your leftist uncle must be kept far away from your rightist aunt.
- Your uncle has two young children which should be seated very close to him – ideally, adjacent.
- Your grandmother is deaf in her right ear, so if she is seated next to an adult she likes to chat with, she must have that person to her *left.*

And everyone else has their own priorities as well. How do you come up with a seating everyone can live with, but without too much fuss?

SCENARIO 1.2. A large faculty needs to vote on a rotation for topical course offerings that are given on a two- or three-year rotation. Here are some sample (partial) preferences of faculty in the department.

- The p-adic Hodge theory course should be offered immediately after the basis p-adic analysis course, since they can build on each other.
- Those courses should be offered far apart, so that there is always a p-adic course available.
- The Computational Social Choice course should only be offered the semester immediately *before* the Cooperative Game Theory course, so that most students in Cooperative Game Theory will be able to do advanced projects on NP-completeness.

[1]More exotic ballot types include truncated rankings (such as in voting for several college football rankings in the United States) and even more interesting options, such as a partially ordered set of the candidates in an election ([**1, 11, 15, 28**]). In the more general aggregation problems that come up in data analysis (in everything from genetics to search engines), one could even have just binary relations as both ballot and output, as in the classic paper [**5**].

- The only person qualified to teach Computational Social Choice is also the same one as must teach Cooperative Game Theory, and she does not want to prepare both of them in the same academic or calendar year.

Even if your whole department votes on a rotation, what method should you choose to do so?

SCENARIO 1.3. Poker is a gambling game of both luck and skill. A long poker game can consist of many rounds, where the starting player shifts around a table after each round, so that in the long run everyone bids first about the same number of times. In addition, it is well-known[2] that it is advantageous to be in bidding order *after* a strong player, and to be placed immediately *before* a weak player.

Suppose that for a particular live-streaming poker championship, the audience of thousands is allowed to place bets[3] as well on who will win – and is allowed to *vote* on what order the players sit in! Clearly an advantageous seating arrangement for your chosen winner will help your odds of winning. How should the organizers of this tournament create ballots, and how should they tally them?

It should be clear that such scenarios could be multiplied indefinitely. In each of them, the common factor is that there is some set of agents with preferences on some non-linear, 'cyclic' domain. However, the most important aspects to agents seem to differ – is it proximity, order, or some combination? As above, submitting an entire ranking of all possible such orderings could be unwieldy, while simply using plurality on the set of all the orderings is too likely to give a massive tie.

1.3. Cyclic orders. The intent of this paper is to examine voting on these cyclic orders, as well as to begin examining some alternate ballots of interest. First, let us give a proper definition.

DEFINITION 1.4. Let $\mathcal{A} = \{A_1, A_2, \ldots, A_n\}$ be a finite set and S_n be the symmetric group. Consider the set of permutations (or complete linear orders) $\mathcal{L}(\mathcal{A})$ of $\mathcal{A}$, and let $\sigma = (1234\cdots n) \in S_n$ be a natural n-cycle. Then construct the equivalence relation where

$$A_{i_1} \succ A_{i_2} \succ \cdots \succ A_{i_n} \sim A_{i_{\sigma(1)}} \succ A_{i_{\sigma(2)}} \succ \cdots \succ A_{i_{\sigma(n)}}$$

for any element of $\mathcal{L}(A)$ (and hence also $A_{i_{\sigma^k(1)}} \succ A_{i_{\sigma^k(2)}} \succ \cdots \succ A_{i_{\sigma^k(n)}}$ for any k). A (total) *cyclic order* (sometimes *cyclic ordering*) on $\mathcal{A}$ is an equivalence class of $\mathcal{L}(A)$ under this equivalence relation.

We will occasionally use the notation CO_n for this set for a given n, regardless of $\mathcal{A}$. There are necessarily $n!/n = (n-1)!$ cyclic orders for a set of cardinality n.

NOTATION 1.5. Most of the time, it will be easier to think of a cyclic order as a single permutation in cycle notation, beginning with a common element. If $\mathcal{A} = \{1, 2, 3, 4, 5\}$, then we might write (13524) to represent the cyclic order where $1 \succ 3 \succ 5 \succ 2 \succ 4 \succ 1$, where it is implied that (35241), (52413), and other analogous cyclic reorderings are equivalent. Sometimes, to emphasize the cyclic nature, we may instead write (135241). However, we will usually use a set of the

[2]See for example, `https://www.pokernews.com/strategy/10-hold-em-tips-08-25417.html`.

[3]Although we do not endorse gambling as such in this paper, such meta-gambling certainly occurs. One can place bets on the winner of the World Series of Poker final round courtesy of the owner of the tournament, Caesar's.

form $\mathcal{A} = \{A, B, C, D, \ldots\}$ because this is the traditional one to use in a voting context, though to be clear our methods will not be voting on them as candidates!

REMARK 1.6. The above action is a right action of S_n. But there is also an important natural *left* action on the set of cyclic orders by the symmetric group S_n. Given $\sigma \in S_n$, the action takes

$$A_{i_1} \succ A_{i_2} \succ \cdots \succ A_{i_n} \succ A_{i_1} \xrightarrow{\sigma} A_{\sigma(i_1)} \succ A_{\sigma(i_2)} \succ \cdots \succ A_{\sigma(i_n)} \succ A_{\sigma(i_1)}$$

which one may think of as permuting the *elements* of $\mathcal{A}$, rather than their *placement* as in Definition 1.4. For example, $(13)(A_1A_2A_3A_4) = (A_3A_2A_1A_4) \sim (A_1A_4A_3A_2)$. Again, for convenience we may write this action as $(AB)(ABCD) = (ACDB)$, with $(AB) \in S_n$ but $(ABCD)$, $(ACDB) \in CO_4$, when there should be no danger of confusion.

REMARK 1.7. One can also define a cyclic order in terms of *ternary* relations with certain properties. While cyclic orders have been studied in terms of set properties and certain graphs ([**38**] is a notable paper along these lines), they are almost always treated in the literature either as relating to *partial* cyclic orders, or as a means to some other end. See Section 6 for some additional references.

We will consider several types of ballots related to cyclic orders. In addition to ‘plurality’ type ballots which ask for just the voter’s favorite cyclic order (which will be addressed in Sections 2 and 5), we will also consider ballots which request certain additional pieces of information from voters. For instance, in Scenario 1.1 we could ask for ballots which allow a voter to select someone to placed immediately to their right as well as a person as far away from them as possible, or in Scenario 1.3 one could submit a ballot placing optimal players to either side of the player you are actually betting on. We will formalize notions like this as TRAD and ROLO ballots in Section 4.

The remainder of the paper will unfold as follows:

- We will start in Section 2 by investigating procedures for $n = 4$ spots in the cyclic order, given that the ballot is a single order.
- Then we will recall some representation theory in Section 3, which will allow us to be far more systematic in our approach in the remainder of the paper.
- Next, in Section 4 we will introduce and analyze various methods for $n = 4$ which allow for more flexibility in balloting, but do not require a full ranking of all 6 possible cyclic orders.
- We will then use the same techniques in Section 5 to examine the case $n = 5$, once again given that the ballot is a single cyclic order.

2. Cyclic orders for $n = 4$

It can easily be checked that there are only two possible cyclic orders with $n = 3$, so we begin by examining the situation where $n = 4$. In this case there are exactly six possible cyclic orders. In Figure 1 we will set a particular order of these outcomes for the purposes of the paper, and include two types of representations which are convenient in different situations.

$(ACBD)$	$(ADBC)$	$(ABCD)$	$(ADCB)$	$(ABDC)$	$(ACDB)$
AC DB	AD CB	AB DC	AD BC	AB CD	AC BD

FIGURE 1. Cyclic orders for $n = 4$

Note that the latter representations are meant to be read clockwise. There are three pairs of orders, each of which is evidently the reversal of the other.

For convenience, in this section we may sometimes think of cyclic orders as seats at a table, as in Scenario 1.1. Think of some set of agents as voting on cyclic orders or seating arrangements, where in this section we allow each agent to select just one cyclic order (a 'plurality' ballot, in some sense). Then we formally extend the notion of a (necessarily anonymous) voting profile to this context.

DEFINITION 2.1. Given n and an ordering of the $(n-1)!$ cyclic orders on a set of cardinality n, a *profile* is a vector $\mathbf{p} \in \mathbb{Q}^{(n-1)!}$. We write $\mathbf{p}[(ABCD)]$ to indicate the value of the profile for a particular cyclic order. (If there are negative entries, especially if the sum of entries is zero, sometimes we call it a *profile differential*.)

REMARK 2.2. The S_n-action of Remark 1.6 extends to profiles, via $\sigma(\mathbf{p})[x] = \mathbf{p}[\sigma^{-1}x]$. Writing elements of S_4 for convenience as (AB), we see that if $\mathbf{p} = (2,1,1,0,0,0)^T$ then $(AB)\mathbf{p} = (1,2,0,0,0,1)^T$.

Although there might be many ways to vote on cyclic orders, in this paper we confine ourselves to neutral points-based procedures[4], where neutrality should have the same connotation as it typically does in voting theory.

DEFINITION 2.3. Suppose we have a *ballot-scoring function*[5] $s : CO_n \times CO_n \to \mathbb{Q}$. Then a *points-based voting rule* f, with ballot and outcome from the set of cyclic orders (for a given n), is a function from the set of all profiles on CO_n to a nonempty subset of CO_n, such that $f(\mathbf{p})$ is the (set of) cyclic order(s) h which maximizes $\sum_{g \in CO_n} \mathbf{p}[g]s(g,h)$. The rule f is *neutral* if s is neutral, in the sense that for any $\sigma \in S_n$, $s(g,h) = s(\sigma(g), \sigma(h))$.

Informally, the function s represents how many points are allocated to the cyclic order h for a single vote for the cyclic order g. It is easier to envision this as applying a matrix M (indexed by the elements of CO_n) with entries $M_{h,g} = s(g,h)$ to a profile $\mathbf{p}$, and then taking the argmax of the resulting vector, and we will do so in the sequel. When, as is usual, there is no danger of ambiguity, we will very often refer to the outcome vector as the result of the voting rule, rather than the argmax.

EXAMPLE 2.4. As an example, suppose we let s be defined so that s gives two points to a ranking for its appearance in a profile, but also awards one point to the

[4]See [**8**] and [**40**] for inspiration and more general related notions.

[5]Indeed, everything in this paper also works over the real numbers, but we restrict ourselves to the rationals for simplicity in certain situations.

reversed cyclic order. In this case,

$$M = \begin{pmatrix} 2 & 1 & 0 & 0 & 0 & 0 \\ 1 & 2 & 0 & 0 & 0 & 0 \\ 0 & 0 & 2 & 1 & 0 & 0 \\ 0 & 0 & 1 & 2 & 0 & 0 \\ 0 & 0 & 0 & 0 & 2 & 1 \\ 0 & 0 & 0 & 0 & 1 & 2 \end{pmatrix}.$$

With $\mathbf{p} = (2,1,0,0,0,1)^T$ we get $M\mathbf{p} = (5,4,0,0,1,2)^T$, so $(ACBD)$ is the most popular ballot in $\mathbf{p}$ as well as the winner under this rule.

Note that this rule is also neutral; for example, the second column can be obtained from the first by applying (AB) (or (CD)), and the third from the first by applying (BC).

EXAMPLE 2.5. The previous example's scoring function s implied that a voter would consider a cyclic order to be related to its reversal, but not to others. What if instead we wanted a model which implied the *worst* thing that could happen for a voter for $(ACBD)$ was for the seating arrangement to be $(ADBC)$? Then we might enjoy this matrix instead.

$$M = \begin{pmatrix} 2 & 0 & 1 & 1 & 1 & 1 \\ 0 & 2 & 1 & 1 & 1 & 1 \\ 1 & 1 & 2 & 0 & 1 & 1 \\ 1 & 1 & 0 & 2 & 1 & 1 \\ 1 & 1 & 1 & 1 & 2 & 0 \\ 1 & 1 & 1 & 1 & 0 & 2 \end{pmatrix}$$

With $\mathbf{p} = (2,1,0,0,0,1)^T$ we get $M\mathbf{p} = (5,3,4,4,3,5)^T$, with the perhaps surprising result that $(ACBD)$ and $(ACDB)$ share the victory. Then again, three-quarters of the electorate wished for A to sit to the right of C, so perhaps this is appropriate.

The matrices in question already seem to have a very high degree of symmetry. This is not a coincidence.

PROPOSITION 2.6. *Given the order of elements of CO_4 in Figure 1, every neutral points-based voting rule for $n = 4$ comes from a matrix as follows, with $a, b, c \in \mathbb{Q}$.*

$$\begin{pmatrix} a & b & c & c & c & c \\ b & a & c & c & c & c \\ c & c & a & b & c & c \\ c & c & b & a & c & c \\ c & c & c & c & a & b \\ c & c & c & c & b & a \end{pmatrix}$$

The proof follows by elementary methods; see the end of the section. A corollary follows by simple computation.

COROLLARY 2.7. *With everything as in the previous proposition, every neutral points-based voting rule for $n = 4$ can be described in terms of the linear algebra of M as follows.*

- *Every profile vector in the subspace spanned by $(1,1,1,1,1,1)^T$ must go to $a + b + 4c$ times itself.*

- *Every profile vector in the span of the vectors* $(2,2,-1,-1,-1,-1)^T$, $(-1,-1,2,2,-1,-1)^T$, *and* $(-1,-1,-1,-1,2,2)^T$ *(which is two dimensional) must go to* $a+b-2c$ *times itself.*
- *Every profile vector in the span of the set of vectors* $\{(1,-1,0,0,0,0)^T$, $(0,0,1,-1,0,0)^T$, $(0,0,0,0,1,-1)^T\}$ *must go to* $a-b$ *times itself.*

Finally, since these subspaces span $\mathbb{Q}^6$, *any linear combination of the vectors above goes to the same linear combination of their images.*

Each of these subspaces also has an interpretation in terms of cyclic orders[6].

- The first component is a trivial profile where all orders are equally likely.
- The third space is comprised of linear combinations of profiles that strongly support one cyclic order and do not support its reversal.
- The second component may seem more mysterious, but is not if we recall Figure 1. Each of the three vectors may be thought of as supporting one of the three options of who (of B, C, D) is not adjacent to A in a given cyclic order. Note also that these three vectors sum to zero, just like the Basic/Borda vectors in Saari's work.

EXAMPLE 2.8. Let's examine Example 2.4 from this standpoint. We can write $\mathbf{p} = (2,1,0,0,0,1)^T$ as

$$\frac{2}{3}(1,1,1,1,1,1)^T + \frac{1}{3}(2,2,-1,-1,-1,-1)^T + \frac{-1}{6}(-1,-1,2,2,-1,-1)^T + \frac{1}{2}(1,-1,0,0,0,0)^T + \frac{-1}{2}(0,0,0,0,1,-1)^T .$$

Since $a+b+4c=3$, $a+b-2c=3$, and $a-b=1$, we compute $f(\mathbf{p})$ from

$$2(1,1,1,1,1,1)^T + (2,2,-1,-1,-1,-1)^T + \frac{-1}{2}(-1,-1,2,2,-1,-1)^T + \frac{1}{2}(1,-1,0,0,0,0)^T + \frac{-1}{2}(0,0,0,0,1,-1)^T,$$

which is indeed $(5,4,0,0,1,2)^T$.

What is remarkable about this decomposition is that any neutral points-based rule can be fully described by three numbers, each of which simply dilate a given subspace of the profile space, and each of which (subspaces) has a natural interpretation. One might say that the rule is really picking which spaces to emphasize, rather than which cyclic orders.

As a result, we can analyze any method. The one in Examples 2.4 and 2.8 preserves non-adjacency more than it emphasizes the difference between a cyclic order and its reversal. This isn't immediately intuitive, since the columns of the matrix are permutations of $(2,1,0,0,0,0)^T$, which would seem to definitely emphasize that reversals are not the same. However, we can actually break this vector into weighted pieces as well[7], so it is truly a combination of all emphases:

$$(2,1,0,0,0,0)^T = \frac{1}{3}(1,1,1,1,1,1)^T + \frac{1}{2}(2,2,-1,-1,-1,-1)^T + \frac{1}{2}(1,-1,0,0,0,0)^T .$$

[6]All this should be strongly reminiscent of decompositions in [**32, 33, 39**], not to mention the treatments mentioned in the introduction to Section 3.

[7]Again, recall papers like [**33**] and [**12**], as we will see below.

We can also create rules that have desired properties, just as with analyzing other voting procedures. If we want a method that ignores full ties, we should ensure that $a+b+4c=0$. Here is a more interesting one.

EXAMPLE 2.9. Suppose we want a vote using a points-based rule that only takes into account the part of a profile that contrasts a cyclic order with its reversal. One can think of this as requiring that the first and second subspaces in question span the kernel of the matrix.

So if we let $a+b+4c=0$ and $a+b-2c=0$, then by solving we see that $c=0$ and $a=-b$. A sample matrix might be:

$$\begin{pmatrix} 1 & -1 & 0 & 0 & 0 & 0 \\ -1 & 1 & 0 & 0 & 0 & 0 \\ 0 & 0 & 1 & -1 & 0 & 0 \\ 0 & 0 & -1 & 1 & 0 & 0 \\ 0 & 0 & 0 & 0 & 1 & -1 \\ 0 & 0 & 0 & 0 & -1 & 1 \end{pmatrix}.$$

An observant reader will note this is just the matrix for 2.5 but with every entry lowered by one.

It is evident from Corollary 2.7 that there are only a few possible kernels and images (the spans of the subspaces in question). This is a sign of the underlying symmetry.

PROOF OF PROPOSITION 2.6. Recall that our matrix has $M_{h,g}=s(g,h)$ for a scoring function s, which by hypothesis is neutral. So our goal would be to show:

- That $s(g,g)$ is the same for all g.
- That $s(g,g')$ is the same for all reversal pairs g,g'.
- All other values of s are the same as each other.

The first requirement is easy to see, because neutrality says $s(g,g)=s(\sigma(g),\sigma(g))$ and we certainly have σ which change any cyclic order to another one. Similarly, $s(g,h)=s(h,g)$, as it is easy to see by inspection that for any pair g,h there is a 2-cycle $\sigma_{g,h}$ such that $\sigma_{g,h}(g)=h$ and vice versa, so our matrix is symmetric.

This shows that $s(g,g')=s(g',g)$ for *each* set of reversal pairs g,g'. To see that they are *all* the same, note that $\sigma=(BC)$ sends the first reversal *pair* $(ACBD),(ADBC)$ to the second pair, and $\sigma=(AC)$ does likewise to the third pair (and apply neutrality).

Now consider $s(g,h)$ where $h\neq g,g'$. Apply a 2-cycle σ such that $\sigma(g)=g'$ (such as (AB) or (CD) for $(ACBD)$). Then $s(g,h)=s(g',\sigma(h))$, where $\sigma(h)\neq g,g',h$. There are two possible choices for σ, which are disjoint, so their product (such as $(AB)(CD)$) sends h to h'. So if we do this twice we have that $s(g,h)=s(g,h')$ for any g,h that are not a reversal pair! By symmetry the same is true for $s(g,h)=s(g',h)$, and so all of the other entries are equal to each other as desired. □

3. Representation theory

There is one loose end in the previous section. Consider the subspaces in Corollary 2.7. Post-hoc it's easy to see they behave nicely, and with some thought about the structure of the cyclic orders one could come up with them 'by hand'; still, it seems a little mysterious where they came from. Likewise, although it's not a loose end, the proof of Proposition 2.6 is quite ad-hoc, and one might despair of generalizing it.

In principle, it is nearly always *possible* to analyze voting systems without any tools of higher power. However, what is possible is not always doable in practice, as the sheer number of options demands more. We choose, as this more powerful tool for analyzing both the systems and the spaces of profiles involved, the representation theory of the symmetric group. This section will recall the information we need, with [**36**] as one of a number of handy references.

In doing so we follow the introduction of these techniques in mathematical social sciences, notably by [**20**] in cooperative game theory and [**12**] in voting theory. More powerful techniques do need some justification, of course. One of the advantages of this approach is the ability to provide very general results; as an example, [**4**] and [**23**] do this with some of the questions asked in [**30**]. Another advantage is to provide a unifying framework, as with [**12**] and [**9**] vis-a-vis the decompositions in the magisterial [**34, 35**]. See [**10**] for a recent overview.

Although we began our analysis with the case where the ballot for a voting procedure is the same as a potential outcome, from Subsection 1.1 and the comments after Remark 1.7 we know our motivation is to be able to handle a wider variety of possible ballots, such as ones asking for partial information on adjacency or non-adjacency. With that in mind, we might as well generalize Definition 2.3 of a points-based rule somewhat (to a generalized scoring rule, in the terminology of [**40**]).

DEFINITION 3.1. Given n, suppose we have a *ballot set* B, the set of *profiles*[8] $\mathbb{Q}^{\text{card}(B)}$ on B, and a *ballot-scoring function* $s : B \times CO_n \to \mathbb{Q}$. Then a *points-based voting rule* f, with *ballot set* B and *outcome set* the set of cyclic orders (for a given n), is a function from the set of all profiles on B to a nonempty subset of CO_n, such that $f(\mathbf{p})$ is the (set of) cyclic order(s) h which maximizes $\sum_{g\in B} \mathbf{p}[g]s(g, h)$.

Further, assume that there is an action of S_n on B (extended to the set of profiles). Then the rule f is *neutral* if s is neutral, in the sense that for any $\sigma \in S_n$, $s(g, h) = s(\sigma(g), \sigma(h))$.

As before, the function s represents how many points are allocated to the cyclic order h for a single vote for the ballot g. The matrix M we visualize now would be indexed by the elements of CO_n in rows and elements of B in columns, so that $M_{h,g} = s(g, h)$ as before; likewise, the system takes the argmax of the resulting vector $M\mathbf{p}$ to obtain the result for any profile $\mathbf{p}$. Assuming that B does have an S_n-action, we can also define $\sigma(\mathbf{p})[x] = \mathbf{p}[\sigma^{-1}x]$ on the space of profiles, as in 2.2.

In that case, we can call the ballot space $\mathbb{Q}B$ and the outcome space $\mathbb{Q}CO_n$. We make the following observations:

- By definition, our profile and outcome spaces have the S_n-action described, with a $\mathbb{Q}$-vector space structure compatible with S_n. That is, both spaces have a $\mathbb{Q}S_n$-module structure.

[8](Again, if there are negative entries, especially if the sum of entries is zero, sometimes we call it a *profile differential*.)

- Since the scoring functions are neutral ($s(g,h) = s(\sigma(g), \sigma(h))$), the S_n-action propagates from profiles to outcomes. Given $\sigma \in S_n$ and $h \in CO_n$,

$$\sigma\left(\sum_{g\in B)} \mathbf{p}[g]s(g,h)\right) = \sum_{g\in B)} \mathbf{p}[(\sigma^{-1}(g)]s(\sigma^{-1}(g), \sigma^{-1}(h))\ ,$$

  which is the same as the effect of σ on the voting rule. (This is the usual way neutrality would be defined.)
- Hence[9]

$$f : \mathbb{Q}B \to \mathbb{Q}CO_n$$

  is a $\mathbb{Q}S_n$-module homomorphism.

Since we know f is a $\mathbb{Q}S_n$-module homomorphism, we can use representation theory to analyze it, via decomposition of $\mathbb{Q}B$ and $\mathbb{Q}CO_n$ into irreducible submodules, and the following key results.

PROPOSITION 3.2. *Let G be a group. Any (finitely generated) $\mathbb{Q}G$-module has a decomposition as a direct sum of irreducible $\mathbb{Q}G$-modules which is unique up to order.*

SCHUR'S LEMMA. *Let G be a group. If M and N are irreducible $\mathbb{Q}G$-modules and $g : M \to N$ is a $\mathbb{Q}G$-module homomorphism, then either $g = 0$ or g is an isomorphism ([**36**] Theorem 1.6.5).*

*Moreover, if $G = S_n$, then $g = 0$ or g is actually multiplication by an element of $\mathbb{Q}$ ([**36**] Corollary 1.6.8, along with [**19**] Theorem 2.1.12 that $\mathbb{Q}$ is a splitting field for S_n).*

PROPOSITION 3.3. *Any irreducible representation of S_n (irreducible $\mathbb{Q}S_n$-module) is isomorphic[10] to one of a finite set of (nonisomorphic) modules. These modules are indexed by the partitions λ of n (denoted $\lambda \vdash n$), and are called the* Specht modules *([**36**] Theorem 2.4.6), denoted S^λ.*

EXAMPLE 3.4. The standard example of this in voting theory is when $n = 3$ and the ballot space is $\mathcal{L}(\{A, B, C\})$. In this case $(3), (2, 1), (1, 1, 1) \vdash 3$ and

$$\mathbb{Q}B \simeq \mathbb{Q}S_3 \simeq S^{(3)} \oplus S^{(2,1)^{\oplus 2}} \oplus S^{(1,1,1)}\ .$$

We can identify $S^{(3)}$ as the subspace of complete ties generated by $(1, 1, 1, 1, 1, 1)^T$, and $S^{(1,1,1)}$ is the subspace generated by $(1, -1, 1, -1, 1, -1)^T$, where permutations coming from the cyclic order $(ABCA)$ have the opposite value of those coming from $(ACBA)$. This is harnessed in [**32, 33, 39**], by name in [**9, 12**].

Note that in the previous example we had the fortune that the ballots in B actually could be considered as elements of a group. That does not happen with CO_n, unfortunately, but something *nearly* as good is true which will allow us to decompose $\mathbb{Q}CO_n$ in practical cases. We'll need a bit of additional terminology, but it will be worth it. First, we need to identify $\mathbb{Q}CO_n$ more precisely in group-theoretic terms.

[9] Abusing notation slightly as alluded to after 2.3.

[10] This is true over both $\mathbb{Q}$ and $\mathbb{C}$ for the same reason as in the previous lemma.

DEFINITION 3.5. Let $H = \langle(123\cdots n)\rangle \leqslant S_n$. We can restate 1.4 by saying that CO_n is the set of (left) cosets S_n/H.

Further, the $\mathbb{Q}S_n$-module $\mathbb{Q}CO_n$ is the *permutation representation* of S_n/H. This is defined by taking a vector space over $\mathbb{Q}$ with one basis vector b_{gH} for each coset in S_n/H, and then letting $g'(b_{gH}) = b_{g'gH}$ (and extending linearly).

Finally, we need to recall the prime tool for *computing* decomposition of representations.

DEFINITION 3.6. Given an $\mathbb{Q}G$-module V, for each $g \in G$ there is a linear transformation on V (as a finite-dimensional vector space) induced by applying g to each basis element. We can create the matrix $X(g)$ to represent this map. We call its trace $\operatorname{tr}(X(g))$ the *character* of V on $g \in G$, usually denoted χ_V.

PROPOSITION 3.7. *The characters χ_λ of the irreducible modules V_λ for G are particularly nice, in that there is an inner product (χ, χ') on the set of* all *functions*[11] *$G \to \mathbb{Q}$ such that (χ_λ, χ_V) is a nonnegative integer where there are exactly (χ_λ, χ_V) 'copies' of V_λ in the decomposition of V. (See* **[36]** *Section 1.9.)*

EXAMPLE 3.8. In Example 3.4, we have $(\chi_{(2,1)}, \chi_{\mathbb{Q}B}) = 2$, $(\chi_{(1,1,1)}, \chi_{\mathbb{Q}B}) = 1$.

PROPOSITION 3.9. *The permutation representation $\mathbb{Q}CO_n$ has character χ where*[12]

$$\begin{aligned}\chi(g) = &\ \text{the number of fixed points of } g \text{ acting on } S_n/H \\ &= \text{the number of cyclic orders left unchanged by } g\end{aligned}$$

This is something we can compute! Recall that characters are a *class function*, that is to say the value of a character on a group element g depends only on its conjugacy class in G. Further recall that the classes of S_n are precisely determined by their cycle decomposition (also indexed by the partitions of n).

THEOREM 3.10. *Let D be the set of (positive) divisors d of n, with divisor complement $e = \frac{n}{d}$. Each divisor corresponds to a conjugacy class of S_n of C_d^e,*

$$\chi(g) = \begin{cases} \frac{e!d^e}{n}\phi(d) & g \text{ of form } C_d^e \\ 0 & \text{otherwise} \end{cases}.$$

COROLLARY 3.11. *If $n = p$ is prime then* $\chi(g) = \begin{cases} (n-1)! & g \text{ the identity} \\ n-1 & g \text{ an } n\text{-cycle} \\ 0 & \text{otherwise} \end{cases}$.

PROOF OF THEOREM 3.10. This is an easy exercise using standard induced character facts, but it is worth making an explicit calculation of the number of fixed points.

For a cyclic order x to remain fixed under an element of S_n, all A_i must be moved by the same amount, modulo n, in one of the underlying permutations representing x. This immediately rules out any g with more than one cycle type in the decomposition fixing any x, so assume g has type C_d^e.

[11] In general from $G \to \mathbb{C}$, but over S_n we may once again restrict to $\mathbb{Q}$.

[12] See e.g. **[36]** Exercise 1.13.3b for this standard fact; see also formula (1.28) since this is actually an induced representation.

Without loss of generality let g have the form

$$(1\cdots\ell_{1,d})(2\cdots\ell_{2,d})\cdots(e\cdots\ell_{e,d}) = g_1g_2\cdots g_e\ .$$

A typical cyclic order fixed by this has the form $(12\cdots e\ell_{1,2}\ell_{2,2}\cdots\ell_{e,d})$ so that each of the d-cycles are intertwined. However, it is evident that other cyclic orders may be fixed, such as $(213\cdots e\ell_{2,2}\ell_{1,2}\ell_{3,2}\cdots\ell_{e,d})$. In addition, the same cycle will be fixed if we replace a d-cycle $g_1 = (1\cdots\ell_{1,d})$ by $(1\ell_{1,d}\ell_{1,d-1}\cdots\ell_{1,2})$, or any of the $\phi(d)$ powers of the cycle which preserve the cycle structure, so *mutatis mutandis* there are $\phi(d)$ (cyclic!) orders of the elements of g_1 fixed by g_1 itself. We now use a basic counting argument based on these observations.

Keeping the order of each d-cycle, we have $n = de$ choices of where to put the first element of the d-cycle g_1 among the elements of x. In addition, we have $\phi(d)$ orderings of the elements of g_1 to choose from in constructing x. For each of the remaining d-cycles $\{g_i\}_{i=2}^{e}$, we have $n - d(i-1) = d(e-i+1)$ positions of x left to put the first element of g_i. However, whichever of the $\phi(d)$ orderings of g_1 was chosen must now be chosen for g_i as well, or g will not *fix* x, only the parts of x corresponding to each g_i. Finally, we have of course overcounted, because each of these x is equivalent to n other cyclic orders with the same properties. Factoring out the powers of d finishes the computation. □

We can reprove Proposition 2.6 and Corollary 2.7 from this point of view, which will prepare us for similar results in Section 5. We first compute directly and with character tables for this proposition:

PROPOSITION 3.12. *The character* $\chi_{\mathbb{Q}CO_4} = \chi_4$ *has values* $\chi_4(e) = 6$, $\chi_4((AB)(CD)) = 2$, $\chi_4((ABCD)) = 2$, *and is zero otherwise. Further,*

$$\mathbb{Q}CO_4 \simeq S^{(4)} \oplus S^{(2,2)} \oplus S^{(2,1,1)}$$

of dimensions 1, 2, *and* 3, *respectively.*

The subspaces in Corollary 2.7 we have already seen to be invariant under S_4, so by uniqueness of decomposition we can restate that result (using Schur's Lemma) as follows.

COROLLARY 3.13. *Any neutral points-based rule* f *on* CO_4 *is determined by three scalars* t, u, v *such that if* $\mathbf{p} \in S^{(4)} \oplus S^{(2,2)} \oplus S^{(2,1,1)}$ *is* $\mathbf{p} = \mathbf{p}_1 + \mathbf{p}_2 + \mathbf{p}_3$, *then*

$$f(\mathbf{p}) = t\mathbf{p}_1 + u\mathbf{p}_2 + v\mathbf{p}_3\ .$$

Further, we would have in Proposition 2.6 that $t = a + b + 4c$, $u = a + b - 2c$, and $v = a - b$; solving for a, b, c yields

$$a = \frac{1}{6}t + \frac{1}{3}u + \frac{1}{2}v, b = \frac{1}{6}t + \frac{1}{3}u - \frac{1}{2}v, c = \frac{1}{6}t - \frac{1}{6}u$$

if one wanted to recreate the matrix in terms of these scalars instead.

Finally, we wish to recall one final very useful concept from [**12**]. In any $\mathbb{Q}G$-module homomorphism, we have seen that the kernel is itself a $\mathbb{Q}G$-module. However, for the purposes of voting theory this simply corresponds to the space of profiles that goes to a complete tie of zero points – certainly a useful space, but we might be interested in what point totals are *possible*, not just impossible.

DEFINITION 3.14. Given a neutral points-based voting rule (or indeed a $\mathbb{Q}G$-module homomorphism) f with (profile) domain $\mathbb{Q}B$, let the kernel be K_f. Then call the orthogonal complement (using the usual inner product on $\mathbb{Q}^{\text{card}(B)}$) to K_f the *effective space of* f, and denote it by E_f.

While any profile vector not in the kernel will go to some output vector other than a complete zero tie, any such profiles $\mathbf{p}$ can be written as $\mathbf{p} = \mathbf{k} + \mathbf{e}$, where $\mathbf{k} \in K_f$ and $\mathbf{e} \in E_f$ are orthogonal. Then $f(\mathbf{p}) = f(\mathbf{e})$, so in some very real sense E_f contains all the information there is to know. Moreover, as we have seen in Corollary 3.12/Corollary 2.7, any element of E_f can be written as a sum of vectors which are then sent by scalar multiplication to vectors in the outcome space. Identifying this space would therefore seem to be highly useful in constructing examples or proving theorems about any given f, or family of f.

4. Meaningful ballots for $n = 4$

We now harness this symmetry to analyze more complex ballots. Because the general results are pretty abstract, we will motivate them with some much more explicit computations.

4.1. An in-depth example. To start us off, consider Scenario 1.3. There is a particularly nice setup here, because we can clearly identify a small subset of the many possible pieces of a cyclic order of greatest relevance to a voter.

First, each voter has a specific player A_v he desires to win. In addition, it seems reasonable that this voting public will be knowledgeable enough that each voter will be able to identify a strongest competitor A_s and weakest competitor A_w (neither the same as A_v). In that case, allowing for a ballot that specifies the two adjacent players to a desired winner seems very reasonable (and certainly requires much less information than a full ranking of all cyclic orders).

DEFINITION 4.1. We define a *ROLO ballot* to be a ballot that specifies a player, along with one to the Right Of and one to the Left Of this player. We use the notation $A\{D, C\}$, where we interpret this as desiring A to have D to its right and C to its left, so that a generic cyclic order would look something like $\begin{smallmatrix} DAC \\ X \cdots Y \end{smallmatrix}$.

In the poker example, a good ballot to submit would be $A_v\{A_s, A_w\}$. However, it should be evident that the same ballot could be profitably used in many settings. Note that there are $n(n-1)(n-2)$ ballots compared to $(n-1)!$ cyclic orders, which means that other than for $n \leq 6$ there are many more cyclic orders than possible ballots. In this paper we will restrict ourselves to $n = 4$, where there are already 24 ballot types, so things are not dull.

Let's apply Definition 3.1 to this situation. Here, B is the set of ROLO ballots, so $\mathbb{Q}B \simeq \mathbb{Q}^{24}$ as vector spaces. Then once we devise a neutral scoring function $s : B \times CO_n \to \mathbb{Q}$, we will automatically have a new neutral points-based system to explore.

EXAMPLE 4.2. Here is a system simple to describe informally, which is perhaps analogous to Example 2.4. Since an element of B in this case completely describes a favorite cyclic order, given a ballot $b \in B$ we could give two points to that order, *one* point to any other order that shares either the right or the left, and then zero points to everything else[13]. For example, a vote for $A\{D, C\}$ would count two

[13]Another way to think of this is that one point is awarded from $A\{D, C\}$ to any order with AC in that order, and one point to any order with DA in that order.

points to $(ACBD)/\frac{AC}{DB}$ and one point each for $(ACDB)/\frac{AC}{BD}$ and $(ABCD)/\frac{AB}{DC}$. We will call this system ROLO(2,1).

Before analyzing this further, we will need to fix an order for the ballots. Recall from Figure 1 the order given to CO_4 is[14] $(ACBD)$, $(ADBC)$, $(ABCD)$, $(ADCB)$, $(ABDC)$, $(ACDB)$. We will give the ROLO ballots the following order[15], which has the four ballots corresponding to each cyclic order placed together.

$$\begin{array}{cccc} A\{D,C\} & B\{C,D\} & D\{B,A\} & C\{A,B\} \\ C\{B,A\} & D\{A,B\} & A\{C,D\} & B\{D,C\} \\ A\{D,B\} & C\{B,D\} & B\{A,C\} & D\{C,A\} \\ B\{C,A\} & D\{A,C\} & C\{D,B\} & A\{B,D\} \\ D\{B,C\} & A\{C,B\} & B\{A,D\} & C\{D,A\} \\ C\{A,D\} & B\{D,A\} & D\{C,B\} & A\{B,C\} \end{array}$$

FIGURE 2. ROLO ballots for $n = 4$

With this in mind, here is the full matrix for the points-based system ROLO(2,1).

$$\begin{pmatrix} 2 & 2 & 2 & 2 & 0 & 0 & 0 & 0 & 1 & 0 & 0 & 1 & 1 & 0 & 1 & 0 & 1 & 0 & 1 & 0 & 1 & 0 & 0 & 1 \\ 0 & 0 & 0 & 0 & 2 & 2 & 2 & 2 & 0 & 1 & 1 & 0 & 0 & 1 & 0 & 1 & 0 & 1 & 0 & 1 & 0 & 1 & 1 & 0 \\ 1 & 0 & 1 & 0 & 1 & 0 & 0 & 1 & 2 & 2 & 2 & 2 & 0 & 0 & 0 & 0 & 0 & 1 & 1 & 0 & 1 & 0 & 1 & 0 \\ 0 & 1 & 0 & 1 & 0 & 1 & 1 & 0 & 0 & 0 & 0 & 0 & 2 & 2 & 2 & 2 & 1 & 0 & 0 & 1 & 0 & 1 & 0 & 1 \\ 0 & 1 & 1 & 0 & 1 & 0 & 1 & 0 & 1 & 0 & 1 & 0 & 0 & 1 & 1 & 0 & 2 & 2 & 2 & 2 & 0 & 0 & 0 & 0 \\ 1 & 0 & 0 & 1 & 0 & 1 & 0 & 1 & 0 & 1 & 0 & 1 & 1 & 0 & 0 & 1 & 0 & 0 & 0 & 0 & 2 & 2 & 2 & 2 \end{pmatrix}$$

FIGURE 3. ROLO(2,1) matrix for $n = 4$

The reader will immediately note a high degree of symmetry. That symmetry comes from the comment, 'the four ballots corresponding to each cyclic order,' and will be fully utilized momentarily. For now, let's see two representative results for ROLO(2,1). First, a 'paradox.'

EXAMPLE 4.3. Suppose we have an electorate of about four thousand, distributed in the following profile:

$$(141, 141, 141, 141, 73, 313, 133, 253, 133, 159, 99,$$
$$193, 223, 9, 103, 129, 193, 159, 133, 219, 163, 9, 9, 163)^T.$$

About forty percent of the voters would like $(ACDB)$ or its reversal, and another third prefer $(ABCD)$ or its reversal. However, under the ROLO(2,1) system, *all four* of those options tie for last place, and $(ACBD)$ is the winning cyclic order by nearly one hundred scoring points!

---

[14]To keep the right and left in view easier one may recall $\frac{AC}{DB}, \frac{AD}{CB}, \frac{AB}{DC}, \frac{AD}{BC}, \frac{AB}{CD}, \frac{AC}{BD}$.

[15]The cognoscenti will later point out, when we see the group action explicitly, that the order could be even more symmetric, but redoing the various vectors and matrices involved is very error-prone, so we are sticking with this order!

To be sure, we do not know of any axioms for voting on cyclic orders that would cause this to be 'paradoxical' in the usual voting sense. At the same time, granted that we are currently only assuming neutrality and anonymity (not even a Pareto or unanimity axiom for the purposes of this paper!), this outcome seems troubling. This is especially so since we went out of our way with ROLO ballots to not simply use a voting system equivalent to plurality.

On the positive side of the ledger, we have results like this.

EXAMPLE 4.4. Pick a cyclic order. A profile differential over ROLO ballots with three voters for each ballot corresponding to that cyclic order, and negative three for each one corresponding to its reversal, yields an outcome vector of 24 points for the order and -24 for its reversal.

This is unsurprising. Moreover, we get a similar profile (differential), with an identical result, if we simply subtract the row of the matrix in Figure 3 corresponding to the reversal of the chosen cyclic order from the row corresponding to the chosen one.

However, linearity immediately then implies that all profiles of forms like

$$(1,1,1,1,-1,-1,-1,-1,-1,1,1,-1,1,-1,1,-1,1,-1,1,-1,-1,1,1,-1)^T$$

go to a complete zero tie. That is to say, ROLO(2,1) considers a profile (differential), preferring ballots coming from a given cyclic order exactly as much as it *dislikes* ballots sharing only one adjacency with that same cyclic order, to be irrelevant to the scores.

In some sense, this profile cancels out the points ROLO(2,1) gives to the two adjacencies of each possible ballot, as there are exactly the same number of each adjacency, positive and negative. Again, without intuition about which profiles *should* cancel out, this may or may not be controversial to assume as an axiom. At the very least, though, it is a positive statement of something that is only obvious in retrospect about the rule.

4.2. The regular representation. Let's return to the symmetry of Figure 3. What is it? There are four elements of S_4 which fix each cyclic order, and we have already seen that these are the cosets in question defining cyclic orders. That means that the action of S_4 on B is isomorphic to that of S_4 *on itself*, which is the definition (when linearized to vector spaces) of the regular representation. Now let's apply Proposition 3.3 in this well-known case.

PROPOSITION 4.5. *The space $\mathbb{Q}B$ for ROLO ballots decomposes as*

$$S^{(4)} \oplus S^{(3,1)^{\oplus 3}} \oplus S^{(2,2)^{\oplus 2}} \oplus S^{(2,1,1)^{\oplus 3}} \oplus S^{(1,1,1,1)} .$$

From Proposition 3.12 we then have a complete characterization of *all* neutral points-based procedures on ROLO ballots!

PROPOSITION 4.6. *The scoring part of any neutral points-based voting rule for $n=4$ from ROLO ballots to CO_4 is a $\mathbb{Q}S_4$-homomorphism*

$$S^{(4)} \oplus S^{(3,1)^{\oplus 3}} \oplus S^{(2,2)^{\oplus 2}} \oplus S^{(2,1,1)^{\oplus 3}} \oplus S^{(1,1,1,1)} \to S^{(4)} \oplus S^{(2,2)} \oplus S^{(2,1,1)}.$$

COROLLARY 4.7. *Thus the kernel of* any *such system is a superspace of the ten dimensions of $S^{(3,1)^{\oplus 3}} \oplus S^{(1,1,1,1)}$ direct summed with an eight-dimensional submodule isomorphic to $S^{(2,2)} \oplus S^{(2,1,1)^{\oplus 2}}$, while the effective space is isomorphic* *to an invariant subspace of* $S^{(4)} \oplus S^{(2,2)} \oplus S^{(2,1,1)}$. *We may describe it fully with six scalars, which represent the scaling factors for the subspaces possibly not in the kernel* $S^{(4)} \oplus {S^{(2,2)}}^{\oplus^2} \oplus {S^{(2,1,1)}}^{\oplus^3}$.

There are several advantages to this point of view. First off, it is much easier to find specific subspaces of interest once you know what you are looking for. Secondly, we can not only describe how profiles will behave, but also more easily create 'paradoxes', or at least unusual results.

Before going in more depth, it will be helpful to examine the structure of Figures 2 and 3 a little more closely. As we pointed out, each ballot can be obtained from $A\{D,C\}$ by a specific element of S_4 (considered as symmetries on the set $\{A,B,C,D\}$). Less evidently, the matrix can also be obtained in the same way, by our neutrality requirement that $s(\sigma(b),\sigma(x)) = \sigma(s(b,x))$. This is evident in that each group of four columns corresponds to a coset of $\langle (ABCD) \rangle$ in S_4, which does not change the cyclic order that a given ballot favors, but *does* cycle through the cyclic orders that share only the right-of or left-of characteristic.

We can now give far more explicit constructions both of any similar procedure (Proposition 4.8) *and* of the space $\mathbb{Q}B$ (Figure 4).

PROPOSITION 4.8. *Any neutral points-based voting rule on a ballot of twenty-four options which is a transitive* S_4*-set to the set of cyclic orders must have a matrix of the following form:*

$$\begin{pmatrix}
a & a & a & a & b & b & b & b & c & d & e & f & c & d & f & e & c & d & f & e & c & d & e & f \\
b & b & b & b & a & a & a & a & d & c & f & e & d & c & e & f & d & c & e & f & d & c & f & e \\
c & d & f & e & c & d & e & f & a & a & a & a & b & b & b & b & e & f & c & d & f & e & c & d \\
d & c & e & f & d & c & f & e & b & b & b & b & a & a & a & a & f & e & d & c & e & f & d & c \\
e & f & c & d & f & e & c & d & f & e & c & d & e & f & c & d & a & a & a & a & b & b & b & b \\
f & e & d & c & e & f & d & c & e & f & d & c & f & e & d & c & b & b & b & b & a & a & a & a
\end{pmatrix}.$$

PROOF. Consider the permutations of the first column of weights which give the other columns. The first four columns are generated by the 4-cycle (3546), and the next four by (12)(56) in addition, forming a dihedral subgroup of S_6. The next eight columns are achieved by a coset of that subgroup by (135)(246) and the final eight columns are the coset generated by $[(135)(246)]^2 = (153)(264)$. The set of all these permutations is a subgroup of S_6 isomorphic to S_4, as can be easily checked. □

We can see that ROLO(2,1) has $a = 2$ and $c = 1 = f$, otherwise zero.

In Figure 4 we give the decomposition as well as brief voting-theoretic commentary. There is *not* a canonical way to distinguish among the infinitely many ways to write ${S^{(2,2)}}^{\oplus^2}$ or ${S^{(2,1,1)}}^{\oplus^3}$ as a direct sum of invariant subspaces, but our ordering of vectors shows one useful way, as we will see. We used **[37]** (in a non-automatic way) as well as computing characters to find this decomposition into orthogonal subspaces. Bases are arranged in such a way to make it evident the symmetries are indeed the ones in question.

The decomposition of $\mathbb{Q}B$ as S_4-invariant subspaces.

- The trivial subspace $S^{(4)} = T$:

$$S^{(4)} = \mathrm{span}(\{(1,1,1,1,1,1,1,1,1,1,1,1,1,1,1,1,1,1,1,1,1,1,1,1)^T\})$$

- These four vectors span $S^{(2,2)^{\oplus 2}}$, and $\mathrm{span}(\{v_1, v_2\})$ is an $S^{(2,2)}$ emphasizing one cyclic order/reversal pair and de-emphasizing the others.

$$v_1 = (2,2,2,2,2,2,2,2,-1,-1,-1,-1,-1,-1,-1,-1,-1,-1,-1,-1,-1,-1,-1,-1)^T,$$
$$v_2 = (-1,-1,-1,-1,-1,-1,-1,-1,2,2,2,2,2,2,2,2,-1,-1,-1,-1,-1,-1,-1,-1)^T,$$
$$v_3 = (2,2,-2,-2,2,2,-2,-2,1,1,-1,-1,1,1,-1,-1,-1,-1,1,1,-1,-1,1,1)^T,$$
$$v_4 = (1,1,-1,-1,1,1,-1,-1,2,2,-2,-2,2,2,-2,-2,1,1,-1,-1,1,1,-1,-1)^T$$

- These vectors span $S^{(2,1,1)^{\oplus 3}}$. We will call each set $\{w_{i1}, w_{i2}, w_{i3}\}$ a *triplet*; the spaces spanned by each set of three are both S_4-invariant and mutually orthogonal. Note $\mathrm{span}(\{w_{11}, w_{12}, w_{13}\}) \simeq S^{(2,1,1)}$ is a subspace of vectors favoring a cyclic order strongly, and disfavoring its reversal strongly.

$$w_{11} = (1,1,1,1,-1,-1,-1,-1,0,0,0,0,0,0,0,0,0,0,0,0,0,0,0,0)^T,$$
$$w_{12} = (0,0,0,0,0,0,0,0,1,1,1,1,-1,-1,-1,-1,0,0,0,0,0,0,0,0)^T,$$
$$w_{13} = (0,0,0,0,0,0,0,0,0,0,0,0,0,0,0,0,1,1,1,1,-1,-1,-1,-1)^T,$$
$$w_{21} = (0,0,0,0,0,0,0,0,1,-1,0,0,1,-1,0,0,1,-1,0,0,1,-1,0,0)^T,$$
$$w_{22} = (1,-1,0,0,1,-1,0,0,0,0,0,0,0,0,0,0,0,0,1,-1,0,0,1,-1)^T,$$
$$w_{23} = (0,0,1,-1,0,0,1,-1,0,0,1,-1,0,0,1,-1,0,0,0,0,0,0,0,0)^T,$$
$$w_{31} = (0,0,0,0,0,0,0,0,0,0,-1,1,0,0,1,-1,0,0,1,-1,0,0,-1,1)^T,$$
$$w_{32} = (0,0,1,-1,0,0,-1,1,0,0,0,0,0,0,0,0,-1,1,0,0,1,-1,0,0)^T,$$
$$w_{33} = (-1,1,0,0,1,-1,0,0,1,-1,0,0,-1,1,0,0,0,0,0,0,0,0,0,0)^T$$

- The sign subspace $S^{(1,1,1,1)}$ gives a ± 1 depending on whether the underlying permutation is even or odd; presumably this should be ignored by a system voting on cyclic orders, which will have equal amounts of each.

$$S^{(1,1,1,1)} = \mathrm{span}\left(\{(1,1,-1,-1,1,1,-1,-1,-1,-1,1,1,-1,-1,1,1,1,1,-1,-1,1,1,-1,-1)^T\}\right)$$

- The $S^{(3,1)^{\oplus 3}}$ subspaces, spanned by the following triplets of vectors, all are in the kernel as well.

$$u_{11} = (1,1,-1,-1,-1,-1,1,1,0,0,0,0,0,0,0,0,0,0,0,0,0,0,0,0)^T,$$
$$u_{12} = (0,0,0,0,0,0,0,0,-1,-1,1,1,1,1,-1,-1,0,0,0,0,0,0,0,0)^T,$$
$$u_{13} = (0,0,0,0,0,0,0,0,0,0,0,0,0,0,0,0,1,1,-1,-1,-1,-1,1,1)^T,$$
$$u_{21} = (0,0,0,0,0,0,0,0,-1,1,0,0,-1,1,0,0,1,-1,0,0,1,-1,0,0)^T,$$
$$u_{22} = (1,-1,0,0,1,-1,0,0,0,0,0,0,0,0,0,0,0,0,-1,1,0,0,-1,1)^T,$$
$$u_{23} = (0,0,-1,1,0,0,-1,1,0,0,1,-1,0,0,1,-1,0,0,0,0,0,0,0,0)^T,$$
$$u_{31} = (0,0,0,0,0,0,0,0,0,0,1,-1,0,0,-1,1,0,0,1,-1,0,0,-1,1)^T,$$
$$u_{32} = (1,-1,0,0,-1,1,0,0,1,-1,0,0,-1,1,0,0,0,0,0,0,0,0,0,0)^T,$$
$$u_{33} = (0,0,-1,1,0,0,1,-1,0,0,0,0,0,0,0,0,-1,1,0,0,1,-1,0,0)^T$$

FIGURE 4. List of invariant subspaces for ROLO ballots

EXAMPLE 4.9. Let's use Figure 4 to explain our examples. Recall that a generic[16] procedure will have a kernel and effective space component of $S^{(2,2)}$ and similarly for $S^{(2,1,1)}$ (twice as large of a kernel in the latter case), the effective parts of which will go directly to the corresponding space in $\mathbb{Q}CO_4$ in Corollary 2.7. We can compute that the $S^{(2,1,1)}$ piece of the effective space of ROLO(2,1) is the space spanned by three vectors, each of which is the sum $2w_{1i} + w_{2i} + w_{3i}$. (The kernel is the orthogonal complement to this, within that nine-dimensional subspace.) This explains how Example 4.4 was created.

Similarly, we can now see how Example 4.3 was created. We started with a large piece of the $S^{(2,1,1)}$ effective space going to multiples of $(2, 2, -1, -1, -1, -1)$ and added to it some large vectors in $S^{(2,2)}$ going to $(1, -1, 0, 0, 0, 0)$, so that $(ACBD)$ would be the winner. Then, in order to obfuscate this reality, we found vectors in the kernel $S^{(2,1,1)}$ component (using linear algebra and Figure 4) which strongly favored $(ACBD)$ and *subtracted those.* Adding enough complete tie ballots to give positive entries finished things up.

For completeness, note that the span of $\{v_3, v_4\}$ in the $S^{(2,2)}$ bullet of Figure 4 is the $S^{(2,2)}$ piece of the kernel of ROLO(2,1), while the span of $\{v_1, v_2\}$ is the effective space.

4.3. Everything you can say. Once we have these computations, one can make as general of statements as necessary for an application. Although we hope people will consider voting/aggregating ballots related to cyclic orders, in some sense this is also an advertisement for the power of these methods.

As a concrete example, we might consider more general ROLO systems with the same ballot, such as ROLO(x,1) where $a = x$ rather than $a = 2$ as before. This is a system that values more (or less) having exactly the 'favorite' cyclic order over other cyclic orders compatible with the adjacencies in a ROLO ballot.

PROPOSITION 4.10. *The effective and kernel $S^{(2,2)}$ subspaces for all ROLO(x,1) procedures (except when $x = 1$) are the same as in Example 4.9. The $S^{(2,1,1)}$ effective space for ROLO(x,1) is spanned by the vectors $xw_{1i} + w_{2i} + w_{3i}$.*

PROOF. One can directly compute by a suitably modified matrix that the $S^{(2,2)}$ kernel for ROLO(2,1) also goes to zero for ROLO(x,1), and likewise that the other two $S^{(2,2)}$ basis vectors (as in the statement) go where indicated. Schur's Lemma and the orthogonality of spaces spanned by $\{v_1, v_2\}$ and $\{v_3, v_4\}$, respectively, in Figure 4 complete the proof of the first statement. When $x = 1$ the kernel $S^{(2,2)}$ component is the whole four-dimensional subspace. The verification of the second statement is very similar to Example 4.9. □

PROPOSITION 4.11. *We can obtain more explicit results with raw computation.*

- *Vectors fully supporting one cyclic order and its reversal over against a different reversal pair, such as*

$$(1, 1, 1, 1, 1, 1, 1, 1, -1, -1, -1, -1, -1, -1, -1, -1, 0, 0, 0, 0, 0, 0, 0, 0)^T$$

  are sent to outcomes of the form $4(x-1, x-1, 1-x, 1-x, 0, 0)$. (Verify by adding the x and 1 points for each cyclic order, without using the matrix.)

[16]Of course, some, such as the zero matrix, will be trivial enough to have a bigger kernel and smaller effective space!

- *As in Example 4.4, in the matrix for ROLO(x,1) we subtract the second row from the first (and likewise with the others) to get a basis. The orthogonal complement to this is the kernel.*
- *On the other hand, vectors fully supporting one cyclic order over against its own reversal, such as*

$$(0,0,0,0,0,0,0,0,0,0,0,0,0,0,0,0,1,1,1,1,-1,-1,-1,-1),$$

are sent to outcomes of the form $4(0,0,0,0,x,-x)$ *fully supporting in the same way. Note that when* $x = 0$ *these are in the kernel, so they are clearly* not *the effective space.*

More generally, we have some much more powerful statements, applicable to *any neutral points-based rule* for this ballot space along the lines of Proposition 4.8. Call the matrix in that Proposition M, the rule f_M, and recall the constants a through f.

THEOREM 4.12. *The image of* T *is* $4(a+b+c+d+e+f)(1,1,1,1,1,1)$, *so* T *is in the kernel precisely when the weighting vector is sum-zero.*

The $S^{(2,2)}$ *effective space is the span of the following two vectors*[17]*:*

$$\begin{aligned}(&2a+2b-c-d-e-f, 2a+2b-c-d-e-f, 2a+2b-c-d-e-f,\\ &2a+2b-c-d-e-f, 2a+2b-c-d-e-f, 2a+2b-c-d-e-f,\\ &2a+2b-c-d-e-f, 2a+2b-c-d-e-f, -a-b+2c+2d-e-f,\\ &-a-b+2c+2d-e-f, -a-b-c-d+2e+2f, -a-b-c-d+2e+2f,\\ &-a-b+2c+2d-e-f, -a-b+2c+2d-e-f, -a-b-c-d+2e+2f,\\ &-a-b-c-d+2e+2f, -a-b+2c+2d-e-f, -a-b+2c+2d-e-f,\\ &-a-b-c-d+2e+2f, -a-b-c-d+2e+2f, -a-b+2c+2d-e-f,\\ &-a-b+2c+2d-e-f, -a-b-c-d+2e+2f, -a-b-c-d+2e+2f)\end{aligned}$$

and

$$\begin{aligned}(&-a-b+2c+2d-e-f, -a-b+2c+2d-e-f, -a-b-c-d+2e+2f,\\ &-a-b-c-d+2e+2f, -a-b+2c+2d-e-f, -a-b+2c+2d-e-f,\\ &-a-b-c-d+2e+2f, -a-b-c-d+2e+2f, 2a+2b-c-d-e-f,\\ &2a+2b-c-d-e-f, 2a+2b-c-d-e-f, 2a+2b-c-d-e-f,\\ &2a+2b-c-d-e-f, 2a+2b-c-d-e-f, 2a+2b-c-d-e-f,\\ &2a+2b-c-d-e-f, -a-b-c-d+2e+2f, -a-b-c-d+2e+2f,\\ &-a-b+2c+2d-e-f, -a-b+2c+2d-e-f, -a-b-c-d+2e+2f,\\ &-a-b-c-d+2e+2f, -a-b+2c+2d-e-f, -a-b+2c+2d-e-f)\end{aligned}$$

The image is

$$\begin{aligned}&4a^2+8ab+4b^2-4ac-4bc+4c^2-4ad-4bd+8cd+4d^2\\ &\qquad -4ae-4be-4ce-4de+4e^2-4af-4bf-4cf-4df+8ef+4f^2\end{aligned}$$

times the second space of Corollary 2.7.

[17] Because of the symmetries involved, it does not seem possible to find a basis for the $S^{(2,2)\oplus 2}$ space that both decomposes into two *orthogonal* pieces, but also looks natural for the general case.

Finally, the $S^{(2,1,1)}$ effective space for f_M has a basis comprised of the weighted vector sums of corresponding vectors in each 'triplet', namely

$$(a-b)w_{1i} + (c-d)w_{2i} + (f-e)w_{3i}$$

with image $4(a-b)^2 + 4(c-d)^2 + 4(f-e)^2$ times the third space of Corollary 2.7.

PROOF. In some sense, this is all banal computation with the matrix M. A particularly efficient way to observe these is to note that we desire $M\mathbf{p} = k\mathbf{v}$ for the various vectors $\mathbf{v}$ in the bases of Corollary 2.7. Then consider the real symmetric (and hence orthogonally diagonalizable) matrix MM^T; an eigenvector $\mathbf{v}$ of this yields a solution $M^T\mathbf{v} = \mathbf{p}$ to our problem, and the eigenvalues coming from MM^T are the multiplicative factors above. □

We can apply this result easily, if desired. The following example extends Proposition 4.11:

EXAMPLE 4.13. The basis vector in the $S^{(2,1,1)}$ effective space for ROLO(x,1) supporting a particular cyclic order goes to $4x^2 + 8$ times[18] the outcome supporting an order and not its reversal. This recovers the result for ROLO(2,1) in Example 4.4, as $4 \cdot 2^2 + 8 = 24$.

4.4. Even more ballots. Importantly, there is no reason to presume that the ballots in question are actually ROLO ballots. To use this analysis, *we only need a set of twenty-four ballots which are a transitive S_4-set.* As a perhaps less intuitive example, imagine we are in the setting of Scenario 1.1. Suppose each voter can submit a ballot that consists not of a preferred cyclic order or adjacency, but two people who should definitely be *far* apart from each other (which immediately gives the other pair which are opposite), and a pair who should be adjacent in a particular direction (compatible with the other information, of course).

For example, there are four such ballots which would define $(ACBD)/\frac{AC}{DB}$, where $XY - ZW$ means X and Y are opposite and Z is to the right of W:

$$AB - DA \quad AB - CB \quad AB - BD \quad AB - AC$$

We may call such ballots TRAD[19] ballots, and a system which gives two points to a cyclic order for fulfilling both conditions, one for fulfilling one, might be called TRAD(2,1). Indeed, we obtain this from the matrix in Proposition 4.8 when $a = 2$, $b = 1 = c$, and zero otherwise. Here are two sample results, analogous to earlier ones for ROLO, and using the same techniques.

EXAMPLE 4.14. In TRAD(2,1), one emphasizes the diagonal a fair bit. Nonetheless, by adding elements from the effective $S^{(2,1,1)}$ subspace of TRAD(2,1) to the $S^{(2,2)}$ part of the *kernel* of TRAD(2,1), we can create the following profile, which strongly emphasizes cyclic orders where A is across from B, and somewhat less emphasizes those where A is across from D.

$$(218, 198, 128, 128, 218, 198, 128, 128, 26, 26, 186, 186,$$
$$6, 6, 166, 166, 0, 0, 250, 230, 0, 0, 250, 230)$$

[18] Although the algebra is linear, this shows that the way linear procedures *vary* need not be linear.

[19] To the Right, Across Diagonally.

Applying the appropriate matrix yields $(ABCD)$ as the winning cyclic order, which not very many people wanted!

EXAMPLE 4.15. The effective space $S^{(2,1,1)}$ component for TRAD(2,1) is spanned by vectors of the form

$$(1,1,1,1,-1,-1,-1,-1,1,-1,0,0,1,-1,0,0,1,-1,0,0,1,-1,0,0)$$

which are the sums $w_{1i}+w_{2i}$ in the $S^{(2,1,1)}$ in Figure 4.

Moreover, this effective space is in the kernel of ROLO(-1,1)! In some sense, a rule which wants to promote either of two compatible directional adjacencies, but *disadvantage* a cyclic order which has both, must a fortiori ignore a large part of diagonal information.

Although we will not pursue this here, the same analysis could be done for other ballot spaces. An relatively easy one would be the set of ROLO ballots where $A\{CD\}$ is identified with $A\{DC\}$, since we could use our previous analysis – but not directly, since we can't just have 'half as big' of everything! Doing this is a good exercise for anyone interested in using these techniques – either by computing a character as in Theorem 3.10, or by creating matrices. Note that the outcome space is likely to be isomorphic to a quotient of $\mathbb{Q}CO_4$, since reversals can no longer be distinguished!

5. The case of five

In this paper we do not attempt the characterizations for all n which often appear in the literature. This is largely because (as we have mentioned above) we don't have axiomatics for which subspaces are of most importance; contrast the situation with the $S^{(n-1,1)}$ and $S^{(n-2,1,1)}$ components in other related settings, such as voting on full orderings, or values for cooperative games. The complexity of the character computations also grow substantially with n and the number of partitions of n.

To give a taste of what lies in store, we end with a set of characterizations of neutral points-based procedures where voters pick a single cyclic order, as in Definition 2.3 with $n=5$; that is, where $B=CO_5$. The earlier-mentioned combinatorial explosion will already be evident, but so will intriguing consistency with our $n=4$ observations.

5.1. The decomposition. First, Corollary 3.11 gives us the character of $\mathbb{Q}CO_5$, which we can then apply Proposition 3.7 to in order to find[20] the decomposition of $\mathbb{Q}CO_5$ in terms of irreducible $\mathbb{Q}S_5$-modules.

PROPOSITION 5.1. *We can decompose $\mathbb{Q}CO_5$ as a sum of irreducible $\mathbb{Q}S_5$-modules as follows:*

$$\mathbb{Q}CO_5 \simeq S^{(5)} \oplus S^{(3,2)} \oplus {S^{(3,1,1)}}^{\oplus 2} \oplus S^{(2,2,1)} \oplus S^{(1,1,1,1,1)}$$

where these modules have dimensions 1, 5, $2\cdot 6$, 5, *and* 1, *respectively.*

[20]The interactive calculator at `https://www.jgibson.id.au/articles/characters/` is helpful for doing these computations, as well as being remarkably addictive.

PROOF. This is a standard calculation, so we just give $S^{(3,1,1)}$ as an example. The only nonzero values of χ are for the identity and for the conjugacy class of 5-cycles, and the values of $\chi_{(3,1,1)}$ for those are 6 and 1. There is one element in the identity's class, and 24 in the 5-cycle class, so we compute:

$$24 \cdot 1 \cdot 6 + 4 \cdot 1 \cdot 24 = 240 = 2 \cdot 120$$

which verifies this irreducible has multiplicity two in the decomposition. □

COROLLARY 5.2. *Any neutral points-based rule on CO_5 where ballots are a single cyclic order only depends on eight parameters.*

PROOF. We have one parameter each by Schur's Lemma for $S^{(5)}$, $S^{(3,2)}$, $S^{(2,2,1)}$, and $S^{(1,1,1,1,1)}$. Further, each component of $S^{(3,1,1)^{\oplus 2}}$ (in any particular way to write it as a direct sum) needs two parameters to describe what linear combination of vectors in the (same two) components of $S^{(3,1,1)^{\oplus 2}}$ it goes to. □

This means that even though there are many more possible irreducibles for $n = 5$ than $n = 4$, the degrees of freedom are not much more than for the ROLO ballots for $n = 4$.

More interesting is identifying which concrete subspaces actually correspond to the irreducibles, and whether these identify new voting-theoretic ideas. For this purpose, we fix the following ordering[21] of CO_5. Each cyclic order is paired with its reversal.

$(ABCDE)$	$(AEDCB)$	$(ABCED)$	$(ADECB)$	$(ABDCE)$	$(AECDB)$
$(ABDEC)$	$(ACEDB)$	$(ABECD)$	$(ADCEB)$	$(ABEDC)$	$(ACDEB)$
$(ACBDE)$	$(AEDBC)$	$(ACDBE)$	$(AEBDC)$	$(ACEBD)$	$(ADBEC)$
$(ADBCE)$	$(AECBD)$	$(AEBCD)$	$(ADCBE)$	$(ACBED)$	$(ADEBC)$

FIGURE 5. Cyclic orders for $n = 5$

In Figure 6 we have the full decomposition of the vector space $\mathbb{Q}^{24}$ with respect to the S_5 action. In this figure, it is helpful to think of the 24 entries in each vector as corresponding to twelve sets of reversal pairs. Some subspaces are easier to identify than others! For instance, $S^{(5)}$ will clearly be the space spanned by $(1, 1, \ldots, 1)^T$. Since when $n = 5$ any 5-cycle is already an even permutation, we can likewise ignore the 'cyclic' nature of the cyclic orders and see that the sign representation $S^{(1,1,1,1,1)}$ must simply have ± 1 as according to whether the underlying permutation is even or odd. The other irreducibles need some additional ingenuity (and computational assistance) to find, but once found then it is once again a standard character computation to verify they are correct.

EXAMPLE 5.3. As an example, consider the $S^{(3,1,1)^{\oplus 2}}$ component in Figure 6 and compare to the ordering in Figure 5. It might be surprising that the space generated by all profiles (or differentials) which support a given cyclic order and oppose its reversal would have only one type of irreducible, one type of symmetry

[21] As with the ROLO ballots, there are several other possible useful orderings, some even more symmetric, but this one works well for our purpose.

– and, despite the same thing happening in Corollary 2.7, this does not generalize even to $n = 6$ or $n = 7$.

Yet the computation is clear. The space is clearly completely fixed under the identity. A given 5-cycle will fix *two* of the reversal pairs; for instance, $(ABCDE)$ fixes the pairs $(ABCDE), (AEDCB)$ and $(ACEBD), (ADBEC)$. Finally, while most other cycle classes obviously cannot fix anything, a product of two-cycles, such as $(BE)(CD)$ can fix a pair, but swapping the places of a cyclic order and its reversal. A given element does so with four of the pairs[22], so the trace of the action on this subspace for such an element is -4. This is exactly twice the character for $S^{(3,1,1)}$ for all classes, so the space is isomorphic (not canonically, to be sure) to $S^{(3,1,1)} \oplus S^{(3,1,1)}$.

Before we look at specifics of the voting rules, let's now use these spaces (and the ones from $n = 4$) to propose some potential voting ideas, with brief commentary. The reader does not have to agree with the proposals! They are meant solely as food for thought. We do think they are reasonable, and in most cases the decompositions merely confirm what one might have thought of anyway.

PROPOSAL 5.4. The following should be considered as useful data on any voting system on cyclic orders, points-based or not:

- The number of total voters in a profile should be recorded.
  - This corresponds to the trivial subspace, as usual. One could choose to kill it or to use it when adding profiles, as in normal voting.
- When n is odd, since the number of simple transpositions needed to convert a cyclic order into another one is invariant, this number and its parity may be taken into account in evaluating a system.
  - For instance, one could ask for a sign subspace to go to zero.
- There is a fundamental distinction between systems that consider there to be no difference between a cyclic order and its reversal, and those which distinguish between them.
  - One doesn't need the subspaces which are generated by profile differentials of the type $\{1, -1, 0, \ldots, 0\}$ supporting an order and denigrating its reversal to assert this, but their consistency certainly helps.
- Reversal pairs should be considered to behave in some ways like candidates in ordinary voting theory, and procedures should be designed/analyzed with this in mind.
  - The $S^{(2,2)}$, $S^{(3,2)}$, and $S^{(2,2,1)}$ spaces support this line of thought.

Here is a less evident proposal. Recall the computation of the character of $S^{(3,1,1)^{\oplus 2}}$ in Figure 6 and observe the distinction between $S^{(3,2)}$ and $S^{(2,2,1)}$.

PROPOSAL 5.5. There is meaningful voting information involved in the distinction between 'seating' people in a given cyclic order and instead seating them k 'seats' apart (in the same order) for some $k > 1$, when this is possible number-theoretically.

This is a generalization of reversal, which may be considered to be the case $k = n - 1$. For $n = 4$ no others are possible, but for $n = 5$ several of these

[22]For this element, it would be the pairs $(ABCDE), (AEDCB)$, $(ACEBD), (ADBEC)$, $(ABDCE), (AECDB)$, and $(ACBED), (ADEBC)$ which have B, E or C, D surrounding A in the cyclic order.

The decomposition $\mathbb{Q}CO_5$ as S_5-invariant subspaces.

- The trivial subspace $S^{(5)} = T$:

$$S^{(5)} = \text{span}\left\{(1,1,1,1,1,1,1,1,1,1,1,1,1,1,1,1,1,1,1,1,1,1,1,1)^T\right\}$$

- The sign subspace $S^{(1,1,1,1,1)}$ gives ± 1 as outlined above, where

$$S^{(1,1,1,1)} = \text{span}\left\{(1,1,-1,-1,-1,-1,1,1,1,1,-1,-1,-1,-1,1,1,-1,-1,1,1,-1,-1,1,1)^T\right\}$$

- We then have two *non*-isomorphic, but similar, subspaces. $S^{(3,2)}$ is the span of the $\{y_i\}$ and $S^{(2,2,1)}$ is the span of the y_i.

$y_1 = (5,5,-1,-1,-1,-1,-1,-1,-1,-1,-1,-1,-1,-1,-1,-1,5,5,-1,-1,-1,-1,-1,-1)^T,$

$y_2 = (-1,-1,5,5,-1,-1,-1,-1,-1,-1,-1,-1,-1,-1,-1,-1,5,5,-1,-1,-1,-1,-1,-1)^T,$

$y_3 = (-1,-1,-1,-1,5,5,-1,-1,-1,-1,-1,-1,-1,-1,-1,-1,-1,-1,-1,-1,-1,-1,5,5)^T,$

$y_4 = (-1,-1,-1,-1,-1,-1,5,5,-1,-1,-1,-1,-1,-1,-1,-1,-1,-1,-1,-1,5,5,-1,-1)^T,$

$y_5 = (-1,-1,-1,-1,-1,-1,-1,-1,5,5,-1,-1,5,5,-1,-1,-1,-1,-1,-1,-1,-1,-1,-1)^T$

$z_1 = (5,5,1,1,1,1,-1,-1,-1,-1,1,1,1,1,-1,-1,-5,-5,-1,-1,1,1,-1,-1)^T,$

$z_2 = (1,1,5,5,-1,-1,1,1,1,1,-1,-1,-1,-1,-5,-5,-1,-1,1,1,-1,-1,1,1)^T,$

$z_3 = (1,1,-1,-1,5,5,1,1,1,1,-1,-1,-1,-1,1,1,-1,-1,1,1,-1,-1,-5,-5)^T,$

$z_4 = (-1,-1,1,1,1,1,-1,-1,5,5,1,1,-5,-5,-1,-1,1,1,-1,-1,1,1,-1,-1)^T,$

$z_5 = (1,1,-1,-1,-1,-1,1,1,1,1,5,5,-1,-1,1,1,-1,-1,-5,-5,-1,-1,1,1)^T$

- The final (reducible) subspace $S^{(3,1,1)^{\oplus 2}}$ is the span of twelve very basic vectors.

$$(1,-1,0,0,0,0,0,0,0,0,0,0,0,0,0,0,0,0,0,0,0,0,0,0)^T,$$
$$(0,0,1,-1,0,0,0,0,0,0,0,0,0,0,0,0,0,0,0,0,0,0,0,0)^T,$$
$$(0,0,0,0,1,-1,0,0,0,0,0,0,0,0,0,0,0,0,0,0,0,0,0,0)^T,$$
$$(0,0,0,0,0,0,1,-1,0,0,0,0,0,0,0,0,0,0,0,0,0,0,0,0)^T,$$
$$(0,0,0,0,0,0,0,0,1,-1,0,0,0,0,0,0,0,0,0,0,0,0,0,0)^T,$$
$$(0,0,0,0,0,0,0,0,0,0,1,-1,0,0,0,0,0,0,0,0,0,0,0,0)^T,$$
$$(0,0,0,0,0,0,0,0,0,0,0,0,1,-1,0,0,0,0,0,0,0,0,0,0)^T,$$
$$(0,0,0,0,0,0,0,0,0,0,0,0,0,0,1,-1,0,0,0,0,0,0,0,0)^T,$$
$$(0,0,0,0,0,0,0,0,0,0,0,0,0,0,0,0,1,-1,0,0,0,0,0,0)^T,$$
$$(0,0,0,0,0,0,0,0,0,0,0,0,0,0,0,0,0,0,1,-1,0,0,0,0)^T,$$
$$(0,0,0,0,0,0,0,0,0,0,0,0,0,0,0,0,0,0,0,0,1,-1,0,0)^T,$$
$$(0,0,0,0,0,0,0,0,0,0,0,0,0,0,0,0,0,0,0,0,0,0,1,-1)^T$$

FIGURE 6. List of invariant subspaces for CO_5 ballots

reorderings are possible since 5 is prime. For example, note the distinction between $(ABCDE)$ and $(ACEBD)$ via the *right* action of (1254), which is $k = 3$ in this setting.

See Example 5.7 for evidence in favor of this. We strongly suspect that *additional* ideas will come with further study of the subspaces – see also Section 6.

5.2. Voting rules. Although we will give the most general form of the matrix for $s : CO_5 \to CO_5$ shortly, one 24×24 matrix will suffice! But we can give a voting rule simply from s itself, as we have now practiced several times.

EXAMPLE 5.6. Let $s : CO_5 \to CO_5$ be the scoring function which gives points as follows:

- It awards four points if the two cyclic orders are the same.
- If they differ by a transposition, score three points.
- If they differ by two transpositions, score two points
- If by three, score one point.
- The reversal (which differs by four transpositions) scores zero points.

To be concrete, we have scores such as $s((ABCDE),(ABDCE)) = 3$, $s((ABCDE),(AEDCB)) = 0$, and $s((ABCDE),(ABECD)) = 2$. The way in which it is defined makes it clear it is neutral, and one could perhaps consider this to be a 'Borda-like' system.

Now, what can we say about this system? Even without a matrix (but see Proposition 5.8) it is pretty easy, if tedious, to compute that *none* of the vectors in the decomposition of $\mathbb{Q}CO_5$ are killed. In fact, most are simply dilated by a factor of two.

However, the twelve-dimensional space $S^{(3,1,1)^{\oplus 2}}$ has a more complicated combination where a profile vector supporting a cyclic order (and opposing its reversal) is sent to an outcome vector giving:

- Four points to the order itself,
- Negative four points to its reversal,
- Two points to one differing by a transposition,
- Negative two points to *their* reversals, and
- Zero to everything else.

That's a lot more like what we should be used to seeing from more-studied voting systems!

We can create a system which *only* performs the latter transformation by taking the hint to have a sum-zero set of weights, along with one more change.

EXAMPLE 5.7. Let $s : CO_5 \to CO_5$ be the scoring function which gives points as follows:

- It awards two points if the two cyclic orders are the same.
- If they differ by a transposition, score one point.
- The reversal of one differing by a transposition scores negative one point.
- The reversal will score negative two points.
- All other scores are two points.

All scoring is the same as before (minus two points each), except that now $s((ABCDE),(ACEBD)) = 0$, which otherwise would have been -1 if we had just subtracted two (and its reversal would have also garnered -1 before). (Recall Proposal 5.5.)

This system behaves *exactly the same* as the one in Example 5.6 on the 12-dimensional subspace, but completely ignores the information from the other subspaces! Which, to be honest, is a lot like what the Borda Count does; it amplifies one very specific piece of information (profiles which emphasize one particular candidate, symmetrically) and intentionally ignores any other information. This is an appropriate system to end our initial investigations on!

It is possible to construct lengthier justifications of all this, but the easiest way to do so is for us to finally provide a matrix that represents any such voting rule. The justification that this works is simply matrix multiplication on each of the subspaces in question, though of course *finding* this matrix required ingenuity like in Proposition 2.6. Our hope is that some of the ideas in Section 6 will help in providing general results of this type in the future.

PROPOSITION 5.8. *Any neutral points-based voting rule on CO_5 must have a matrix of scoring function values of the following form:*

$$\begin{pmatrix}
a & b & c & d & c & d & e & f & e & f & d & c & c & d & e & f & g & h & e & f & c & d & f & e \\
b & a & d & c & d & c & f & e & f & e & c & d & d & c & f & e & h & g & f & e & d & c & e & f \\
c & d & a & b & e & f & d & c & c & d & e & f & f & e & g & h & e & f & c & d & e & f & c & d \\
d & c & b & a & f & e & c & d & d & c & f & e & e & f & h & g & f & e & d & c & f & e & d & c \\
c & d & e & f & a & b & c & d & d & c & e & f & e & f & d & c & e & f & c & d & f & e & h & g \\
d & c & f & e & b & a & d & c & c & d & f & e & f & e & c & d & f & e & d & c & e & f & g & h \\
e & f & d & c & c & d & a & b & e & f & c & d & c & d & f & e & d & c & f & e & h & g & f & e \\
f & e & c & d & d & c & b & a & f & e & d & c & d & c & e & f & c & d & e & f & g & h & e & f \\
e & f & c & d & d & c & e & f & a & b & c & d & h & g & e & f & d & c & f & e & d & c & e & f \\
f & e & d & c & c & d & f & e & b & a & d & c & g & h & f & e & c & d & e & f & d & c & f & e \\
d & c & e & f & e & f & c & d & c & d & a & b & f & e & d & c & f & e & h & g & f & e & c & d \\
c & d & f & e & f & e & d & c & d & c & b & a & e & f & c & d & e & f & g & h & e & f & d & c \\
c & d & f & e & e & f & c & d & g & h & f & e & a & b & c & d & e & f & d & c & f & e & c & d \\
d & c & e & f & f & e & d & c & h & g & e & f & b & a & d & c & f & e & c & d & e & f & d & c \\
e & f & h & g & d & c & f & e & e & f & d & c & c & d & a & b & d & c & e & f & d & c & e & f \\
f & e & g & h & c & d & e & f & f & e & c & d & d & c & b & a & c & d & f & e & c & d & f & e \\
h & g & e & f & e & f & d & c & d & c & f & e & e & f & d & c & a & b & d & c & e & f & c & d \\
g & h & f & e & f & e & c & d & c & d & e & f & f & e & c & d & b & a & c & d & f & e & d & c \\
e & f & c & d & c & d & f & e & f & e & g & h & d & c & e & f & d & c & a & b & d & c & f & e \\
f & e & d & c & d & c & e & f & e & f & h & g & c & d & f & e & c & d & b & a & c & d & e & f \\
c & d & e & f & f & e & g & h & c & d & f & e & f & e & d & c & e & f & d & c & a & b & d & c \\
d & c & f & e & e & f & h & g & d & c & e & f & e & f & c & d & f & e & c & d & b & a & c & d \\
f & e & c & d & g & h & f & e & e & f & c & d & c & d & e & f & c & d & f & e & d & c & a & b \\
e & f & d & c & h & g & e & f & f & e & d & c & d & c & f & e & d & c & e & f & c & d & b & a
\end{pmatrix}$$

This transformation scales each irreducible subspace in Figure 6 as follows:

- *The trivial subspace is scaled by* $a + b + 5c + 5d + 5e + 5f + g + h$.
- *The sign subspace is scaled by* $a + b - 5c - 5d + 5e + 5f - g - h$.
- *The five-dimensional subspaces* $S^{(3,2)}$ *and* $S^{(2,2,1)}$ *are scaled by* $a+b-c-d-e-f+g+h$ *and* $a+b+c+d-e-f-g-h$*, respectively.*
- *Any one of the vectors in the twelve-dimensional subspace* $S^{(3,1,1)^{\oplus 2}}$ *corresponding to a cyclic order* x *is sent to a vector with*
  - $a-b$ *points for* x *(and* $b-a$ *points for its reversal),*
  - $c-d$ *points for the orders differing by one transposition (and* $d-c$ *for their reversals),*
  - $e-f$ *points for the orders differing by a 3-cycle (and* $f-e$ *for their reversals, which differ by two disjoint transpositions),*

– *$g - h$ points for the (unique!) order differing as in Proposal 5.5, and $h - g$ points for its reversal.*

6. Further work and acknowledgments

It should be clear that this paper is just scratching the surface of a larger research program in voting on cyclic orders under various ballots. In particular, we have only dealt with points-based procedures based on the scoring functions. While successfully *completely* characterizing all systems with ballot space CO_n seems unlikely, future possible directions include the following.

- Deciding which, if any, of the ideas in Proposals 5.4 or 5.5 are worthy of further investigation.
- Hence, an important extension is to analyze voting when we do *not* worry about reversal. This halves the number of cyclic orders, and one may note that the decompositions already provided in this paper clearly have some subspaces which would remain invariant if reversed orders were identified. There are evident generalizations of Theorem 3.10 which would immediately yield further detail.
- Unlike some alternate ballots, the ROLO systems are easily extended to $n \geq 4$. As noted above, the situation of more ballots than cyclic orders is a low-dimensional phenomenon.
- Finally, there are important graphs that are associated with the various sets of cyclic orders (see also below). They have additional symmetry which may lead to a finer-grained analysis.

In addition (or perhaps more importantly), using the algebra to come up with axiomatics of the usual voting variety would be a very good goal. It seems very unlikely that there are not analogues to the Borda or Kemeny rules (for instance, as maximum likelihood estimators). Alternately to raw axiomatics, using the methods of [**5**] or [**40**] might be a good place to start. This would be fairly necessary to address ballots consisting of full or truncated rankings of cyclic orders, analysis of which otherwise would look pretty much like similar ballots in 'ordinary' voting theory.

A different direction for further research is the set of cyclic orders[23]. There is quite a bit of work on cyclic orders considered as ternary relations, not focusing at all on overall structure of the set of cyclic orders. For example, one may ask whether a partial cyclic order (as a relation) can be completed to a total cyclic order (like the ones under discussion in this paper); this problem is actually NP-complete. See among many others [**18, 24, 29**] and especially many papers by Novák, such as [**26, 27**]. They also appear in many graph-theoretic contexts, such as graph crossing numbers; the paper [**38**] proves some very interesting things about a graph[24] with cyclic orders as vertices. Nonetheless there is a lot more which could be asked about them[25], particularly regarding distances on the set. We welcome any references we are not aware of for this interesting object.

[23]Also called circular orders, circular permutations, etc. These objects have often been rediscovered.

[24]For graphs, see also [**14**], [**25**], and [**21**].

[25]Cyclic orders are also used in defining so-called ribbon graphs in mathematical physics, but there they simply fulfill the role of a topological datum.

The authors wish to particularly thank the Gordon College Provost's Summer Undergraduate Research Fellowship for support, and the Gordon College Department of Mathematics and Computer Science, Mike Orrison, and Bill Zwicker for helpful early feedback. We also wish to thank the organizers of the AMS Special Session on The Mathematics of Decisions, Elections and Games for inviting a talk based on this research, for organizing this volume of papers, and for helpful comments from the editors and a referee. We finally want to recognize the teams behind Sage Math ([**37**]) and GAP ([**17**]) for software that greatly eased the exploration of these topics, particularly with respect to computing characters and kernels.

References

[1] Michael Ackerman, Sul-Young Choi, Peter Coughlin, Eric Gottlieb, and Japheth Wood, *Elections with partially ordered preferences*, Public Choice **157** (2013), no. 1-2, 145–168 (English).

[2] José Carlos R. Alcantud and Annick Laruelle, *Dis&approval voting: a characterization*, Soc. Choice Welf. **43** (2014), no. 1, 1–10, DOI 10.1007/s00355-013-0766-7. MR3217234

[3] Kenneth J. Arrow, *A difficulty in the concept of social welfare*, The Journal of Political Economy **58** (1950), no. 4, 328–346.

[4] Hélène Barcelo, Megan Bernstein, Sarah Bockting-Conrad, Erin McNicholas, Kathryn Nyman, and Shira Viel, *Algebraic voting theory & representations of $S_m \wr S_n$*, Adv. in Appl. Math. **120** (2020), 102077, 23, DOI 10.1016/j.aam.2020.102077. MR4121950

[5] Jean-Pierre Barthélémy and Bernard Monjardet, *The median procedure in cluster analysis and social choice theory*, Math. Social Sci. **1** (1980/81), no. 3, 235–267, DOI 10.1016/0165-4896(81)90041-X. MR616379

[6] Steven J. Brams and Peter C. Fishburn, *Approval voting*, The American Political Science Review **72** (1978), no. 3, 831–847.

[7] Steven J. Brams and D. Marc Kilgour, *Satisfaction approval voting*, Voting power and procedures, Stud. Choice Welf., Springer, Cham, 2014, pp. 323–346, DOI 10.1007/978-3-319-05158-1_18. MR3290332

[8] Vincent Conitzer, Matthew Rognlie, and Lirong Xia, *Preference functions that score rankings and maximum likelihood estimation*,Proceedings of the 21st International Joint Conference on Artifical Intelligence (San Francisco, CA, USA), IJCAI'09, Morgan Kaufmann Publishers Inc., 2009, pp. 109–115.

[9] Karl-Dieter Crisman, *The Borda count, the Kemeny rule, and the permutahedron*, The mathematics of decisions, elections, and games, Contemp. Math., vol. 624, Amer. Math. Soc., Providence, RI, 2014, pp. 101–134, DOI 10.1090/conm/624/12475. MR3308273

[10] Karl-Dieter Crisman and Michael E. Orrison, *Representation theory of the symmetric group in voting theory and game theory*, Algebraic and geometric methods in discrete mathematics, Contemp. Math., vol. 685, Amer. Math. Soc., Providence, RI, 2017, pp. 97–115, DOI 10.1090/conm/685. MR3625572

[11] John Cullinan, Samuel K. Hsiao, and David Polett, *A Borda count for partially ordered ballots*, Soc. Choice Welf. **42** (2014), no. 4, 913–926, DOI 10.1007/s00355-013-0751-1. MR3196998

[12] Zajj Daugherty, Alexander K. Eustis, Gregory Minton, and Michael E. Orrison, *Voting, the symmetric group, and representation theory*, Amer. Math. Monthly **116** (2009), no. 8, 667–687, DOI 10.4169/193009709X460796. MR2572103

[13] Matt Davis, Michael E. Orrison, and Francis Edward Su, *Voting for committees in agreeable societies*, The mathematics of decisions, elections, and games, Contemp. Math., vol. 624, Amer. Math. Soc., Providence, RI, 2014, pp. 147–157, DOI 10.1090/conm/624/12477. MR3308275

[14] Duane DeTemple and Jack Robertson, *Permutations associated with billiard paths*, Discrete Math. **47** (1983), no. 2-3, 211–219, DOI 10.1016/0012-365X(83)90091-2. MR724659

[15] Ronald Fagin, Ravi Kumar, Mohammad Mahdian, D. Sivakumar, and Erik Vee, *Comparing partial rankings*, SIAM J. Discrete Math. **20** (2006), no. 3, 628–648, DOI 10.1137/05063088X. MR2272220

[16] Piotr Faliszewski, Piotr Skowron, Arkadii Slinko, and Nimrod Talmon, *Multiwinner voting: a new challenge for social choice theory*, Trends in computational social choice, AI Access, [2017] ©2017, pp. 27–47. MR4481023

[17] The GAP Group, *GAP – Groups, Algorithms, and Programming, Version 4.11.1*, 2021.

[18] Stefan Haar, *Cyclic ordering through partial orders*, J. Mult.-Valued Logic Soft Comput. **27** (2016), no. 2-3, 209–228. MR3532465

[19] Gordon James and Adalbert Kerber, *The representation theory of the symmetric group*, Encyclopedia of Mathematics and its Applications, vol. 16, Addison-Wesley Publishing Co., Reading, Mass., 1981. With a foreword by P. M. Cohn; With an introduction by Gilbert de B. Robinson. MR644144

[20] Norman L. Kleinberg and Jeffrey H. Weiss, *Algebraic structure of games*, Math. Social Sci. **9** (1985), no. 1, 35–44, DOI 10.1016/0165-4896(85)90005-8. MR795555

[21] Daniel J. Kleitman, *The crossing number of* $K_{5,n}$, J. Combinatorial Theory **9** (1970), 315–323. MR280403

[22] Jean-François Laslier and M. Remzi Sanver (eds.), *Handbook on approval voting*, Studies in Choice and Welfare, Springer, Heidelberg, 2010, DOI 10.1007/978-3-642-02839-7. MR2742437

[23] Stephen Lee, *Understanding voting for committees using wreath products*, 2010.

[24] Nimrod Megiddo, *Partial and complete cyclic orders*, Bull. Amer. Math. Soc. **82** (1976), no. 2, 274–276, DOI 10.1090/S0002-9904-1976-14020-7. MR404065

[25] Nagi H. Nahas, *On the crossing number of* $K_{m,n}$, Electron. J. Combin. **10** (2003), Note 8, 6, DOI 10.37236/1748. MR2014536

[26] Vítězslav Novák and Miroslav Novotný, *On determination of a cyclic order*, Czechoslovak Math. J. **33(108)** (1983), no. 4, 555–563. MR721087

[27] Miroslav Novotný and Vítězslav Novák, *Dimension theory for cyclically and cocyclically ordered sets*, Czechoslovak Math. J. **33(108)** (1983), no. 4, 647–653. MR721091

[28] Maria Silvia Pini, Francesca Rossi, Kristen Brent Venable, and Toby Walsh, *Aggregating partially ordered preferences*, J. Logic Comput. **19** (2009), no. 3, 475–502, DOI 10.1093/logcom/exn012. MR2506683

[29] Alain Quilliot, *Cyclic orders*, European J. Combin. **10** (1989), no. 5, 477–488, DOI 10.1016/S0195-6698(89)80022-8. MR1014556

[30] Thomas C. Ratliff, *Selecting commitees*, Public Choice **126** (2006), no. 3, 343–355.

[31] Thomas C. Ratliff and Donald G. Saari, *Complexities of electing diverse committees*, Soc. Choice Welf. **43** (2014), no. 1, 55–71, DOI 10.1007/s00355-013-0773-8. MR3217238

[32] Donald G. Saari, *Basic geometry of voting*, Springer-Verlag, Berlin, 1995, DOI 10.1007/978-3-642-57748-2. MR1410265

[33] Donald G. Saari, *Explaining all three-alternative voting outcomes*, J. Econom. Theory **87** (1999), no. 2, 313–355, DOI 10.1006/jeth.1999.2541. MR1713824

[34] Donald G. Saari, *Mathematical structure of voting paradoxes. I. Pairwise votes*, Econom. Theory **15** (2000), no. 1, 1–53, DOI 10.1007/s001990050001. MR1731508

[35] Donald G. Saari, *Mathematical structure of voting paradoxes. II. Positional voting*, Econom. Theory **15** (2000), no. 1, 55–102, DOI 10.1007/s001990050001. MR1731509

[36] Bruce E. Sagan, *The symmetric group*, 2nd ed., Graduate Texts in Mathematics, vol. 203, Springer-Verlag, New York, 2001. Representations, combinatorial algorithms, and symmetric functions, DOI 10.1007/978-1-4757-6804-6. MR1824028

[37] W. A. Stein, M. Hansen, et al., *Sage Mathematics Software (Version 9.0)*, The Sage Development Team, 2020, `http://www.sagemath.org`.

[38] D. R. Woodall, *Cyclic-order graphs and Zarankiewicz's crossing-number conjecture*, J. Graph Theory **17** (1993), no. 6, 657–671, DOI 10.1002/jgt.3190170602. MR1244681

[39] William S. Zwicker, *The voters' paradox, spin, and the Borda count*, Math. Social Sci. **22** (1991), no. 3, 187–227, DOI 10.1016/0165-4896(91)90023-K. MR1149504

[40] William S. Zwicker, *Consistency without neutrality in voting rules: When is a vote an average?*, Math. Comput. Modelling **48** (2008), no. 9-10, 1357–1373, DOI 10.1016/j.mcm.2008.05.031. MR2459706

Department of Mathematics and Computer Science, Gordon College, Wenham, Massachusetts 01902

Email address: karl.crisman@gordon.edu

Department of Mathematics and Computer Science, Gordon College, Wenham, Massachusetts 01902

Department of Mathematics and Computer Science, Gordon College, Wenham, Massachusetts 01902

Department of Mathematics and Computer Science, Gordon College, Wenham, Massachusetts 01902

Contemporary Mathematics
Volume **795**, 2024
https://doi.org/10.1090/conm/795/15968

Impossibility theorems involving weakenings of expansion consistency and resoluteness in voting

Wesley H. Holliday, Chase Norman, Eric Pacuit, and Saam Zahedian

Abstract. A fundamental principle of individual rational choice is Sen's γ axiom, also known as *expansion consistency*, stating that any alternative chosen from each of two menus must be chosen from the union of the menus. Expansion consistency can also be formulated in the setting of social choice. In voting theory, it states that any candidate chosen from two fields of candidates must be chosen from the combined field of candidates. An important special case of the axiom is *binary* expansion consistency, which states that any candidate chosen from an initial field of candidates and chosen in a head-to-head match with a new candidate must also be chosen when the new candidate is added to the field, thereby ruling out *spoiler effects*. In this paper, we study the tension between this weakening of expansion consistency and weakenings of *resoluteness*, an axiom demanding the choice of a single candidate in any election. As is well known, resoluteness is inconsistent with basic fairness conditions on social choice, namely anonymity and neutrality. Here we prove that even significant weakenings of resoluteness, which are consistent with anonymity and neutrality, are inconsistent with binary expansion consistency. The proofs make use of SAT solving, with the correctness of a SAT encoding formally verified in the Lean Theorem Prover, as well as a strategy for generalizing impossibility theorems obtained for special types of voting methods (namely majoritarian and pairwise voting methods) to impossibility theorems for arbitrary voting methods. This proof strategy may be of independent interest for its potential applicability to other impossibility theorems in social choice.

Contents

2020 *Mathematics Subject Classification.* 91B12, 91B14.
Key words and phrases. Social choice theory, voting, expansion consistency, impossibility theorems, SAT solving, formal verification.

References

1. Introduction

Since the inception of modern social choice theory, a key question has been to what extent principles of individual rational choice can be imposed on social choice. For his famous Impossibility Theorem, Arrow [**1**] assumed that society's choice function obeys the same rationality constraints as a rational individual's choice function. In combination with his axioms of Independence of Irrelevant Alternatives (IIA) and universal domain, Arrow's social rationality assumptions leave room for only undemocratic social choice functions (see, e.g., [**57**, Thm. 5]). In fact, in the presence of IIA and universal domain, even much weaker social rationality assumptions still lead to a slew of Arrow-style impossibility theorems (see, e.g., [**54**, Ch. 3]). More important for our purposes here is that even without IIA, principles of individual rational choice can cause problems in social choice.

On the conception of individual rational choice as a matter of *maximization*, a rational agent is representable as choosing from any menu of options the maximal elements according to some underlying relation of preference between options. As shown by Sen [**50**], assuming finitely many options, an individual's choices can be represented in this way if and only if their choices satisfy two axioms, known as α and γ, or *contraction consistency* and *expansion consistency*.

When applied to social choice—in particular to voting—contraction consistency states that a candidate chosen from some field of candidates must still be chosen in any smaller field in which the candidate remains, holding fixed the preferences of the voters. For example, if b is among the winning candidates in an election when the field is $\{a, b, c\}$, then b must still be among the winning candidates when the field is $\{a, b\}$. Yet this contradicts basic principles of social choice—in particular, anonymity and neutrality (see Section 2.2) plus majority rule in two-candidate elections. Consider an election in which three voters submit the following rankings of the candidates: *abc*, *bca*, *cab*. In this case, anonymity and neutrality require that a, b, and c are all tied for winning the election. Then by contraction consistency, b must be among the winners when the field is $\{a, b\}$. But this contradicts majority rule in two-candidate elections, since two voters rank a over b, while only one ranks b above a.[1] Thus, for one of the central problems in voting theory—namely, devising anonymous and neutral voting methods that agree with majority-rule in two candidate elections—contraction consistency is unacceptable.

By contrast, expansion consistency appears more benign. When applied to voting, expansion consistency states that any candidate chosen from two fields of candidates must be chosen from the combined field of candidates, holding fixed the preferences of the voters. For example, if a is chosen from $\{a, b, c\}$ and from $\{a, d, e\}$, then a must also be chosen from $\{a, b, c, d, e\}$. Now there are anonymous and neutral

[1] In fact, the same style of reasoning shows that contraction consistency is inconsistent with anonymity, neutrality, and the requirement that a single winner be chosen in any two-candidate election with an odd number of voters. For anonymity, neutrality, and the requirement together imply that for the election described in the main text, either majority rule is applied to its two-candidate subelections or minority rule is applied to its two-candidate subelections. In the first case, we are done by the argument in the main text. In the second case, contraction consistency implies that b must be among the winners when the field is $\{b, c\}$, which contradicts minority rule.

voting methods that agree with majority-rule in two candidate elections while also satisfying expansion consistency, as studied systematically in [**8**]. Examples include Split Cycle [**30, 31**], Top Cycle [**49, 52**], Uncovered Set [**21, 39**], and Weighted Covering [**23, 45**].

However, we will show that expansion consistency still imposes significant constraints on social choice. A striking feature of the voting methods satisfying expansion consistency listed above is that they are "less resolute" than some other well-known voting methods that violate expansion consistency, such as Beat Path [**47, 48**], Minimax [**35, 51**], and Ranked Pairs [**55, 58**]. A voting method is *resolute* if it selects a unique winner in any election. While resoluteness is inconsistent with anonymity and neutrality, as shown by considering the three-voter election above, one can weaken resoluteness to avoid such an inconsistency. For example, one may require a unique winner only in elections without the kind of symmetries that force a tie in the election above (technically, in any election that is *uniquely weighted*, as defined in Section 2.2). Such weaker forms of resoluteness are satisfied by some of the voting methods that violate expansion consistency but violated by the voting methods that satisfy expansion consistency.[2] In this paper, we will prove that this is no coincidence: we identify weakenings of resoluteness that are inconsistent with expansion consistency.

In response, one may turn to weaker forms of expansion consistency, such as *binary* expansion consistency. Binary expansion consistency states that any candidate chosen from an initial field of candidates and (uniquely) chosen in a head-to-head match with a new candidate must also be chosen when the new candidate is added to the field. Some voting methods that violate expansion consistency still satisfy binary expansion consistency, which is good enough to rule out *spoiler effects* in elections (see Section 2.2). However, we will prove that the relevant weakenings of resoluteness are even inconsistent with binary expansion consistency. Thus, there appears to be a basic tension between axioms in the spirit of expansion consistency and axioms in the spirit of resoluteness, assuming anonymity and neutrality.

The rest of the paper is organized as follows. In Section 2, we motivate and precisely state our main impossibility theorems, and we prove special cases of the theorems. The rest of the paper is devoted to proving the theorems in full generality. Our strategy to do so is outlined in Section 3 and carried out in Sections 4–6. We conclude in Section 7 with some suggestions for future research. Appendices A-C contain details of computer-aided techniques used in the proofs of our theorems.

2. The impossibility theorems

2.1. Origins in individual choice. In this paper, we consider impossibility theorems arising from the attempt to transfer axioms on individual choice to social choice. We briefly recall the necessary background on individual choice. Given a finite set X, a *choice function* on X is a map $C : \wp(X) \setminus \{\varnothing\} \to \wp(X) \setminus \{\varnothing\}$ such that for all $S \in \wp(X) \setminus \{\varnothing\}$, we have $\varnothing \neq C(S) \subseteq S$. The intuitive interpretation is that when offered a choice from the menu S, the agent considers all the elements in $C(S)$ choiceworthy. To force the choice of a single alternative, one can impose

[2]The two-stage majoritarian rule proposed by Horan and Spurmont [**33**] satisfies expansion consistency and resoluteness but at the expense of violating neutrality. In this paper, we are interested only in anonymous and neutral methods.

the following axiom on C:

$$\textbf{resoluteness}: |C(S)| = 1 \text{ for all } S \in \wp(X) \setminus \{\varnothing\}.$$

The classic question concerning choice functions is a representation question: under what conditions can we view the agent as if she chooses from any given menu the greatest elements according to a *preference relation*? Where R is a binary relation on X, we say that C is *represented by* R if for all $S \in \wp(X) \setminus \{\varnothing\}$,

$$C(S) = \{x \in S \mid \text{for all } y \in S,\ xRy\}.$$

Sen [**50**] identified the following axioms on C that are necessary and sufficient for representability by a binary relation:

$$\alpha\text{: if } S \subseteq S'\text{, then } S \cap C(S') \subseteq C(S);$$
$$\gamma\text{: } C(S) \cap C(S') \subseteq C(S \cup S').$$

THEOREM 2.1 (Sen). *Let C be a choice function on a finite set X.*

(1) *C is representable by a binary relation on X iff C satisfies α and γ.*
(2) *if C is resolute, then the following are equivalent:*
 (a) *C is representable by a binary relation on X;*
 (b) *C is representable by a linear order on X;*
 (c) *C satisfies α.*

One can also characterize choice functions that satisfy only α (resp. γ) in terms of different notions of representability by a family of relations instead of a single relation (see [**22**, Thm. 6] for α and [**56**, Thm. 6B], [**22**, Thm. 7], and [**8**, Thm. 1] for γ).[3]

In this paper, we shall see that there are difficulties transferring even weaker versions of the above axioms from individual choice to social choice. First, consider the following weakening of γ:

$$\textbf{binary } \gamma\text{: if } x \in C(A) \text{ and } C(\{x,y\}) = \{x\}\text{, then } x \in C(A \cup \{y\}).$$

As Bordes [**5**] puts it (in the context of choice between candidates in voting, to which we turn in Section 2.2), "if x is a winner in A... one cannot turn x into a loser by introducing new alternatives to which x does not lose in duels" (p. 125). In fact, Bordes' formulation suggests replacing $C(\{x,y\}) = \{x\}$ by $x \in C(\{x,y\})$, resulting in a slightly stronger axiom, but we will work with the weaker version, as impossibility theorems with weaker axioms are stronger results. Binary γ is an obvious consequence of γ, setting $S = A$ and $S' = \{x,y\}$.

We will also consider a weakening of α with the same antecedent as binary γ:

$$\textbf{binary } \boldsymbol{\alpha}\text{: if } x \in C(A) \text{ and } C(\{x,y\}) = \{x\}\text{, then } C(A \cup \{y\}) \subseteq C(A).$$

In other words, adding to the menu an alternative that loses to an initially chosen alternative does not lead to any new choices from the menu. That binary α follows from α is not quite as immediate as the analogous implication for γ—but almost.

PROPOSITION 2.2. *Any choice function satisfying α also satisfies binary α.*

---

[3] One can also consider a notion of representability by a binary relation between subsets of X, rather than between elements of X, and there are analogues of α and γ in this setting of set-based rationalizability [**13**].

PROOF. Suppose $x \in C(A)$ and $C(\{x,y\}) = \{x\}$. Since $\{x,y\} \subseteq A \cup \{y\}$, we have $\{x,y\} \cap C(A \cup \{y\}) \subseteq C(\{x,y\})$ by α. The since $y \notin C(\{x,y\})$, we conclude $y \notin C(A \cup \{y\})$. Hence $C(A \cup \{y\}) \subseteq A$. By α again, $A \cap C(A \cup \{y\}) \subseteq C(A)$, which with the previous sentence implies $C(A \cup \{y\}) \subseteq C(A)$. □

Finally, we will consider a weakening of resoluteness. This weakening says that while some non-singleton choice sets may be allowed, we should never enlarge the choice set when adding an alternative that loses to some initially chosen alternative:

α-resoluteness: if $x \in C(A)$ and $C(\{x,y\}) = \{x\}$, then $|C(A \cup \{y\})| \leq |C(A)|$.

Clearly α-resoluteness is implied by binary α and by resoluteness.

2.2. From individual choice to social choice. We now explain how the axioms on individual choice from Section 2.1 can be applied to social choice.

Fix infinite sets $\mathcal{X}$ and $\mathcal{V}$ of *candidates* and *voters*, respectively. A *profile* $\mathbf{P}$ is a function $\mathbf{P} : V \to \mathcal{L}(X)$ where V is a finite subset of $\mathcal{V}$, X is a finite subset of $\mathcal{X}$, and $\mathcal{L}(X)$ is the set of all strict linear orders on X.[4] We write $V(\mathbf{P})$ for V and $X(\mathbf{P})$ for X. Let prof be the set of all profiles. Given a set A of candidates, the profile $\mathbf{P}|_A$ assigns to each voter their ranking of the candidates in $\mathbf{P}$ restricted to A, and for a particular candidate y, we write $\mathbf{P}_{-y} = \mathbf{P}|_{X(\mathbf{P})\setminus\{y\}}$.

Given two candidates x, y in a profile $\mathbf{P}$, the *margin of x over y in* $\mathbf{P}$ is the number of voters who rank x above y in $\mathbf{P}$ minus the number who rank y above x in $\mathbf{P}$; if this margin is positive, we say that x is *majority preferred to y in* $\mathbf{P}$. The *majority graph of* $\mathbf{P}$, denoted $M(\mathbf{P})$, is the directed graph whose set of vertices is $X(\mathbf{P})$ with an edge from x to y if x is majority preferred to y. The *margin graph of* $\mathbf{P}$, denoted $\mathcal{M}(\mathbf{P})$, is the weighted directed graph obtained from $M(\mathbf{P})$ by weighting the edge from x to y by the margin of x over y in $\mathbf{P}$. See Figure 1.

v_1	v_2	v_3	v_4	v_5	v_6	v_7	v_8	v_9
b	b	a	a	a	c	c	c	c
a	a	c	c	c	b	b	b	b
c	c	b	b	b	a	a	a	a

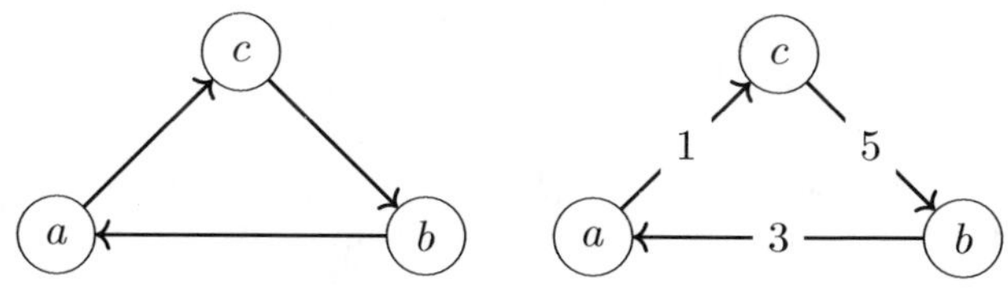

FIGURE 1. A preference profile $\mathbf{P}$ (top), its majority graph $M(\mathbf{P})$ (bottom left), and its margin graph $\mathcal{M}(\mathbf{P})$ (bottom right).

[4]This linearity assumption strengthens our impossibility theorems. That is, our impossibility theorems for voting methods defined only on profiles of linear orders immediately imply corresponding impossibility theorems for voting methods defined on profiles of weak orders; for if there were a voting method satisfying the relevant axioms on profiles of weak orders, then restricting the voting method to profiles of linear orders would yield the kind of voting method we will prove does not exist. By contrast, proving that no voting method defined on profiles of weak orders satisfies certain axioms does not immediately imply that no voting method defined only on linear orders satisfies the axioms. Selecting winners in profiles of weak orders may be more difficult.

REMARK 2.3. Since voters submit linear orders of the candidates, it follows that all weights in $\mathcal{M}(\mathbf{P})$ must have the same parity, which is even if there are any distinct candidates with a margin of zero between them.

McGarvey [**38**] proved that any asymmetric directed graph is isomorphic to the majority graph of a profile. Debord [**20**] proved that any asymmetric directed graph with weighted edges satisfying the parity constraint in Remark 2.3 is isomorphic to the margin graph of a profile. We will review the proofs of these results in Section 4.2.

A *voting method* F assigns to each profile $\mathbf{P}$ in some nonempty set $\mathrm{dom}(F)$ of profiles a nonempty set $F(\mathbf{P})$ of candidates who are tied for winning the election. All of the axioms on individual choice functions from the previous section corresponds to axioms on voting methods as follows, where all profiles are assumed to belong to $\mathrm{dom}(F)$:

resoluteness: $|F(\mathbf{P})| = 1$ for all profiles $\mathbf{P} \in \mathrm{dom}(F)$;

α: if $Y \subseteq X(\mathbf{P})$, then $Y \cap F(\mathbf{P}) \subseteq F(\mathbf{P}|_Y)$;

γ: if $X(\mathbf{P}) = Y \cup Z$, then $F(\mathbf{P}|_Y) \cap F(\mathbf{P}|_Z) \subseteq F(\mathbf{P})$;

binary γ: if $x \in F(\mathbf{P}_{-y})$ and $F(\mathbf{P}|_{\{x,y\}}) = \{x\}$, then $x \in F(\mathbf{P})$;[5]

binary α: if $x \in F(\mathbf{P}_{-y})$ and $F(\mathbf{P}|_{\{x,y\}}) = \{x\}$, then $F(\mathbf{P}) \subseteq F(\mathbf{P}_{-y})$;

α-**resoluteness:** if $x \in F(\mathbf{P}_{-y})$ and $F(\mathbf{P}|_{\{x,y\}}) = \{x\}$, then $|F(\mathbf{P})| \leq |F(\mathbf{P}_{-y})|$.

Let us consider three examples illustrating how binary γ, binary α, and α-resoluteness compare in strength to γ and α. We begin by showing that binary γ is weaker than γ.

EXAMPLE 2.4. The Banks voting method [**3**] is defined as follows.[6] Say that a *chain* in $M(\mathbf{P})$ is a subset of $X(\mathbf{P})$ linearly ordered by the edge relation of $M(\mathbf{P})$. Then $a \in \mathit{Banks}(\mathbf{P})$ if a is the maximum element with respect to the edge relation of some maximal chain in $M(\mathbf{P})$. For convenience, we repeat here a proof that Banks satisfies binary γ but not γ from [**30**].

For binary γ, suppose $a \in \mathit{Banks}(\mathbf{P}_{-b})$ and $a \in \mathit{Banks}(\mathbf{P}|_{\{a,b\}})$. Hence a is the maximum element of some maximal chain R in $M(\mathbf{P}_{-b})$. Let R' be a maximal chain in $M(\mathbf{P})$ with $R \subseteq R'$. Since $a \in \mathit{Banks}(\mathbf{P}|_{\{a,b\}})$, b is not the maximum of R', so a is the maximum of R', so $a \in \mathit{Banks}(\mathbf{P})$. Note this establishes the strengthening of binary γ with $x \in F(\mathbf{P}|_{\{x,y\}})$ in place of $F(\mathbf{P}|_{\{x,y\}}) = \{x\}$.

For a violation of γ, consider a profile $\mathbf{P}$ with the majority graph in Figure 2, which exists by McGarvey's [**38**] theorem (Theorem 4.5.1 below). The maximal chains of $M(\mathbf{P})$, ordered by the edge relation, are as follows:

$$(c,a,g,b) \quad (c,g,d,b) \quad (c,e,a,g) \quad (c,d,b,e) \quad (c,d,e,a)$$
$$(f,c,g) \quad (e,a,f,g) \quad (d,b,e,f) \quad (d,a,b,f) \quad (d,e,a,f).$$

[5] Note that binary γ for choice functions can be written equivalently as: if $x \in C(B \setminus \{y\})$ and $C(\{x,y\}) = \{x\}$, then $x \in C(B)$.

[6] We thank Felix Brandt for providing this example.

Then $Banks(\mathbf{P}|_{\{a,b,c,e,f\}}) \cap Banks(\mathbf{P}|_{\{a,d,f,g\}}) \ni a \notin Banks(\mathbf{P}) = \{c, d, e, f\}$.

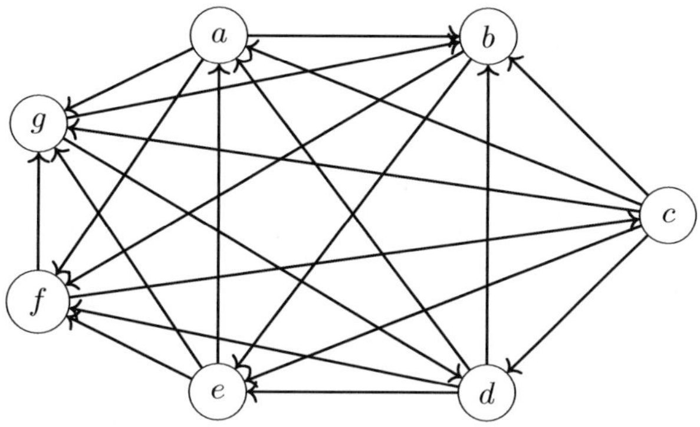

FIGURE 2

Yet we cannot use Banks to show that binary α is weaker than α, as it violates even α-resoluteness, as shown by a profile with the following majority graph in Figure 3. Note that $Banks(\mathbf{P}_{-y}) = \{x, z, w\}$, $Banks(\mathbf{P}|_{\{x,y\}}) = \{x\}$, and $Banks(\mathbf{P}) = \{x, y, z, w\}$, violating α-resoluteness.

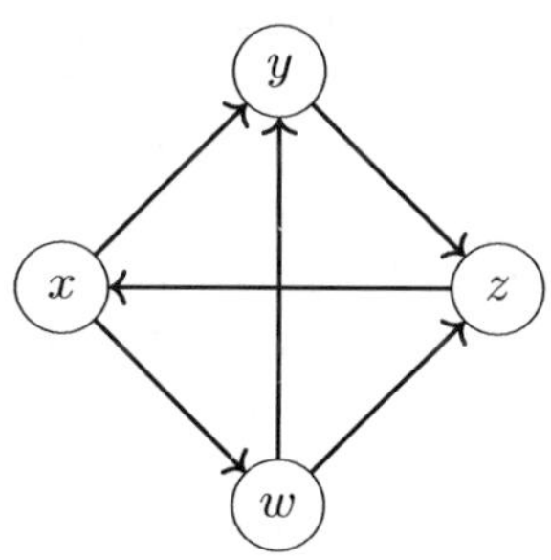

FIGURE 3

The following example shows both that binary γ is weaker than γ and that binary α is weaker than α for voting methods.

EXAMPLE 2.5. A candidate x *Pareto dominates* a candidate y in profile $\mathbf{P}$ if all voters rank x above y. Say that the *Pareto score* of a candidate x is the number of candidates whom x Pareto dominates. Then define the Pareto scoring method as follows: $x \in PS(\mathbf{P})$ if x is Pareto undominated and x has maximal Pareto score among candidates who are Pareto undominated. To see that Pareto scoring satisfies binary γ and binary α, suppose $x \in PS(\mathbf{P}_{-y})$ and $PS(\mathbf{P}|_{\{x,y\}}) = \{x\}$. The latter implies that x Pareto dominates y. Hence $y \notin PS(\mathbf{P})$, and x's Pareto score in $\mathbf{P}$ is one greater than x's Pareto score in $\mathbf{P}_{-y}$. Since the most a candidate's Pareto score can increase from $\mathbf{P}_{-y}$ to $\mathbf{P}$ is one, it follows that $x \in PS(\mathbf{P})$ and $PS(\mathbf{P}) \subseteq PS(\mathbf{P}_{-y})$. However, Pareto scoring violates γ and α; in fact, it violates even the weakenings of γ and α obtained from binary γ and binary α by replacing $F(\mathbf{P}|_{\{x,y\}}) = \{x\}$ with $x \in F(\mathbf{P}|_{\{x,y\}})$, as shown by the profile in Figure 4. Note that x and y are Pareto undominated, while y (but not x) Pareto dominates z. Then $x \in PS(\mathbf{P}_{-y})$ and $x \in PS(\mathbf{P}|_{\{x,y\}})$, but $PS(\mathbf{P}) = \{y\}$.

1	1	1
y	y	x
x	z	y
z	x	z

FIGURE 4

Finally, we show that α-resoluteness is weaker than binary α over a restricted domain of profiles.

EXAMPLE 2.6. Consider profiles with an odd number of voters and at most four candidates. With respect to this restricted domain, the Copeland voting method [**19**] satisfies α-resoluteness, as well as binary γ, but violates binary α. The *Copeland score* of a candidate x in a profile $\mathbf{P}$ is defined using the majority graph $M(\mathbf{P})$ as the outdegree of x minus the indegree of x, i.e., the number of head-to-head wins minus the number of head-to-head losses. The Copeland voting method selects as the winner in $\mathbf{P}$ the candidates with maximal Copeland score. Since this method is anonymous and neutral (see below for definitions), it violates α by the argument given in Section 1. To see that it satisfies α-resoluteness, as well as binary γ, over the restricted domain, suppose $x \in Copeland(\mathbf{P}_{-y})$ and $Copeland(\mathbf{P}|_{\{x,y\}}) = \{x\}$. We consider two cases. First, suppose x has an arrow to every other candidate in $M(\mathbf{P}_{-y})$. Then since $Copeland(\mathbf{P}|_{\{x,y\}}) = \{x\}$, x has an arrow to every other candidate in $M(\mathbf{P})$, so $Copeland(\mathbf{P}) = \{x\}$. Now suppose that x has an incoming arrow from some other candidate in $M(\mathbf{P}_{-y})$. Then since $x \in Copeland(\mathbf{P}_{-y})$, it follows that $M(\mathbf{P}_{-y})$ is a cycle $x \to a \to b \to x$, so each candidate has a Copeland score of 0 and hence $Copeland(\mathbf{P}_{-y}) = \{x, a, b\}$. Then $M(\mathbf{P})$ is of the form in Figure 5, where the dashed lines represent arrows that could point in either direction.

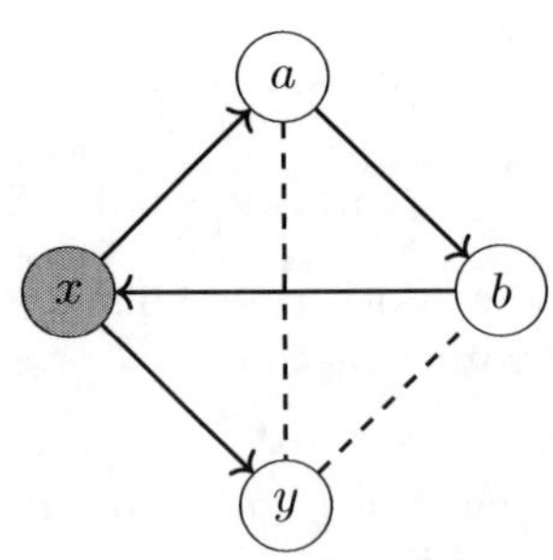

FIGURE 5

No matter the choice of direction for the arrows replacing the dashed lines in Figure 5, $x \in Copeland(\mathbf{P})$ and $|Copeland(\mathbf{P})| \leq 3$. This shows that Copeland satisfies binary γ and α-resoluteness for profiles in the restricted domain. However, it is easy to see that Copeland violates these axioms when moving from 4 to 5 candidates.

We will be interested in the interaction of the above axioms borrowed from individual choice with some standard axioms on voting methods that are distinctively social-choice theoretic. In particular, we are interested in basic conditions of fairness with respect to voters and candidates. For the fairness condition with respect to voters, given $i, j \in \mathcal{V}$, let $\tau^{i \leftrightarrows j}$ be the permutation of $\mathcal{V}$ that swaps only i and j. Given a profile $\mathbf{P}$, let $\mathbf{P}^{i \leftrightarrows j}$ be the profile with $V(\mathbf{P}^{i \leftrightarrows j}) = \tau^{i \leftrightarrows j}[V(\mathbf{P})]$ and $X(\mathbf{P}^{i \leftrightarrows j}) = X(\mathbf{P})$ such that for $k \in V(\mathbf{P}^{i \leftrightarrows j})$, we have $\mathbf{P}^{i \leftrightarrows j}(k) = \mathbf{P}(\tau^{i \leftrightarrows j}(k))$.

anonymity: for any profile $\mathbf{P} \in \text{dom}(F)$ and $i, j \in \mathcal{V}$, if $\mathbf{P}^{i \leftrightarrows j} \in \text{dom}(F)$, then $F(\mathbf{P}^{i \leftrightarrows j}) = F(\mathbf{P})$.[7]

One can equivalently state anonymity with an arbitrary permutation τ of $\mathcal{V}$, since the transpositions $\tau^{i \leftrightarrows j}$ generate the full permutation group. Another equivalent statement is that if two profiles $\mathbf{P}$ and $\mathbf{P}'$ are such that for each linear order L, the same number of voters submit L as their ballot in $\mathbf{P}$ as in $\mathbf{P}'$, then $F(\mathbf{P}) = F(\mathbf{P}')$.

For the fairness condition with respect to candidates, given $a, b \in \mathcal{X}$, let $\pi_{a \leftrightarrows b}$ be the permutation of $\mathcal{X}$ that swaps only a and b. Given a linear order L of some finite subset of $\mathcal{X}$, define $L_{a \leftrightarrows b}$ such that for all $x, y \in \mathcal{X}$, $(x, y) \in L_{a \leftrightarrows b}$ iff $(\pi_{a \leftrightarrows b}(x), \pi_{a \leftrightarrows b}(y)) \in L$. Given $X \subseteq \mathcal{X}$, let $X_{a \leftrightarrows b} = \pi_{a \leftrightarrows b}[X]$. Given a profile $\mathbf{P}$, let $\mathbf{P}_{a \leftrightarrows b}$ be the profile with $V(\mathbf{P}_{a \leftrightarrows b}) = V(\mathbf{P})$ and $X(\mathbf{P}_{a \leftrightarrows b}) = X(\mathbf{P})_{a \leftrightarrows b}$ such that for $i \in V(\mathbf{P}_{a \leftrightarrows b})$, we have $\mathbf{P}_{a \leftrightarrows b}(i) = \mathbf{P}(i)_{a \leftrightarrows b}$.

neutrality: for any profile $\mathbf{P} \in \text{dom}(F)$ and $a, b \in \mathcal{X}$, if $\mathbf{P}_{a \leftrightarrows b} \in \text{dom}(F)$, then $F(\mathbf{P}_{a \leftrightarrows b}) = F(\mathbf{P})_{a \leftrightarrows b}$.[8]

One can equivalently state neutrality with an arbitrary permutation π of $\mathcal{X}$.

One of the most basic facts in voting theory is that there is no voting method satisfying anonymity, neutrality, and resoluteness on all profiles. This can be seen with the three-candidate profile in Section 1, for which anonymity and neutrality force the selection of all three candidates. In fact, the problem already arises with two candidates:

binary resoluteness: $|F(\mathbf{P})| = 1$ for all profiles $\mathbf{P}$ such that $|X(\mathbf{P})| = 2$.

Consider a profile $\mathbf{P}$ with only two candidates x, y and an even number of voters, half of whom rank x above y and half of whom rank y above x. Then anonymity and neutrality imply $F(\mathbf{P}) = \{x, y\}$, violating resoluteness.

PROPOSITION 2.7. *There is no voting method defined on all two-candidate profiles satisfying anonymity, neutrality, and binary resoluteness.*

A natural response is to weaken binary resoluteness to the following:

binary quasi-resoluteness: $|F(\mathbf{P})| = 1$ for all profiles $\mathbf{P}$ such that $|X(\mathbf{P})| = 2$ and for $x, y \in X(\mathbf{P})$ with $x \neq y$, the margin of x over y is nonzero.

[7] In our variable-election setting, this axiom could be called *global anonymity* to distinguish it from an axiom of *local anonymity* that only quantifies over $i, j \in V(\mathbf{P})$. Local anonymity permits swapping the ballots assigned to two voters in $\mathbf{P}$ but does not permit changing the set of voters.

[8] In our variable-election setting, this axiom could be called *global neutrality* to distinguish it from an axiom of *local neutrality* that only quantifies over $a, b \in X(\mathbf{P})$. Local neutrality permits swapping the positions of two candidates in $\mathbf{P}$ but does not permit changing the set of candidates.

Unlike binary resoluteness, binary quasi-resoluteness is consistent with anonymity and neutrality. Indeed, almost all standard voting methods satisfy binary quasi-resoluteness, including, e.g., the Plurality voting method that selects in a given profile the candidates with the most first-place votes.

In this paper, we are interested in impossibility theorems that arise by adding axioms to the initial package of anonymity, neutrality, and binary quasi-resoluteness. First, adding binary α to this package results in an inconsistency by the argument in Section 1 (see Footnote 1). Thus, we move on to binary γ. As argued in [**30**], binary γ can be viewed as an axiom that mitigates *spoiler effects* in elections.[9] Consider one of the most famous U.S. elections involving a spoiler effect: the 2000 Presidential Election in Florida. Though the election did not collect ranked ballots from voters, polling suggests [**28, 37**] that Gore would have beaten Bush in the two-candidate election $\mathbf{P}|_{\{\text{Bush, Gore}\}}$, and Gore would have beaten Nader in the two-candidate election $\mathbf{P}|_{\{\text{Gore, Nader}\}}$, yet Gore was not a winner in the three-candidate election $\mathbf{P}|_{\{\text{Bush, Gore, Nader}\}}$ according to Plurality voting.[10] Thus, Plurality voting violates binary γ.

In the example above, Gore is the *Condorcet winner*, a candidate who is majority preferred to every other candidate. Thus, the example also shows that Plurality voting violates the property of *Condorcet consistency*, which requires that a method selects the Condorcet winner whenever one exists. However, binary γ and Condorcet consistency are independent properties. For example, the Condorcet consistent methods Beat Path [**47, 48**], Minimax [**35, 51**], and Ranked Pairs [**55, 58**] all violate binary γ. Conversely, the trivial voting method that always declares all candidates tied for the win satisfies binary γ but not Condorcet consistency. However, there is a connection between binary γ and the concept of the Condorcet winner: if a voting method satisfies binary γ and reduces to majority voting in two-candidate elections, then the Condorcet winner, if one exists, is always *among* the winners of the election. In fact, the examples of voting methods in the literature satisfying binary γ, including Banks [**3**], Split Cycle [**30, 31**], Top Cycle [**49, 52**], Uncovered Set [**21, 39**], and Weighted Covering [**23, 45**], are all Condorcet consistent. Thus, Condorcet consistency and binary γ are cross-cutting but related properties, as shown by the list of voting methods in Table 1.

TABLE 1. Examples of methods violating or satisfying binary γ. Condorcet consistent voting methods are marked with a $\star$.

methods violating binary γ	methods satisfying binary γ
Beat Path*	Banks*
Borda	Split Cycle*
Instant Runoff	Top Cycle*
Minimax*	Uncovered Set*
Plurality	Weighted Covering*
Ranked Pairs*	

[9]The axiom called *stability for winners* in [**30**] is almost the same as binary γ, but with '$F(\mathbf{P}|_{\{x,y\}}) = \{x\}$' in the statement of binary γ replaced by 'the margin of x over y in $\mathbf{P}$ is positive'. Assuming F reduces to majority voting in two-candidate profiles, F satisfying binary γ is equivalent to F satisfying stability for winners.

[10]Of course there were more than three candidates in this election, but this simplified example contains the essential points.

Though we can consistently add binary γ to our initial package of axioms, adding binary γ together with even the weakest consequence of α that we have discussed—α-resoluteness—leads to an impossibility theorem. Because axioms like binary γ and α-resoluteness are variable-candidate axioms, involving choosing winners from elections involving different sets of candidates, we are especially interested in exactly how many candidates are needed to trigger impossibility results.[11] If there is no voting method satisfying the relevant axioms on all profiles with up to n candidates, then of course there is no voting method satisfying the relevant axioms on all profiles with up to m candidates for $m \geq n$; thus, we are interested in the least number of candidates for which an impossibility theorem holds.

THEOREM 2.8.

(1) *There is no voting method satisfying anonymity, neutrality, binary quasi-resoluteness, binary γ, and α-resoluteness for profiles with up to 6 candidates.*

(2) *There is a voting method satisfying the above axioms for profiles with up to 5 candidates.*[12]

We will prove Theorem 2.8 with the help of a SAT solver, as explained in Section 3. First, however, we will give a human-readable proof of a weaker version of Theorem 2.8.1 that assumes that the voting method satisfies the relevant axioms for profiles having up to 7 candidates.

PROPOSITION 2.9. *There is no voting method satisfying anonymity, neutrality, binary quasi-resoluteness, binary γ, and α-resoluteness for profiles with up to 7 candidates.*

PROOF. Suppose for contradiction that there is a voting method F satisfying the axioms for profiles with up to 7 candidates. By anonymity, neutrality, and binary quasi-resoluteness, for every possible margin $m > 0$ and number k of voters, either (i) for every two-candidate profile with k voters in which one candidate has a margin of m over the other, the unique winner is the majority winner or (ii) for every two-candidate profile with k voters in which one candidate has a margin of m over the other, the unique winner is the minority winner. For $m = 6$ and $k = 18$, suppose that (i) holds; the proof in case (ii) is analogous. Now consider the profile $\mathbf{P}$ in Figure 6, whose margin graph is shown in Figure 7 (all arrows in/out of a rectangle go in/out of every node in the rectangle). By (i), if there is an arrow from x to y in Figure 7, then $F(\mathbf{P}|_{\{x,y\}}) = \{x\}$.

Although the profile in Figure 6 contains 8 candidates, in what follows we will only use the assumption that F satisfies the relevant axioms on subprofiles of $\mathbf{P}$ with up to 7 candidates. We will repeatedly use the fact that in a perfectly symmetrical three-candidate profile realizing a majority cycle, such as $\mathbf{P}|_{\{a,b,c\}}$, $\mathbf{P}|_{\{b,c,d\}}$, and $\mathbf{P}|_{\{a,b,e\}}$, anonymity and neutrality imply that every candidate is chosen. Thus, $c \in F(\mathbf{P}|_{\{a,b,c\}})$, so $c \in F(\mathbf{P}|_{\{a,a',a'',b,c\}})$ by two applications of binary γ. By anonymity and neutrality we also have that

$$\{a, a', a''\} \cap F(\mathbf{P}|_{\{a,a',a'',b,c\}}) \neq \varnothing \text{ if and only if } \{a, a', a''\} \subseteq F(\mathbf{P}|_{\{a,a',a'',b,c\}}).$$

Thus, one of the following three cases holds.

[11] Optimizing these impossibility results for the number of voters is an interesting subject for future research.

[12] Note this improves on the example of Copeland from Example 2.6.

1	1	1	1	1	1	1	1	1	1	1	1	1	1	1	1	1	1
e	e	e	e	e	e	d	d	d	d	d	d	c	c'	c	c'	c	c'
a	a	a'	a'	a''	a''	b	b	b	b	b	b	c'	c	c'	c	c'	c
a'	a'	a''	a''	a	a	e	e	e	e	e	e	d	d	d	d	d	d
a''	a''	a	a	a'	a'	c	c'	c	c'	c	c'	a	a	a'	a'	a''	a''
b	b	b	b	b	b	c'	c	c'	c	c'	c	a'	a'	a''	a''	a	a
c	c'	c	c'	c	c'	a	a	a'	a'	a''	a''	a''	a''	a	a	a'	a'
c'	c	c'	c	c'	c	a'	a'	a''	a''	a	a	b	b	b	b	b	b
d	d	d	d	d	d	a''	a''	a	a	a'	a'	e	e	e	e	e	e

FIGURE 6. A profile whose subprofiles (with 7 or fewer candidates) are used in the proof of Proposition 2.9.

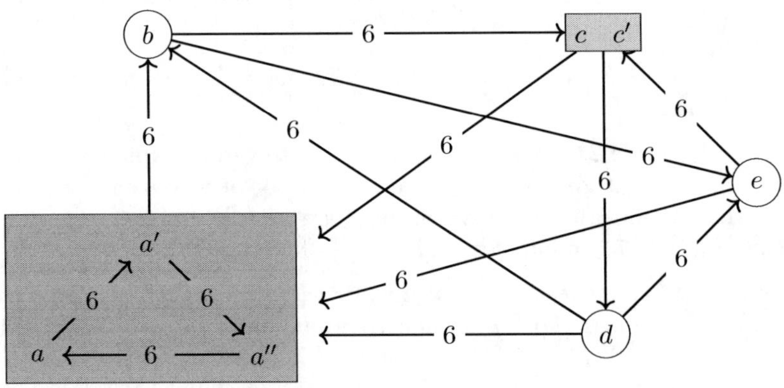

FIGURE 7. Margin graph of the profile in Figure 6.

CASE 1. $F(\mathbf{P}|_{\{a,a',a'',b,c\}}) = \{c\}$. Hence $c \in F(\mathbf{P}|_{\{a,a',a'',b,c,d\}})$ by binary γ. Moreover, by anonymity and neutrality, $d \in F(\mathbf{P}|_{\{b,c,d\}})$, so $d \in F(\mathbf{P}|_{\{a,a',a'',b,c,d\}})$ by three applications of binary γ. But together $F(\mathbf{P}|_{\{a,a',a'',b,c\}}) = \{c\}$, $F(\mathbf{P}|_{\{c,d\}}) = \{c\}$, and $c, d \in F(\mathbf{P}|_{\{a,a',a'',b,c,d\}})$ contradict α-resoluteness.

CASE 2. $F(\mathbf{P}|_{\{a,a',a'',b,c\}}) = \{b, c\}$. Hence $b \in F(\mathbf{P}|_{\{a,a',a'',b,c,c'\}})$ by binary γ. By anonymity and neutrality,

$$\{c, c'\} \cap F(\mathbf{P}|_{\{a,a',a'',b,c,c'\}}) \neq \varnothing \text{ if and only if } \{c, c'\} \subseteq F(\mathbf{P}|_{\{a,a',a'',b,c,c'\}}),$$

and

$$\{a, a', a''\} \cap F(\mathbf{P}|_{\{a,a',a'',b,c,c'\}}) \neq \varnothing \text{ if and only if } \{a, a', a''\} \subseteq F(\mathbf{P}|_{\{a,a',a'',b,c,c'\}}).$$

Then given $F(\mathbf{P}|_{\{a,a',a'',b,c\}}) = \{b, c\}$ and $F(\mathbf{P}|_{\{b,c'\}}) = \{b\}$, α-resoluteness implies $F(\mathbf{P}|_{\{a,a',a'',b,c,c'\}}) = \{b\}$. Next, by anonymity and neutrality, $e \in F(\mathbf{P}|_{\{a,b,e\}})$ and hence $e \in F(\mathbf{P}|_{\{a,a',a'',b,c,c',e\}})$ by four applications of binary γ. Moreover, given $F(\mathbf{P}|_{\{a,a',a'',b,c,c'\}}) = \{b\}$, we have $b \in F(\mathbf{P}|_{\{a,a',a'',b,c,c',e\}})$ by binary γ. But together $F(\mathbf{P}|_{\{a,a',a'',b,c,c'\}}) = \{b\}$, $F(\mathbf{P}|_{\{b,e\}}) = \{b\}$, and $b, e \in F(\mathbf{P}|_{\{a,a',a'',b,c,c',e\}})$ contradict α-resoluteness.

CASE 3. $\{c, a, a', a''\} \subseteq F(\mathbf{P}|_{\{a,a',a'',b,c\}})$. By anonymity and neutrality, we have $F(\mathbf{P}|_{\{a,b,c\}}) = \{a, b, c\}$. Then since $F(\mathbf{P}|_{\{a',c\}}) = \{c\}$, it follows by binary γ and α-resoluteness that $c \in F(\mathbf{P}|_{\{a,a',b,c\}})$ and $|F(\mathbf{P}|_{\{a,a',b,c\}})| \leq 3$, which with

$F(\mathbf{P}|_{\{a'',c\}}) = \{c\}$ and α-resoluteness implies $|F(\mathbf{P}|_{\{a,a',a'',b,c\}})| \leq 3$, which contradicts the assumption of the case. □

REMARK 2.10.

(1) If we drop either anonymity or neutrality, then there are voting methods satisfying the remaining axioms from Theorem 2.8. For an anonymous (resp. neutral) voting method F satisfying the other axioms, fix a linear order L of the set of all possible candidates (resp. for each possible set V of voters, fix a voter $i_V \in V$), and for any profile $\mathbf{P}$, let $F(\mathbf{P}) = \{x\}$ where x is the maximal element in the linear order L restricted to $X(\mathbf{P})$ (resp. in $i_{V(\mathbf{P})}$'s linear order of $X(\mathbf{P})$) such that no candidate ranked below x in this linear order is majority preferred to x. Observe that F satisfies binary γ. Then since F always selects a unique winner, it also satisfies α-resoluteness.
(2) If we drop binary quasi-resoluteness, then there are voting methods satisfying the remaining axioms from Theorem 2.8, e.g., the trivial voting method that always selects all candidates as winners (note that this method trivially satisfies binary γ and α-resoluteness, because the condition $F(\mathbf{P}|_{\{x,y\}}) = \{x\}$ is always false).
(3) It is open whether if we drop binary γ, there are voting methods satisfying the remaining axioms from Theorem 2.8. However, we do know from SAT solving (see Appendix A) that there are voting methods satisfying the remaining axioms at least for profiles having up to 7 candidates.
(4) If we drop α-resoluteness, then all the voting methods in the right column of Table 1 satisfy the remaining axioms from Theorem 2.8.

While binary γ and α-resoluteness for *individual choice functions* are easily satisfied—in particular by any choice function representable by a binary relation—Theorem 2.8.1 shows that binary γ and α-resoluteness for *voting methods* are together inconsistent with some of the most uncontroversial axioms for voting, namely anonymity, neutrality, and binary quasi-resoluteness.

A natural response to Proposition 2.7 and Theorem 2.8.1 is that asking for any kind of resoluteness property with respect to *all* profiles, including highly symmetric profiles, is asking for too much. Thus, α-resoluteness is asking for too much. This motivates a move to more qualified resoluteness properties that only apply to profiles with sufficiently broken symmetries. For example, call a profile $\mathbf{P}$ *uniquely weighted* if for any two distinct pairs a, b and x, y of distinct candidates, the margin of a over b in $\mathbf{P}$ is not equal to the margin of x over y in $\mathbf{P}$. Thus, the profile in Figure 7 is *not* uniquely weighted. By contrast, in elections with many voters, one would expect with high probability to obtain a uniquely weighted profile. Now consider restricting resoluteness only to uniquely weighted profiles:

quasi-resoluteness: $|F(\mathbf{P})| = 1$ for all *uniquely weighted* profiles $\mathbf{P}$.

There are many voting methods satisfying the trio of anonymity, neutrality, and quasi-resoluteness, including Beat Path, Minimax, and Ranked Pairs, as well as any voting method that uses a quasi-resolute voting method as a tiebreaker, e.g., Plurality with Minimax tiebreaking (in the event of a tie for the most first place votes). However, we will show that even this weakening of resoluteness is incompatible with basic principles of individual rational choice applied to voting—in particular, with the binary γ axiom alone—assuming anonymity and neutrality.

THEOREM 2.11. *There is no voting method satisfying anonymity, neutrality, binary γ, and quasi-resoluteness for profiles with up to 4 candidates.*

Theorem 2.11 explains the observation from the literature that voting methods satisfying binary γ, listed in Table 1, all violate quasi-resoluteness, while the quasi-resolute methods mentioned above all violate binary γ. Note that 4 candidates is optimal for the impossibility, since Beat Path, Minimax, Ranked Pairs, and Split Cycle satisfy the axioms in Theorem 2.11 for profiles up to 3 candidates.

We will prove Theorem 2.11 with the help of a SAT solver, as explained in Section 3. First, however, we will give a human-readable proof of a variation on Theorem 2.11. For the variation, we add the following assumptions. Given a profile $\mathbf{P}$, the profile $n\mathbf{P}$ replaces each voter from $\mathbf{P}$ with n copies of that voter with the same ballot.[13]

DEFINITION 2.12. Let F be a voting method.

(1) F is *homogeneous* if for all $\mathbf{P} \in \text{dom}(F)$ and $n \in \mathbb{Z}^+$, if $n\mathbf{P} \in \text{dom}(F)$, then $F(\mathbf{P}) = F(n\mathbf{P})$ [**52**].
(2) F is *cancellative* if for all $\mathbf{P}, \mathbf{P}' \in \text{dom}(F)$, if $\mathbf{P}'$ is obtained from $\mathbf{P}$ by deleting one copy of each linear order of $X(\mathbf{P})$, then $F(\mathbf{P}) = F(\mathbf{P}')$.

Adding these assumptions alongside those of Theorem 2.11 would make for a weaker impossibility theorem. However, because we will be able to construct profiles by hand in our proof, we can weaken quasi-resoluteness to the following property.

DEFINITION 2.13. A profile $\mathbf{P}$ is *uniquely ranked* if for every $Y \subseteq X(\mathbf{P})$ with $|Y| \geq 2$:

(1) for each $n \in \{1, \dots, |Y|\}$, no two candidates have the same number of n-th place votes in $\mathbf{P}|_Y$;
(2) no two candidates have the same *Borda score* in $\mathbf{P}|_Y$, where the Borda score of x in a profile $\mathbf{Q}$ is $\sum_{i \in V(\mathbf{Q})} |\{y \in X(\mathbf{Q}) \mid (x, y) \in \mathbf{Q}(i)\}|$.

A voting method F satisfies *weak quasi-resoluteness* if $|F(\mathbf{P})| = 1$ for every profile $\mathbf{P}$ that is both uniquely weighted and uniquely ranked.

Examples of voting methods that are weakly quasi-resolute but not quasi-resolute include Plurality, Borda, Instant Runoff, Coombs [**18, 27**], Baldwin [**2**], Nanson [**40, 41**], and Bucklin [**29**]. In the following proof, it is tedious to check by hand that all the profiles are uniquely ranked (for code to check this, see `https://github.com/szahedian/binary-gamma`). However, it is easy to check that the profiles are uniquely weighted, which suffices for the version of Proposition 2.14 with quasi-resoluteness in place of weak quasi-resoluteness.

PROPOSITION 2.14. *There is no anonymous, neutral, homogeneous, and cancellative voting method on the domain of profiles with up to 4 candidates satisfying binary γ and weak quasi-resoluteness.*

PROOF. Suppose for contradiction there is such a voting method F.

---

[13] I.e., $V(n\mathbf{P}) = \{(i, m) \mid i \in V(\mathbf{P}), m \in \mathbb{Z}^+, m \leq n\}$ and $(a, b) \in n\mathbf{P}((i, m))$ if $(a, b) \in \mathbf{P}(i)$.

CASE A. In a two-candidate profile $\mathbf{P}_{a,b}$ with 4 voters all of whom rank a above b, a is a winner. Then by anonymity, neutrality, homogeneity, and cancellativity, we have the following fact ($\star$): for any even $m \geq 4$ and $n \in \mathbb{Z}^+$, in any two-candidate $\mathbf{P}'$ profile with m voters and a margin of $4n$ for some candidate x over another, y, x is a winner. For by neutrality, x wins in the profile $\mathbf{P}_{x,y}$ obtained from $\mathbf{P}_{a,b}$ by swapping x for a and y for b. Then by homogeneity, x wins in $n\mathbf{P}_{x,y}$, wherein the margin of x over y is $4n$. Let $\mathbf{P}_0'$ be obtained from $n\mathbf{P}_{x,y}$ by adding $(m-4n)/2$ voters with the linear order xy and $(m-4n)/2$ voters with the linear order yx, so the number of voters is m and the margin of x over y is still $4n$. It follows by cancellativity that x wins in $\mathbf{P}_0'$. Then by anonymity, x wins in $\mathbf{P}'$.

Now consider the following 96-voter profile $\mathbf{P}_0$:

44	12	36	4
a	b	c	c
b	c	a	b
c	a	b	a

$\mathbf{P}_0$

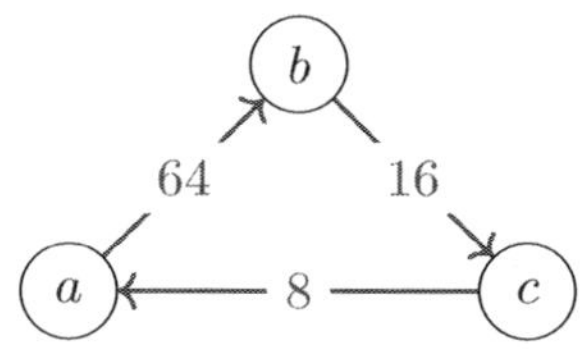

$\mathcal{M}(\mathbf{P}_0)$

CASE A.1. $a \in F(\mathbf{P}_0)$. Then by neutrality and homogeneity, we have $d \in F(\mathbf{Q}_1)$ for the following 48-voter profile $\mathbf{Q}_1$:

22	6	18	2
d	c	a	a
c	a	d	c
a	d	c	d

$\mathbf{Q}_1$

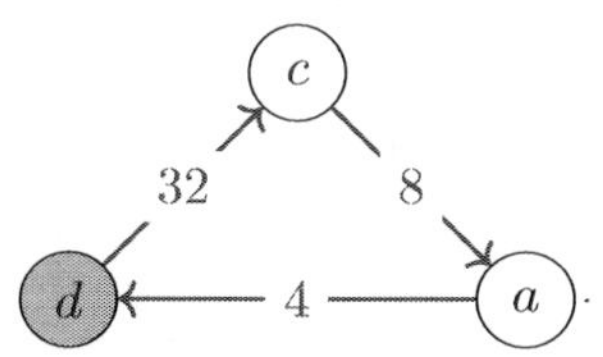

$\mathcal{M}(\mathbf{Q}_1)$

Then by cancellativity, $d \in F(\mathbf{Q}_1^+)$ where $\mathbf{Q}_1^+$ is the 96-voter profile obtained from $\mathbf{Q}_1$ by adding 8 copies of each linear order of $\{a, c, d\}$: .

30	14	26	10	8	8
d	c	a	a	c	d
c	a	d	c	d	a
a	d	c	d	a	c

$\mathbf{Q}_1^+$

Then where $\mathbf{P}_1$ is the following uniquely weighted and uniquely ranked 96-voter profile such that $\mathbf{P}_1|_{\{a,b,c\}} = \mathbf{P}_0$ and $\mathbf{P}_1|_{\{a,c,d\}} = \mathbf{Q}_1^+$, ($\star$) and binary γ imply $\{a, d\} \subseteq F(\mathbf{P}_1)$, violating weak quasi-resoluteness:

30	8	6	26	10	4	2	2	8
d	b	c	a	a	b	c	c	d
c	c	a	d	b	c	d	b	a
a	a	d	b	c	d	b	d	b
b	d	b	c	d	a	a	a	c

$\mathbf{P}_1$

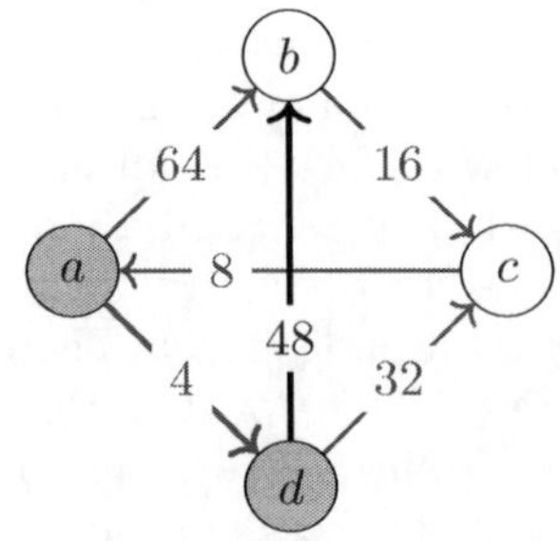

$\mathcal{M}(\mathbf{P}_1)$

CASE A.2. $b \in F(\mathbf{P}_0)$. Then by neutrality and homogeneity, we have $d \in F(\mathbf{Q}_2)$ for the following 768-voter profile $\mathbf{Q}_2$:

352	96	288	32
b	d	a	a
d	a	b	d
a	b	d	b

$\mathbf{Q}_2$

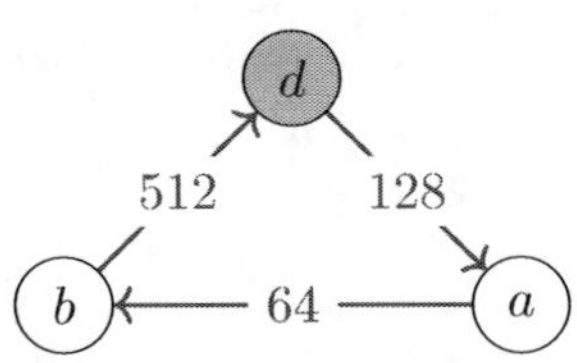

$\mathcal{M}(\mathbf{Q}_2)$

Moreover, by cancellativity, $b \in F(\mathbf{P}_0)$ implies $b \in F(\mathbf{P}_0^+)$ for the 768-voter profile $\mathbf{P}_0^+$ obtained from $\mathbf{P}_0$ by adding 112 copies of each linear order over $\{a, b, c\}$:

156	124	148	116	112	112
a	b	c	c	a	b
b	c	a	b	c	a
c	a	b	a	b	c

$\mathbf{P}_0^+$

Then where $\mathbf{P}_2$ is the following uniquely weighted and uniquely ranked 768-voter profile such that $\mathbf{P}_2|_{\{a,b,c\}} = \mathbf{P}_0^+$ and $\mathbf{P}_2|_{\{a,b,d\}} = \mathbf{Q}_2$, $(\star)$ and binary γ imply $\{b, d\} \subseteq F(\mathbf{P}_2)$, violating weak quasi-resoluteness:

156	124	96	52	116	80	32	112
a	b	d	c	c	a	a	b
b	d	c	a	b	c	d	d
d	c	a	b	d	b	c	a
c	a	b	d	a	d	b	c

$\mathbf{P}_2$

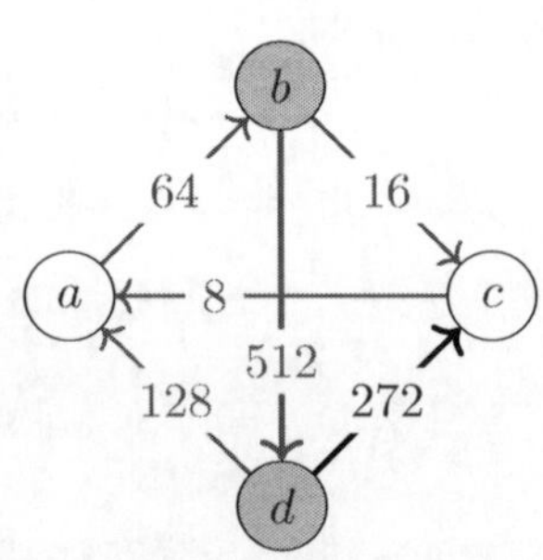

$\mathcal{M}(\mathbf{P}_2)$

CASE A.3. $c \in F(\mathbf{P}_0)$. Then by neutrality and homogeneity, we have $d \in F(\mathbf{Q}_3)$ for the following 24-voter profile $\mathbf{Q}_3$:

11	3	9	1
b	c	d	d
c	d	b	c
d	b	c	b

$\mathbf{Q}_3$

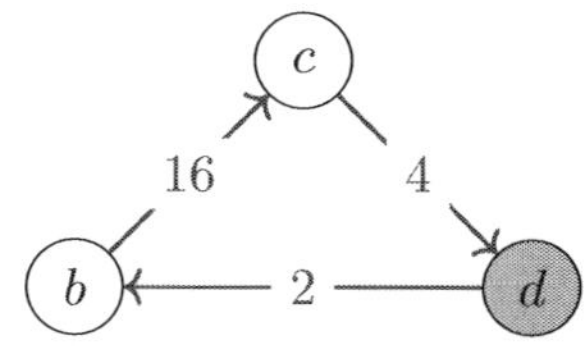

$\mathcal{M}(\mathbf{Q}_3)$

Then by cancellativity, $d \in F(\mathbf{Q}_3^+)$ where $\mathbf{Q}_3^+$ is the 96-voter profile obtained from $\mathbf{Q}_3$ by adding 12 copies of each linear order of $\{b, c, d\}$:

23	15	21	13	12	12
b	c	d	d	b	c
c	d	b	c	d	b
d	b	c	b	c	d

$\mathbf{Q}_3^+$

Then where $\mathbf{P}_3$ is the following uniquely weighted and uniquely ranked 96-voter profile such that $\mathbf{P}_3|_{\{a,b,c\}} = \mathbf{P}_0$ and $\mathbf{P}_3|_{\{b,c,d\}} = \mathbf{Q}_3^+$, ($\star$) and binary γ imply $\{c, d\} \subseteq F(\mathbf{P}_3)$, violating weak quasi-resoluteness:

12	11	15	21	13	12	8	4
b	a	c	d	d	a	c	c
c	b	d	a	c	b	a	b
d	c	a	b	a	d	b	a
a	d	b	c	b	c	d	d

$\mathbf{P}_3$

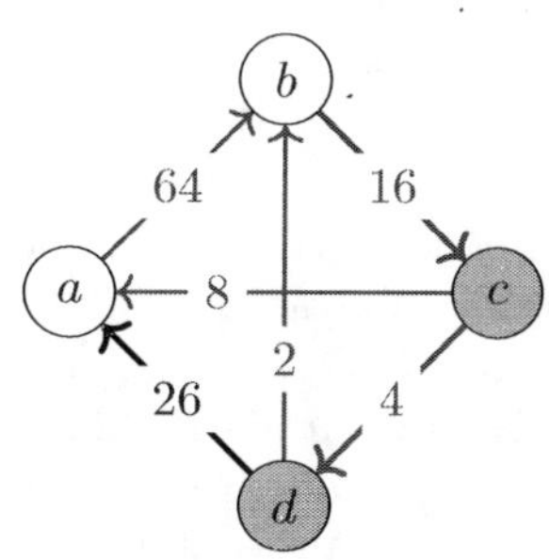

$\mathcal{M}(\mathbf{P}_3)$

CASE B. In a two-candidate profile with 4 voters all of whom rank x above y, x is not a winner, so y is a winner. Then the proof can be obtained from that of Case A by reversing all ballots and all arrows. $\square$

Reflection on the proof of Proposition 2.14 suggests that without homogeneity and cancellativity, there is an explosion in the number of cases one must consider, so we cannot prove Theorem 2.11 in the same manner.

In the rest of the paper, we will build up to proofs of Theorems 2.8.1 and Theorem 2.11. As we have seen, these impossibility theorems show that a core principle of individual rational choice, namely (binary) γ, is inconsistent with weakenings of resoluteness in the setting of anonymous and neutral social choice. Though these impossibility results raise important conceptual questions about which axioms ought to be weakened in response, in this paper we focus only on the proofs of the results. We hope that these proofs may also suggest useful strategies for other impossibility results in social choice.

3. Proof strategy

In this section, we outline our proofs of Theorems 2.8.1 and 2.11. In principle, there is a straightforward proof strategy coming from mathematical logic. Consider a language L_{prof} of propositional logic [**24**, § 1.1] that contains for each profile $\mathbf{P}$ and nonempty $Y \subseteq X(\mathbf{P})$ a propositional variable $A_{\mathbf{P},Y}$, which we interpret to mean that the set of winning candidates for profile $\mathbf{P}$ is Y. Then it is easy to show (cf. the Appendix) that there is a set Σ of formulas of L_{prof} that is satisfiable if and only if there exists a voting method satisfying anonymity, neutrality, binary quasi-resoluteness, binary γ, and α-resoluteness (resp. anonymity, neutrality, binary γ, and quasi-resoluteness). Now if Σ is unsatisfiable, then by the compactness theorem for propositional logic [**24**, § 1.7], there is a finite subset Σ_0 of Σ that is unsatisfiable. Since Σ_0 is finite, there are some maximum numbers n and m of voters and candidates, respectively, appearing in any profile in the set $\{\mathbf{P} \mid A_{\mathbf{P},Y}$ occurs in some formula in $\Sigma_0\}$. Now there is a formula $\varphi_{n,m}$ of L_{prof} that it satisfiable if and only if there is a voting method satisfying the relevant axioms for profiles with up to n voters and m candidates; and the satisfiability of $\varphi_{n,m}$ can be checked using a SAT solver on a computer. Thus, if it is possible to prove a result like Theorem 2.8.1 (resp. 2.11) at all, then it is in principle possible to do so by using a SAT solver to show the unsatisfiability of $\varphi_{n,m}$ for sufficiently large n, m.

For processing by the SAT solver, we write $\varphi_{n,m}$ in *conjunctive normal form* (CNF), i.e., a conjunction of disjunctions of propositional variables or their negations: $(\ell_{1,1} \vee \cdots \vee \ell_{1,k_1}) \wedge \cdots \wedge (\ell_{j,1} \vee \cdots \vee \ell_{j,k_j})$, where each ℓ is of the form $A_{\mathbf{P},Y}$ or $\neg A_{\mathbf{P},Y}$. If the SAT solver finds that $\varphi_{n,m}$ is unsatisfiable, then we may apply algorithms that extract a minimal unsatisfiable set (MUS) of conjuncts from $\varphi_{n,m}$, i.e., an unsatisfiable set of conjuncts such that removing any conjunct results in a satisfiable formula. If the MUS is sufficiently small, we may be able to reconstruct a human-readable proof of its unsatisfiability (as in, e.g., [**6, 10**]). Even if a human-readable proof is infeasible, we may be able to formally verify the SAT result using an interactive theorem prover (as in, e.g., [**6, 15**]). SAT solving has already been used to prove a number of impossibility theorems and other results in computational social choice (see, e.g., [**6, 7, 10–12, 15, 25, 26, 34**]).

The difficulty with the strategy above for our purposes here is that for more than a few candidates and voters, the number of profiles—and hence propositional variables $A_{\mathbf{P},Y}$—is too large to feasibly carry out the strategy to prove Theorems 2.8.1 and 2.11. Of course, since we are interested in voting methods satisfying anonymity, it would suffice to restrict attention to equivalence classes of profiles related by swapping voters as in the anonymity axiom, or equivalently, with anonymous profiles (see Definition 4.1). In fact, since we are interested in voting methods satisfying anonymity and neutrality, it would suffice to work with equivalence classes of profiles related by swapping voters and/or candidates as in the anonymity and neutrality axioms—so-called anonymous and neutral equivalence classes (ANECs) [**43, 44**]. Yet there are still too many ANECs if we have more than a few candidates and voters. Table 2 illustrates this for a mere 6 voters and up to 5 candidates (the relevant number for the possibility result in Theorem 2.8.2).

In fact, we know that profiles with at least 11 voters are needed to prove Theorem 2.11, because the Split Cycle voting method satisfies anonymity, neutrality, binary γ, and quasi-resoluteness on all profiles with fewer than 11 voters. For as shown in [**30**], Split Cycle satisfies anonymity, neutrality, and binary γ on all profiles, and it satisfies quasi-resoluteness on profiles with fewer than 4 candidates. Then since no 4-candidate profile with fewer than 11 voters is uniquely weighted, Split Cycle vacuously satisfies quasi-resoluteness on all 4-candidate profiles with fewer than 11 voters. Thus, profiles with at least 11 voters are indeed needed to prove Theorem 2.11, rendering a purely profile-based SAT strategy infeasible.

TABLE 2. Comparison of the numbers of *unlabeled weak tournaments*, *preference profiles*, *anonymous profiles*, and *anonymous and neutral equivalence classes of profiles* (ANECs). We chose 6 voters because for each $n \in \{3, 4, 5\}$, 6 is the least number of voters such that all n-candidate unlabeled weak tournaments are isomorphic to the majority graph of a profile with at most that number of voters [**53**].

candidates	unlabeled weak tournaments	6-voter ANECs	6-voter anonymous profiles	6-voter profiles
3	7	83	462	46,656
4	42	19,941	475,020	191,102,976
5	582	39,096,565	4,690,625,500	2,985,984,000,000

Fortunately, there is a way to circumvent the problem posed by too many profiles. This involves switching from profiles to abstract majority graphs and abstract margin graphs, not associated with any particular profiles.

An abstract majority graph is simply an asymmetric directed graph, which we will call a *weak tournament* $T = (X, \to)$.[14] A *tournament* is a weak tournament such that for all $a, b \in X$ with $a \neq b$, we have $a \to b$ or $b \to a$.

An abstract margin graph is simply a *weighted weak tournament* $\mathcal{T} = (T, w)$, where T is a weak tournament and w is a function assigning to each edge in T a positive integer weight, satisfying the parity constraint of Remark 2.3. A *weighted tournament* is a weighted weak tournament $\mathcal{T} = (T, w)$ in which T is a tournament. We say that $\mathcal{T}$ is *uniquely weighted* if for all $(a, b), (c, d) \in X(T)^2$ with $a \neq b$, $c \neq d$, and $(a, b) \neq (c, d)$, we have $w(a, b) \neq w(c, d)$.

We now explain our proof strategy in two steps.

Step 1 for Theorem 2.8.1. We prove an a priori weaker version of Theorem 2.8.1 using the additional assumption that the voting method's choice of winners depends only on the majority graph of the profile.

DEFINITION 3.1. A voting method F is *majoritarian* if for all $\mathbf{P}, \mathbf{P}' \in \text{dom}(F)$,

$$M(\mathbf{P}) = M(\mathbf{P}') \text{ implies } F(\mathbf{P}) = F(\mathbf{P}').$$

[14] Here we use the term 'weak tournament' as in [**9**].

If F is majoritarian, then we can think of F as a function that assigns to each weak tournament a subset of the nodes. If F is also neutral, then we can be even more economical, using a strategy from [**10**]. For each isomorphism class of weak tournaments, we fix a canonical representative of the class. Then we regard F as a function that assigns to each canonical weak tournament T a subset of the nodes. In fact, it turns out that our impossibility theorem will hold even if we restrict attention to just canonical *tournaments*. To prove there is no majoritarian voting method satisfying the axioms of Theorem 2.8.1, we can follow the SAT solving strategy sketched above but using a propositional language L_{maj} that contains for each canonical tournament T and subset Y of nodes a propositional variable $A_{T,Y}$, interpreted to mean that in T, the set of winners is Y. Then we write a CNF formula ψ_6 that is satisfiable if and only if there is majoritarian voting method on profiles (whose majority graphs are tournaments) up to 6 candidates satisfying the axioms listed in Theorem 2.8.1. Since there are far fewer canonical tournaments with up to 6 nodes than there are profiles with up to 6 candidates and even a few voters, the formula ψ_6 will have far fewer variables than the formulas $\varphi_{n,6}$ of L_{prof} mentioned above. Indeed, we successfully carry out this strategy in Appendix A to prove the following.

PROPOSITION 3.2. *For any finite $Y \subseteq \mathcal{X}$ with $|Y| \geq 6$, there is no* majoritarian *voting method with domain $\{\mathbf{P} \mid X(\mathbf{P}) \subseteq Y\}$ satisfying anonymity, neutrality, binary quasi-resoluteness, binary γ, and α-resoluteness.*

By contrast, as stated in Theorem 2.8.2, there is such a voting method on the domain $\{\mathbf{P} \mid |X(\mathbf{P})| \leq 5\}$. After obtaining Proposition 3.2 by SAT solving, we extracted an MUS from the unsatisfiable CNF formula, but we were unable to extract from it a proof with few enough cases to make a human-readable proof practical. For details, see Appendix A.5.

Step 1 for Theorem 2.11. Toward proving Theorem 2.11, we will first prove an a priori weaker version using the additional assumption that the voting method's choice of winners depends only on the margin graph of the profile.

DEFINITION 3.3. A voting method F is *pairwise* if for all $\mathbf{P}, \mathbf{P}' \in \text{dom}(F)$,

$$\mathcal{M}(\mathbf{P}) = \mathcal{M}(\mathbf{P}') \text{ implies } F(\mathbf{P}) = F(\mathbf{P}').$$

If F is pairwise, then we can think of F as a function that assigns a set of nodes to each weighted weak tournament satisfying the parity constraint in Remark 2.3. If F is also neutral, then we can fix for each isomorphism class of such weighted weak tournaments a canonical representative and then regard F as a function that assigns to each canonical weighted weak tournament a subset of the nodes. Then following a SAT strategy like the one sketched above for majoritarian methods, we prove the following in Appendix B.

PROPOSITION 3.4. *For any $Y \subseteq \mathcal{X}$ with $|Y| \geq 4$, there is no* pairwise *voting method with domain*

$$\{\mathbf{P} \in \mathsf{prof} \mid X(\mathbf{P}) \subseteq Y, \mathcal{M}(\mathbf{P}) \textit{ is uniquely weighted, and all positive weights belong to } \{2,4,6,8,10,12\}\}$$

satisfying anonymity, neutrality, binary γ, and quasi-resoluteness.

By contrast, as previously noted, there are such voting methods on the domain $\{\mathbf{P} \mid |X(\mathbf{P})| \leq 3\}$. After obtaining Proposition 3.4 by SAT solving, we extracted an MUS from the unsatisfiable CNF formula, but in this case the distance from the MUS to a practical human-readable proof was even greater than for Proposition 3.2. For details, see Appendix B.4. To increase confidence in our SAT-based proof of Proposition 3.4, we also formally verified a SAT encoding in the Lean Theorem Prover, as described in Appendix C.

Step 2 for Theorems 2.8.1 and 2.11. We extend Propositions 3.2 and 3.4 for majoritarian and pairwise voting methods to impossibility theorems for *all* voting methods using the following *transfer lemmas.*

LEMMA 3.5. *For any finite $Y \subseteq \mathcal{X}$, if there is a voting method with domain $\{\mathbf{P} \in \mathsf{prof} \mid X(\mathbf{P}) \subseteq Y\}$ satisfying anonymity, neutrality, binary quasi-resoluteness, binary γ, and α-resoluteness, then there is a* majoritarian *voting method with domain $\{\mathbf{P} \in \mathsf{prof} \mid X(\mathbf{P}) \subseteq Y\}$ satisfying anonymity, neutrality, binary quasi-resoluteness, binary γ, and α-resoluteness.*

LEMMA 3.6. *For any finite $Y \subseteq \mathcal{X}$, if there is a voting method with domain $\{\mathbf{P} \in \mathsf{prof} \mid X(\mathbf{P}) \subseteq Y\}$ satisfying anonymity, neutrality, binary γ, and quasi-resoluteness, then for any $m \in \mathbb{Z}^+$, there is a* pairwise *voting method with domain*

$$\{\mathbf{P} \in \mathsf{prof} \mid X(\mathbf{P}) \subseteq Y,\ m \text{ is the maximum weight in } \mathcal{M}(\mathbf{P})\}$$

satisfying anonymity, neutrality, binary γ, and quasi-resoluteness.

Combining Propositions 3.2 and 3.4 with Lemmas 3.5 and 3.6 establishes Theorems 2.8.1 and 2.11, respectively. Note that to relate Proposition 3.4 to Lemma 3.6, we use the fact that if there is a pairwise voting method F on some domain D satisfying the listed axioms, then for any $D' \subseteq D$, the restriction of F to D' also satisfies the listed axioms.

Proof strategy for the transfer lemmas. Our high-level strategy to prove the transfers lemmas is the following, which we will sketch for Lemma 3.5. Where VM is the class of all voting methods and maj is the class of majoritarian voting methods, we want a map

$$\pi_{\mathsf{maj}} : \mathsf{VM} \to \mathsf{maj}$$

that preserves the axioms in Theorem 2.8.1, i.e., if $F \in \mathsf{VM}$ satisfies the axioms, then so does $\pi_{\mathsf{maj}}(F)$. Our strategy is to obtain π_{maj} as a composition of maps. First, we have the map M that sends each profile to its majority graph. Second, we will choose an appropriate map ϕ that sends each majority graph to a profile $\mathbf{P}$. Then we will define π_{maj} such that for each $F \in \mathsf{VM}$, the voting method $\pi_{\mathsf{maj}}(F)$ is defined by

$$\pi_{\mathsf{maj}}(F)(\mathbf{P}) = F(\phi(M(\mathbf{P}))).$$

It is immediate that if $M(\mathbf{P}) = M(\mathbf{P}')$, then $\pi_{\mathsf{maj}}(F)(\mathbf{P}) = \pi_{\mathsf{maj}}(F)(\mathbf{P}')$, which shows $\pi_{\mathsf{maj}}(F) \in \mathsf{maj}$. The crucial fact to prove, however, is that π_{maj} preserves the axioms of the relevant impossibility theorem. This requires a careful choice of ϕ.

The properties required of ϕ depend on the specific axioms in the impossibility theorem. As an example, let us consider the axiom of neutrality. Recall from

Section 2.2 that a voting method F satisfies neutrality if for any profile $\mathbf{P}$,

$$F(\mathbf{P}_{a\leftrightarrows b}) = F(\mathbf{P})_{a\leftrightarrows b}.$$

Given a majority graph G and vertices a and b, let $G_{a\leftrightarrows b}$ be the isomorphic copy of G with a and b swapped. Then in order to show that ϕ preserves neutrality, we would like to show that for any majority graph G, we have

$$\phi(G_{a\leftrightarrows b}) = \phi(G)_{a\leftrightarrows b}.$$

As we shall see, this commutativity requirement rules out a standard construction of profiles from majority graphs [**38**] as a candidate for ϕ.

Though we have sketched the strategy above for proving the majoritarian transfer lemma, Lemma 3.5, an analogous strategy applies to proving the pairwise transfer lemma, Lemma 3.6, using an appropriate map π_{pairwise}. In the next three sections, we carry out this strategy for proving the transfer lemmas.

4. Profile representation of weighted weak tournaments

In this section, we introduce a method for representing an abstract majority/margin graph as the majority/margin graph of a profile, which will play the role of the ϕ map in our proof strategy outlined at the end of Section 3.

To prove properties of our representation, it will be convenient to use *anonymous profiles*, which simply record how many voters submit each ranking of the candidates, as reviewed in Section 4.1. In Section 4.2, we then review the standard constructions of (anonymous) profiles from majority and margin graphs due to McGarvey and Debord. We introduce our alternative construction in Section 4.3.

4.1. Anonymous profiles. Recall from Section 2.2 that $\mathcal{X}$ is an infinite set of candidates and that for $X \subseteq \mathcal{X}$, $\mathcal{L}(X)$ is the set of all strict linear orders on X.

It will be convenient for certain intermediate calculations to be able to work with a generalization of anonymous profiles allowing the number of voters who submit a certain ballot to be *negative* (cf. [**46**, p. 135]).

Definition 4.1. A *generalized anonymous profile* is a function $\mathbf{P}$ such that for some finite $X(\mathbf{P}) \subseteq \mathcal{X}$, we have $\mathbf{P} : \mathcal{L}(X(\mathbf{P})) \to \mathbb{Z}$.

An *anonymous profile* is a generalized anonymous profile $\mathbf{P} : \mathcal{L}(X(\mathbf{P})) \to \mathbb{N}$.

Given $\mathbf{P}$ and $\mathbf{P}'$ with $X(\mathbf{P}) = X(\mathbf{P}')$, we define $\mathbf{P}+\mathbf{P}'$ with $X(\mathbf{P}+\mathbf{P}') = X(\mathbf{P})$ (resp. $\mathbf{P} - \mathbf{P}'$ with $X(\mathbf{P} - \mathbf{P}') = X(\mathbf{P})$) such that for all $L \in \mathcal{L}(X(\mathbf{P}))$,

$$(\mathbf{P} + \mathbf{P}')(L) = \mathbf{P}(L) + \mathbf{P}(L') \text{ and } (\mathbf{P} - \mathbf{P}')(L) = \mathbf{P}(L) - \mathbf{P}(L').$$

Given $n \in \mathbb{Z}$, we define $n\mathbf{P}$ with $X(n\mathbf{P}) = X(\mathbf{P})$ such that for all $L \in \mathcal{L}(X(\mathbf{P}))$,

$$(n\mathbf{P})(L) = n \times \mathbf{P}(L).$$

For $a, b \in X(\mathbf{P})$, we define the *margin of a over b in* $\mathbf{P}$, denoted $\mu_{a,b}(\mathbf{P})$, by:

$$\mu_{a,b}(\mathbf{P}) = \Bigg(\sum_{L\in\mathcal{L}(X(\mathbf{P}))\colon aLb} \mathbf{P}(L)\Bigg) - \Bigg(\sum_{L\in\mathcal{L}(X(\mathbf{P}))\colon bLa} \mathbf{P}(L)\Bigg).$$

The *majority graph* $M(\mathbf{P})$ and *margin graph* $\mathcal{M}(\mathbf{P})$ of an anonymous profile $\mathbf{P}$ are defined just as for ordinary profiles in Section 2.2. The following is obvious.

Lemma 4.2. *For generalized anonymous profiles $\mathbf{P}, \mathbf{P}'$ with $X(\mathbf{P}) = X(\mathbf{P}')$ and $a, b \in X(\mathbf{P})$, $\mu_{a,b}(\mathbf{P}+\mathbf{P}') = \mu_{a,b}(\mathbf{P})+\mu_{a,b}(\mathbf{P}')$ and $\mu_{a,b}(\mathbf{P}-\mathbf{P}') = \mu_{a,b}(\mathbf{P})-\mu_{a,b}(\mathbf{P}')$.*

4.2. McGarvey and Debord's constructions. The classic constructions of profiles from majority graphs and margin graphs are due to McGarvey [**38**] and Debord [**20**]. We review these constructions for comparison with our alternative construction in Section 4.3.

Henceforth we fix a distinguished strict linear order $L_\star$ on the set $\mathcal{X}$.

DEFINITION 4.3. For any weighted weak tournament $\mathcal{T}$ such that all weights are even numbers, we define the profile $\mathsf{Deb}(\mathcal{T})$ as follows. For each $a, b \in X(\mathcal{T})$ such that $a \to b$ in $\mathcal{T}$ with weight n, where $x_1, \ldots, x_k$ enumerates $X(\mathcal{T}) \setminus \{a, b\}$ according to the order $L_\star$, we add $\frac{n}{2}$ voters with the linear order $abx_1 \ldots x_k$ and $\frac{n}{2}$ voters with the linear order $x_k \ldots x_1 ab$.

For a weak tournament T, let T_2 be the weighted weak tournament based on T in which each edge has weight 2, and define $\mathsf{Mc}(T) = \mathsf{Deb}(T_2)$.

Though it will not play a role in our proofs, we mention how to handle weighted tournaments with weights of odd parity. There is an obvious correspondence between weighted weak tournaments and skew-symmetric matrices. Given weighted weak tournaments $\mathcal{T}_1$ and $\mathcal{T}_2$, let $\mathcal{T}_1 - \mathcal{T}_2$ be the weighted weak tournament corresponding to the matrix obtained by subtracting the matrix for $\mathcal{T}_2$ from that for $\mathcal{T}_1$. In addition, for each finite set X of candidates, let profile $\mathbf{P}_{\star,X}$ be the profile with only one voter who submits the restriction to X of $L_\star$.

DEFINITION 4.4. For a weighted tournament $\mathcal{T}$ with odd weights, we define the profile $\mathsf{Deb}(\mathcal{T})$ as follows. As $\mathcal{T} - \mathcal{M}(\mathbf{P}_{\star,X(\mathcal{T})})$ has even number weights, we set

$$\mathsf{Deb}(\mathcal{T}) = \mathsf{Deb}(\mathcal{T} - \mathcal{M}(\mathbf{P}_{\star,X(\mathcal{T})})) + \mathbf{P}_{\star,X(\mathcal{T})}.$$

Then it is easy to see that $\mathcal{M}(\mathsf{Deb}(\mathcal{T})) = \mathcal{T}$, which yields the following.

THEOREM 4.5.

(1) [**38**] *Every weak tournament is the majority graph of a profile.*
(2) [**20**] *If $\mathcal{T} = (T, w)$ is a weighted weak tournament in which all weights have the same parity, which is even if T is not a tournament, then $\mathcal{T}$ is the margin graph of a profile.*

4.3. An alternative representation. The key to proving the first transfer lemma, Lemma 3.5, is an alternative to McGarvey's representation of weak tournaments by profiles. The basic idea of the representation is simple. To realize a given weak tournament T, start with a profile consisting of one copy of each linear order over the candidates; then if there is an edge in T from candidate a to candidate b, modify the profile such that every ballot with ba on top becomes a ballot with ab on top, thereby creating an edge from a to b in the majority graph of the profile.

In fact, we need to be more careful: in order to ensure that our profile representation commutes with the operation of restricting to a set of candidates, we will realize a given majority graph T *relative to a set* Y of candidates, containing all the candidates in T. Thus, we implement the idea above starting with a profile consisting of one copy of each linear order over the candidates in Y. The following notation will facilitate our proofs about this construction.

DEFINITION 4.6. Given a finite $Y \subseteq \mathcal{X}$, let $\mathbf{L}_Y$ be the profile such that $\mathbf{L}_Y(L) = 1$ for each linear order L on Y. For $a, b \in Y$, let $\mathbf{L}_{ab\text{-}Y}$ be the profile such that $\mathbf{L}_{ab\text{-}Y}(L) = 1$ if L is a linear order on Y of the form $aby_1 \ldots y_n$ for $y_1, \ldots, y_n \in Y \setminus \{a, b\}$ and $\mathbf{L}_{ab\text{-}Y}(L) = 0$ otherwise. We call $\mathbf{L}_{ab\text{-}Y}$ an *ab-Y-block*.

Now we officially define our representation of weak tournaments.

DEFINITION 4.7. Given a finite $Y \subseteq \mathcal{X}$ and weak tournament T with $X(T) \subseteq Y$, we define the *profile representation of T relative to Y*, $\mathfrak{P}_Y(T)$, as follows. Start with the profile $\mathbf{L}_Y$. For each pair $a, b \in X(T)$ with $a \to b$ in T, flip the first two candidates of the *ba-Y-block* to obtain an additional *ab-Y-block*, resulting in a profile $\mathfrak{L}_Y(T)$. More formally, let

$$\mathfrak{L}_Y(T) = \mathbf{L}_Y + \sum_{a \to b \text{ in } T} (\mathbf{L}_{ab\text{-}Y} - \mathbf{L}_{ba\text{-}Y}). \tag{4.1}$$

Then let

$$\mathfrak{P}_Y(T) = \mathfrak{L}_Y(T)|_{X(T)}. \tag{4.2}$$

It is easy to see that like McGarvey's construction, Definition 4.7 allows us to represent any weak tournament as the majority graph of a profile. Let $\mathbb{WT}_Y$ be the set of all weak tournaments T with $X(T) \subseteq Y$.

PROPOSITION 4.8. *For any finite $Y \subseteq \mathcal{X}$, $T \in \mathbb{WT}_Y$, we have $T = M(\mathfrak{P}_Y(T))$.*

This proposition follows from Propositions 4.10 and 4.12 proved below.

For the purposes of the second transfer lemma, Lemma 3.6, we now define a similar construction for certain weighted weak tournaments. The idea is simply to start with a profile that has sufficiently many copies of each linear order, and then to create an edge from a to b with a certain margin, we turn an appropriate number of *ba-Y* blocks into *ab-Y* blocks.

Note that given a finite $Y \subseteq \mathcal{X}$ and $a, b \in Y$, the number of ballots in an *ab-Y-block* is $(|Y| - 2)!$. Later it will be useful to express margins as multiples of

$$\psi_Y = 2 \times (|Y| - 2)!,$$

in which case the number of ballots in an *ab-Y-block* is $\psi_Y/2$. When Y is clear from context, we will write ψ for ψ_Y.

DEFINITION 4.9. Given a finite $Y \subseteq \mathcal{X}$, $m \in \mathbb{Z}^+$, and weighted weak tournament $\mathcal{T}$ with $X(\mathcal{T}) \subseteq Y$ each of whose weights is $k\psi_Y$ for $k \in \mathbb{Z}$ and less than or equal to $m\psi$, we define the *profile representation of $\mathcal{T}$ relative to Y and m*, $\mathfrak{P}_{Y,m}(\mathcal{T})$, as follows. Start with $m\mathbf{L}_Y$, so for each $a, b \in X(\mathcal{T})$, there are m *ab-Y*-blocks in $m\mathbf{L}_Y$. For each pair a, b with $a \to b$ in $\mathcal{T}$ with weight $k\psi_Y$, in k *ba-Y*-blocks in $m\mathbf{L}_Y$, flip the first two candidates to obtain an additional k *ab-Y*-blocks, resulting in a profile $\mathfrak{L}_{Y,m}(\mathcal{T})$. More formally,

$$\mathfrak{L}_{Y,m}(\mathcal{T}) = m\mathbf{L}_Y + \sum_{a \xrightarrow{k\psi} b \text{ in } \mathcal{T}} (k\mathbf{L}_{ab\text{-}Y} - k\mathbf{L}_{ba\text{-}Y}). \tag{4.3}$$

Then let

$$\mathfrak{P}_{Y,m}(\mathcal{T}) = \mathfrak{L}_{Y,m}(\mathcal{T})|_{X(\mathcal{T})}. \tag{4.4}$$

We can regard the construction in Definition 4.7 as a special case of that in Definition 4.9.

PROPOSITION 4.10. *Given a finite $Y \subseteq \mathcal{X}$ and weak tournament T with $X(T) \subseteq Y$, let T_ψ be the weighted tournament based on T with all weights being ψ_Y. Then $\mathfrak{P}_Y(T) = \mathfrak{P}_{Y,1}(T_\psi)$.*

T

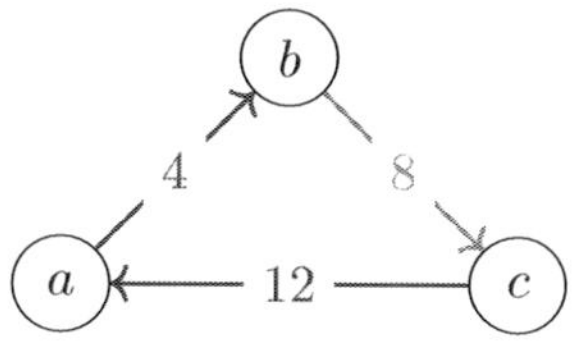

$6\mathbf{L}_Y$

6	6	6	6	6	6	6	6	6	6	6	6	6	6	6	6	6	6	6	6	6	6	6	6
a	a	a	a	a	a	b	b	b	b	b	b	c	c	c	c	c	c	d	d	d	d	d	d
b	b	c	c	d	d	a	a	c	c	d	d	a	a	b	b	d	d	a	a	b	b	c	c
c	d	b	d	b	c	c	d	a	d	a	c	b	d	a	d	a	b	b	c	a	c	a	b
d	c	d	b	c	b	d	c	d	a	c	a	d	b	d	a	b	a	c	b	c	a	b	a

$\mathfrak{L}_{Y,6}(T)$

6	6	3	3	3	3	6	6	1	1	5	5	6	6	6	6	6	6	2	2	4	4	6	6	6	6	6	6	6	6
a	a	c	c	a	a	a	a	a	a	b	b	b	b	b	b	c	c	b	b	c	c	c	c	d	d	d	d	d	d
b	b	a	a	c	c	d	d	b	b	a	a	c	c	d	d	a	a	c	c	b	b	d	d	a	a	b	b	c	c
c	d	b	d	b	d	b	c	c	d	c	d	a	d	a	c	b	d	a	d	a	d	a	b	b	c	a	c	a	b
d	c	d	b	d	b	c	b	d	c	d	c	d	a	c	a	d	b	d	a	d	a	b	a	c	b	c	a	b	a

$\mathfrak{P}_{Y,6}(T)$

6	6	3	3	3	3	6	6	1	1	5	5	6	6	6	6	6	6	2	2	4	4	6	6	6	6	6	6	6	6
a	a	c	c	a	a	a	a	a	a	b	b	b	b	b	b	c	c	b	b	c	c	c	c	a	a	b	b	c	c
b	b	a	a	c	c	b	c	b	b	a	a	c	c	a	c	a	a	c	c	b	b	a	b	b	c	a	c	a	b
c	c	b	b	b	b	c	b	c	c	c	c	a	a	c	a	b	b	a	a	a	a	b	a	c	b	c	a	b	a

$\mathsf{Deb}(T)$

2	2	4	4	6	6
a	c	b	a	c	b
b	a	c	b	a	c
c	b	a	c	b	a

FIGURE 8. Illustration of Definitions 4.7 and 4.3 with $Y = \{a, b, c, d\}$

PROOF. Since all weights in T_ψ are ψ_Y, the k in Definition 4.9 is 1. Then since $m = 1$, (4.3) reduces to (4.1). □

Figures 9 and 8 illustrate our construction from Definition 4.9, contrasting it with Debord's from Definition 4.3. This construction allows us to represent any weighted weak tournament in the following class as the margin graph of a profile.

DEFINITION 4.11. Given a finite $Y \subseteq \mathcal{X}$ and $m \in \mathbb{Z}^+$, let $\mathbb{WT}^w_{Y,m}$ be the set of all weighted weak tournaments $\mathcal{T}$ with $X(\mathcal{T}) \subseteq Y$ such that each weight in $\mathcal{T}$ is of the form $k\psi_Y$ for some $k \in \mathbb{Z}$ and less than or equal to $m\psi_Y$.

PROPOSITION 4.12. *For any finite $Y \subseteq \mathcal{X}$, $m \in \mathbb{Z}^+$, and $\mathcal{T} \in \mathbb{WT}^w_{Y,m}$, we have $\mathcal{T} = \mathcal{M}(\mathfrak{P}_{Y,m}(\mathcal{T}))$.*

PROOF. Recall that $\mathfrak{P}_{Y,m}$ is defined by (4.3)–(4.4). Concerning (4.3), observe that for any $a, b \in X(\mathcal{T})$:

(i) $\mu_{a,b}(m\mathbf{L}_Y) = 0$;

(ii) if $a \overset{k\psi}{\to} b$ in $\mathcal{T}$, then $\mu_{a,b}(k\mathbf{L}_{ab\text{-}Y} - k\mathbf{L}_{ba\text{-}Y}) = k\psi_Y$;

T

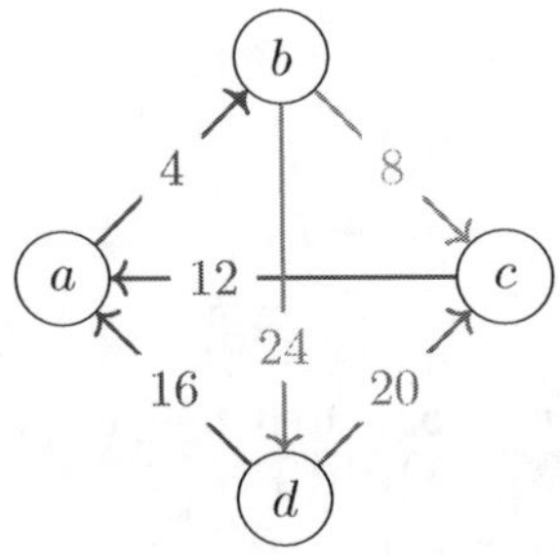

$6\mathbf{L}_Y$

6	6	6	6	6	6	6	6	6	6	6	6	6	6	6	6	6	6	6	6	6	6	6	6
a	a	a	a	a	a	b	b	b	b	b	b	c	c	c	c	c	c	d	d	d	d	d	d
b	b	c	c	d	d	a	a	c	c	d	d	a	a	b	b	d	d	a	a	b	b	c	c
c	d	b	d	b	c	c	d	a	d	a	c	b	d	a	d	a	b	b	c	a	c	a	b
d	c	d	b	c	b	d	c	d	a	c	a	d	b	d	a	b	a	c	b	c	a	b	a

$\mathfrak{L}_{Y,6}(T)$ = $\mathfrak{P}_{Y,6}(T)$

6	6	3	3	3	3	4	4	2	2	1	1	5	5	6	6	6	6	6	6	2	2	4	4	5	5	1	1	6	6	6	6	0	0	6	6
a	a	c	c	a	a	d	d	a	a	a	a	b	b	b	b	b	b	c	c	b	b	c	c	d	d	c	c	d	d	b	b	d	d	d	d
b	b	a	a	c	c	a	a	d	d	b	b	a	a	c	c	d	d	a	a	c	c	b	b	c	c	d	d	a	a	d	d	b	b	c	c
c	d	b	d	b	d	b	c	b	c	c	d	c	d	a	d	a	c	b	d	a	d	a	d	a	b	a	b	b	c	a	c	a	c	a	b
d	c	d	b	d	b	c	b	c	b	d	c	d	c	d	a	c	a	d	b	d	a	d	a	b	a	b	a	c	b	c	a	c	a	b	a

Deb(T)

2	2	4	4	6	6	8	8	10	10	12	12
a	d	b	d	c	d	d	c	d	b	b	c
b	c	c	a	a	b	a	b	c	a	d	a
c	a	a	b	b	c	b	d	a	d	a	b
d	b	d	c	d	a	c	a	b	c	c	d

FIGURE 9. Illustration of Definitions 4.7 and 4.3 with $Y = \{a, b, c, d\}$

(iii) for any $(c, d) \neq (a, b)$ and $(d, c) \neq (a, b)$, if $c \overset{k\psi}{\to} d$, then $\mu_{a,b}(k\mathbf{L}_{cd\text{-}Y} - k\mathbf{L}_{dc\text{-}Y}) = 0$.

Part (i) is immediate from the definition of $\mathbf{L}_Y$. For (ii), we have:

$$\begin{aligned}\mu_{a,b}(k\mathbf{L}_{ab\text{-}Y} - k\mathbf{L}_{ba\text{-}Y}) &= \mu_{a,b}(k\mathbf{L}_{ab\text{-}Y}) - \mu_{a,b}(k\mathbf{L}_{ba\text{-}Y})\\ &= k\mu_{a,b}(\mathbf{L}_{ab\text{-}Y}) - k\mu_{a,b}(\mathbf{L}_{ba\text{-}Y})\\ &= k\frac{\psi_Y}{2} - (-k\frac{\psi_Y}{2})\\ &= k\psi_Y.\end{aligned}$$

For (iii), we have:

$$\mu_{a,b}(k\mathbf{L}_{cd\text{-}Y} - k\mathbf{L}_{dc\text{-}Y}) = k\mu_{a,b}(\mathbf{L}_{cd\text{-}Y}) - k\mu_{a,b}(\mathbf{L}_{dc\text{-}Y}).$$

Recall that $(c, d) \neq (a, b)$ and $(d, c) \neq (a, b)$. If $\{c, d\} \cap \{a, b\} = \varnothing$, so $a, b \in Y$, then $\mu_{a,b}(\mathbf{L}_{cd\text{-}Y}) = \mu_{a,b}(\mathbf{L}_{dc\text{-}Y}) = 0$, so $\mu_{a,b}(k\mathbf{L}_{cd\text{-}Y} - k\mathbf{L}_{dc\text{-}Y}) = 0$. Now suppose $|\{c, d\} \cap \{a, b\}| = 1$. Then without loss of generality suppose $a \in \{c, d\}$ and $b \in Y$. Then $\mu_{a,b}(\mathbf{L}_{cd\text{-}Y}) = \mu_{a,b}(\mathbf{L}_{dc\text{-}Y}) = \psi_Y/2$, so again $\mu_{a,b}(k\mathbf{L}_{cd\text{-}Y} - k\mathbf{L}_{dc\text{-}Y}) = 0$.

It follows from (i)–(iii) that for each pair a, b with $a \to b$ in $\mathcal{T}$ with weight $k\psi_Y$, we have $\mu_{a,b}(\mathfrak{L}_{Y,m}(\mathcal{T})) = k\psi_Y$; moreover, for each pair a, b with neither $a \to b$ nor $b \to a$ in $\mathcal{T}$, we have $\mu_{a,b}(\mathfrak{L}_{Y,m}(\mathcal{T})) = 0$. Thus, $\mathcal{T} = \mathcal{M}(\mathfrak{P}_{Y,m}(\mathcal{T}))$. $\square$

There are two key differences between our construction and those of McGarvey and Debord, namely that representation commutes with restriction to a set of candidates and transposition of candidates, as proved in the following subsections.

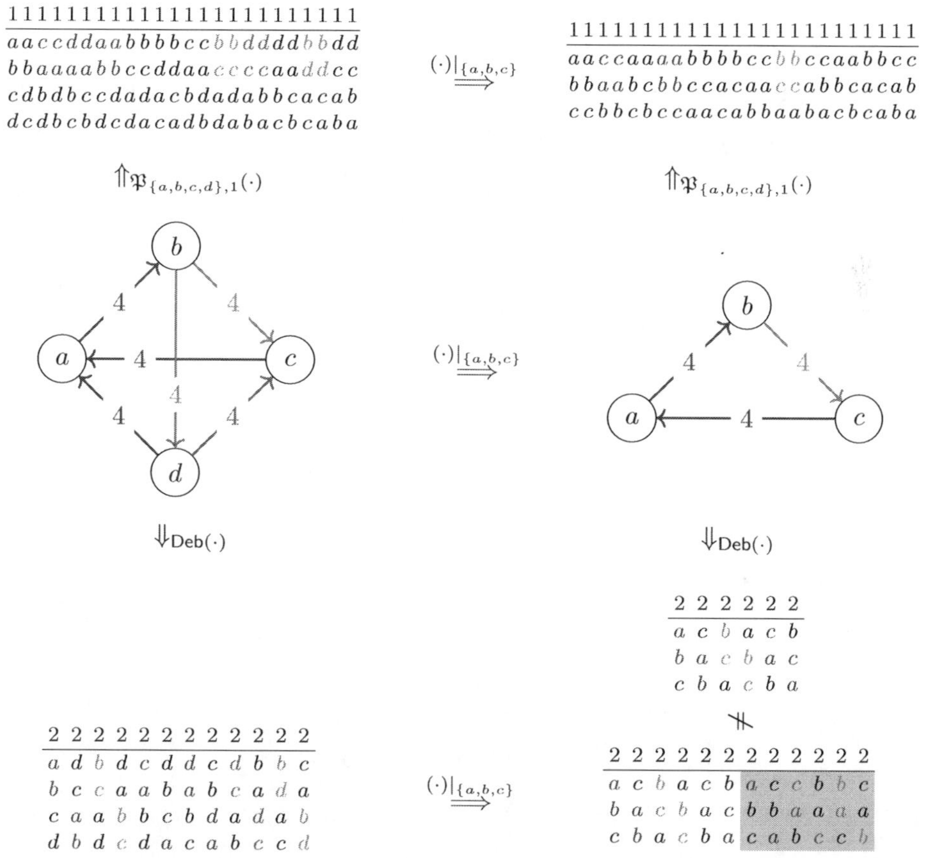

FIGURE 10. Representation commutes with restriction (above), but the McGarvey/Debord construction does not commute with restriction (below).

4.4. Representation commutes with restriction. Given a binary relation R on a set X and $Y \subseteq X$, *the restriction of* R *to* Y is the binary relation $R|_Y$ on Y such that for all $a, b \in Y$, $(a, b) \in R|_Y$ if and only if $(a, b) \in R$.

Given a generalized anonymous profile $\mathbf{P}$ and $Y \subseteq X(\mathbf{P})$, the *restriction of* $\mathbf{P}$ *to* Y is the profile $\mathbf{P}|_Y$ with $X(\mathbf{P}|_Y) = Y$ such that for all $L \in \mathcal{L}(Y)$,

$$\mathbf{P}|_Y(L) = \sum_{L' \in \mathcal{L}(X(\mathbf{P})) :\, L = L'|_Y} \mathbf{P}(L').$$

Given a weak tournament T and $Z \subseteq X(T)$, the *restriction of T to Z* is the directed graph $T|_Z$ whose set of vertices is Y and whose edge relation is $\to|_Z$. The restriction of a weighted weak tournament is defined analogously, with the weight function restricted to $\to|_Z$. Clearly restriction commutes with taking the majority or margin graph of a profile and with multiplication.

LEMMA 4.13. *Let $\mathbf{P}$ be a generalized anonymous profile and $\mathcal{T}$ a weighted weak tournament.*

(1) *For any $\varnothing \neq Z \subseteq X(\mathbf{P})$, $M(\mathbf{P}|_Z) = M(\mathbf{P})|_Z$ and $\mathcal{M}(\mathbf{P}|_Z) = \mathcal{M}(\mathbf{P})|_Z$.*
(2) *For any $\varnothing \neq Z \subseteq X(\mathcal{T})$ and $n \in \mathbb{Z}^+$, $n(\mathcal{T}|_Z) = (n\mathcal{T})|_Z$.*

Next we show that representation commutes with restriction, as in Figure 10.

PROPOSITION 4.14. *For any finite $Y \subseteq \mathcal{X}$, $T \in \mathbb{WT}_Y$, $m \in \mathbb{Z}^+$, $\mathcal{T} \in \mathbb{WT}^w_{Y,m}$:*

(1) *if $\varnothing \neq Z \subseteq X(T)$, then $\mathfrak{P}_Y(T|_Z) = \mathfrak{P}_Y(T)|_Z$;*
(2) *if $\varnothing \neq Z \subseteq X(\mathcal{T})$, then $\mathfrak{P}_{Y,m}(\mathcal{T}|_Z) = \mathfrak{P}_{Y,m}(\mathcal{T})|_Z$.*

PROOF. We give only the proof for (1), as the proof for (2) is essentially the same but with more notation. The key observation is that if $\{a,b\} \not\subseteq Z$, then $(\mathbf{L}_{ab\text{-}Y})|_Z = (\mathbf{L}_{ba\text{-}Y})|_Z$. Then we have

$$\begin{aligned}
\mathfrak{P}_Y(T)|_Z &= (\mathfrak{L}_Y(T)|_{X(T)})|_Z \text{ by (4.2)} \\
&= \mathfrak{L}_Y(T)|_Z \text{ since } Z \subseteq X(T) \\
&= \Big(\mathbf{L}_Y + \sum_{a \to b \text{ in } T} (\mathbf{L}_{ab\text{-}Y} - \mathbf{L}_{ba\text{-}Y})\Big)\Big|_Z \\
&= \Big(\mathbf{L}_Y + \sum_{\substack{\{a,b\}\subseteq Z \\ a \to b \text{ in } T}} (\mathbf{L}_{ab\text{-}Y} - \mathbf{L}_{ba\text{-}Y}) \\
&\qquad + \sum_{\substack{\{a,b\}\not\subseteq Z \\ a \to b \text{ in } T}} (\mathbf{L}_{ab\text{-}Y} - \mathbf{L}_{ba\text{-}Y})\Big)\Big|_Z \\
&= \Big(\mathbf{L}_Y + \sum_{\substack{\{a,b\}\subseteq Z \\ a \to b \text{ in } T}} (\mathbf{L}_{ab\text{-}Y} - \mathbf{L}_{ba\text{-}Y})\Big)\Big|_Z \\
&\qquad + \sum_{\substack{\{a,b\}\not\subseteq Z \\ a \to b \text{ in } T}} ((\mathbf{L}_{ab\text{-}Y})|_Z - (\mathbf{L}_{ba\text{-}Y})|_Z) \\
&= \Big(\mathbf{L}_Y + \sum_{a \to b \text{ in } T|_Z} (\mathbf{L}_{ab\text{-}Y} - \mathbf{L}_{ba\text{-}Y})\Big)\Big|_Z \\
&= \mathfrak{L}_Y(T|_Z)|_Z \\
&= \mathfrak{L}_Y(T|_Z)|_{X(T|_Z)} \text{ since } X(T|_Z) = Z \\
&= \mathfrak{P}_Y(T|_Z) \text{ by (4.2).}
\end{aligned}$$

□

4.5. Representation commutes with transposition. Given $a, b \in \mathcal{X}$, a linear order L, and $X \subseteq \mathcal{X}$, recall the definitions of $L_{a\leftrightarrows b}$ and $X_{a\leftrightarrows b}$ from Section 2.2. Given a generalized anonymous profile $\mathbf{P}$ and $a, b \in \mathcal{X}$, we define $\mathbf{P}_{a\leftrightarrows b}$ such that for any linear order L, $\mathbf{P}_{a\leftrightarrows b}(L) = \mathbf{P}(L_{a\leftrightarrows b})$.

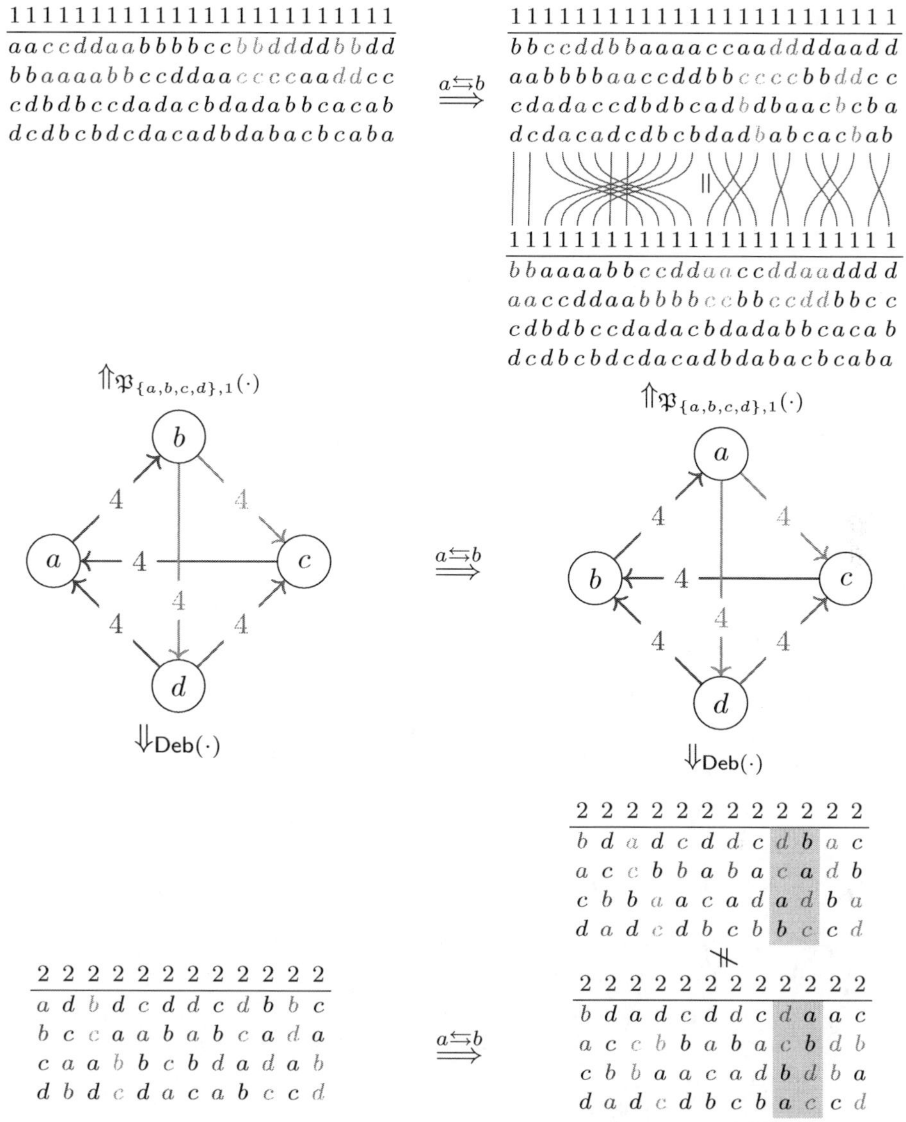

FIGURE 11. Representation commutes with transposition (above), but the McGarvey/Debord construction does not commute with transposition (below).

Given a weak tournament $T = (X(T), \to)$ and $a, b \in \mathcal{X}$, we define the weak tournament $T_{a \leftrightarrows b} = (X(T)_{a \leftrightarrows b}, \to_{a \leftrightarrows b})$ such that for all $x, y \in X(T)$, we have

$$x \to_{a \leftrightarrows b} y \text{ in } T_{a \leftrightarrows b} \text{ if } \pi_{a \leftrightarrows b}(x) \to \pi_{a \leftrightarrows b}(y) \text{ in } T.$$

Similarly, for a weighted weak tournament $\mathcal{T} = (T, w)$, we define $\mathcal{T}_{a \leftrightarrows b} = (T_{a \leftrightarrows b}, w_{a \leftrightarrows b})$ where if $x \to_{a \leftrightarrows b} y$ in $T_{a \leftrightarrows b}$, then

$$w_{a \leftrightarrows b}(x, y) = w(\pi_{a \leftrightarrows b}(x), \pi_{a \leftrightarrows b}(y)).$$

Later we will use the obvious fact that transposition commutes with taking the majority or margin graph of a profile, with restriction, and with multiplication.

LEMMA 4.15. *Let* $\mathbf{P}$ *be a generalized anonymous profile,* $\mathcal{T}$ *a weighted weak tournament, and* $a, b \in \mathcal{X}$.

(1) $M(\mathbf{P}_{a\leftrightarrows b}) = M(\mathbf{P})_{a\leftrightarrows b}$ *and* $\mathcal{M}(\mathbf{P}_{a\leftrightarrows b}) = \mathcal{M}(\mathbf{P})_{a\leftrightarrows b}$.
(2) *For any* $\varnothing \neq Z \subseteq X(\mathbf{P})$, *then* $(\mathbf{P}_{|Z})_{a\leftrightarrows b} = (\mathbf{P}_{a\leftrightarrows b})_{|\pi_{a\leftrightarrows b}[Z]}$.
(3) *For any* $n \in \mathbb{Z}^+$, $n(\mathcal{T}_{a\leftrightarrows b}) = (n\mathcal{T})_{a\leftrightarrows b}$.

As illustrated in Figure 11, unlike the constructions of McGarvey and Debord, representation commutes with transposition.

PROPOSITION 4.16. *For any finite* $Y \subseteq \mathcal{X}$, $T \in \mathbb{WT}_Y$, $m \in \mathbb{Z}^+$, $\mathcal{T} \in \mathbb{WT}^w_{Y,m}$, *and* $a, b \in Y$:

(1) $\mathfrak{P}_Y(T_{a\leftrightarrows b}) = \mathfrak{P}_Y(T)_{a\leftrightarrows b}$;
(2) $\mathfrak{P}_{Y,m}(\mathcal{T}_{a\leftrightarrows b}) = \mathfrak{P}_{Y,m}(\mathcal{T})_{a\leftrightarrows b}$.

PROOF. We give only the proof for (1), as the proof for (2) is essentially the same but with more notation. Define $\rho : Y \to Y$ by $\rho(a) = b$, $\rho(b) = a$, and $\rho(c) = c$ for all $c \in Y \setminus \{a, b\}$. The key observations are that $(\mathbf{L}_Y)_{a\leftrightarrows b} = \mathbf{L}_Y$, since $a, b \in Y$, and for any $c, d \in X(T)$, $(\mathbf{L}_{cd\text{-}Y})_{a\leftrightarrows b} = \mathbf{L}_{\rho(c)\rho(d)\text{-}Y}$. Then we have

$$\begin{aligned}
\mathfrak{P}_Y(T)_{a\leftrightarrows b} &= (\mathfrak{L}_Y(T)|_{X(T)})_{a\leftrightarrows b} \text{ by (4.2)}\\
&= (\mathfrak{L}_Y(T)_{a\leftrightarrows b})|_{X(T)_{a\leftrightarrows b}}\\
&= \Big(\big(\mathbf{L}_Y + \sum_{c \to d \text{ in } T} (\mathbf{L}_{cd\text{-}Y} - \mathbf{L}_{dc\text{-}Y})\big)_{a\leftrightarrows b}\Big)|_{X(T)_{a\leftrightarrows b}}\\
&= \Big((\mathbf{L}_Y)_{a\leftrightarrows b} + \sum_{c \to d \text{ in } T} ((\mathbf{L}_{cd\text{-}Y})_{a\leftrightarrows b} - (\mathbf{L}_{dc\text{-}Y})_{a\leftrightarrows b})\Big)|_{X(T)_{a\leftrightarrows b}}\\
&= \Big(\mathbf{L}_Y + \sum_{c \to d \text{ in } T} (\mathbf{L}_{\rho(c)\rho(d)\text{-}Y} - \mathbf{L}_{\rho(d)\rho(c)\text{-}Y})\Big)|_{X(T)_{a\leftrightarrows b}}\\
&= \Big(\mathbf{L}_Y + \sum_{\rho(c) \to \rho(d) \text{ in } T_{a\leftrightarrows b}} (\mathbf{L}_{\rho(c)\rho(d)\text{-}Y} - \mathbf{L}_{\rho(d)\rho(c)\text{-}Y})\Big)|_{X(T)_{a\leftrightarrows b}}\\
&= \Big(\mathbf{L}_Y + \sum_{c' \to d' \text{ in } T_{a\leftrightarrows b}} (\mathbf{L}_{c'd'\text{-}Y} - \mathbf{L}_{d'c'\text{-}Y})\Big)|_{X(T)_{a\leftrightarrows b}}\\
&= \mathfrak{L}_Y(T_{a\leftrightarrows b})|_{X(T)_{a\leftrightarrows b}}\\
&= \mathfrak{L}_Y(T_{a\leftrightarrows b})|_{X(T_{a\leftrightarrows b})}\\
&= \mathfrak{P}_Y(T_{a\leftrightarrows b}) \text{ by (4.2).}
\end{aligned}$$

□

5. Majoritarian and pairwise projection

Using the representation of weak tournaments in Definition 4.7, the next key step is to associate with each voting method F and finite $Y \subseteq \mathcal{X}$ a voting method that we call its *majoritarian projection* relative to Y. In fact, for convenience we will first do this for *anonymous* voting methods defined on anonymous profiles and then for ordinary voting methods in our proof of the transfer lemma, Lemma 3.5.

DEFINITION 5.1. An *anonymous voting method* is a function F whose domain is a set of anonymous profiles such that for any $\mathbf{P} \in \text{dom}(F)$, $\varnothing \neq F(\mathbf{P}) \subseteq X(\mathbf{P})$.

Every axiom we have defined for voting methods applies in an obvious way to anonymous voting methods, so we will not repeat the definitions here.

DEFINITION 5.2. Given an anonymous voting method F and finite $Y \subseteq \mathcal{X}$, define the *majoritarian projection of F relative to Y*, $F_{maj,Y}$, as the anonymous voting method with

$$\mathrm{dom}(F_{maj,Y}) = \{\mathbf{P} \in \textsf{anon-prof} \mid X(\mathbf{P}) \subseteq Y \text{ and } \mathfrak{P}_Y(M(\mathbf{P})) \in \mathrm{dom}(F)\}$$

and

$$F_{maj,Y}(\mathbf{P}) = F(\mathfrak{P}_Y(M(\mathbf{P}))).$$

We define an analogous notion of pairwise projection using Definition 4.9.

DEFINITION 5.3. Given an anonymous voting method F, finite $Y \subseteq \mathcal{X}$, and $m \in \mathbb{Z}^+$, define the *pairwise projection of F relative to Y and m*, $F_{pair,Y,m}$, as the anonymous voting method with

$$\mathrm{dom}(F_{pair,Y,m}) = \{\mathbf{P} \in \textsf{anon-prof} \mid X(\mathbf{P}) \subseteq Y, \text{ all weights in } \mathcal{M}(\mathbf{P}) \text{ are } \leq m, \\ \text{and } \mathfrak{P}_{Y,m}(\psi\mathcal{M}(\mathbf{P})) \in \mathrm{dom}(F)\}$$

and

$$F_{pair,Y,m}(\mathbf{P}) = F(\mathfrak{P}_{Y,m}(\psi\mathcal{M}(\mathbf{P})))$$

where $\psi = 2 \times (|Y| - 2)!$.

Note that multiplying $\mathcal{M}(\mathbf{P})$ by ψ gives us a weighted tournament suitable for representation by Definition 4.9. Though the resulting profile $\mathfrak{P}_{Y,m}(\psi\mathcal{M}(\mathbf{P}))$ does not represent $\mathcal{M}(\mathbf{P})$, but instead $\psi\mathcal{M}(\mathbf{P})$, this does not matter, since we only use $\mathfrak{P}_{Y,m}(\psi\mathcal{M}(\mathbf{P}))$ to select winners according to F.

It is easy to see that the majoritarian (resp. pairwise) projection of F is indeed majoritarian (resp. pairwise). Moreover, majoritarian and pairwise projection preserve neutrality thanks to the fact in Proposition 4.16 that representation commutes with transposition (in contrast to the McGarvey/Debord construction) plus Lemma 4.15. Recall that given $Y \subseteq \mathcal{X}$, the set $Y_{a\leftrightarrows b}$ is the image of Y under the permutation of $\mathcal{X}$ that only swaps a and b.

LEMMA 5.4. *Given an anonymous voting method F, finite $Y \subseteq \mathcal{X}$, and $m \in \mathbb{Z}^+$:*

(1) *$F_{maj,Y}$ is majoritarian;*
(2) *$F_{pair,Y,m}$ is pairwise;*
(3) *if F is neutral, then $F_{maj,Y}$ and $F_{pair,Y,m}$ are neutral.*

PROOF. Parts 1 and 2 are immediate from Definitions 5.2 and 5.3, respectively. For part 3, assuming F is neutral, given any $\mathbf{P} \in \mathrm{dom}(F_{maj,Y})$ and $\mathbf{P}_{a\leftrightarrows b} \in \mathrm{dom}(F)$, we have $a, b \in Y$, and

$$\begin{aligned}
F_{maj,Y}(\mathbf{P}_{a\leftrightarrows b}) &= F(\mathfrak{P}_Y(M(\mathbf{P}_{a\leftrightarrows b}))) && \text{by Definition 5.2} \\
&= F(\mathfrak{P}_Y(M(\mathbf{P})_{a\leftrightarrows b})) && \text{by Lemma 4.15.1} \\
&= F(\mathfrak{P}_Y(M(\mathbf{P}))_{a\leftrightarrows b}) && \text{by Proposition 4.16.1 since } a, b \in Y \\
&= F(\mathfrak{P}_Y(M(\mathbf{P})))_{a\leftrightarrows b} && \text{since } F \text{ is neutral} \\
&= F_{maj,Y}(\mathbf{P})_{a\leftrightarrows b} && \text{by Definition 5.2,}
\end{aligned}$$

so $F_{maj,Y}$ is neutral. The proof for $F_{pair,Y,m}$ is analogous, using Definition 5.3, Proposition 4.16.2, and also Lemma 4.15.3. □

6. Preservation of axioms

We now turn to the key question of which axioms on voting methods are preserved by majoritarian and pairwise projection, besides neutrality (Lemma 5.4.3). These preservation results will be key to proving the transfer lemmas from Section 3.

6.1. Axioms as sets. To prove general results about the preservation of axioms by majoritarian and pairwise projections, we will make a move that is unusual in social choice theory but standard in mathematical logic, by treating axioms on voting methods as mathematical objects. Thinking syntactically, we could introduce a formal language and define a *universal axiom* as a formula of the form

$$\forall \mathbf{P}_1 \ldots \forall \mathbf{P}_n \varphi(\mathbf{P}_1, \ldots, \mathbf{P}_n)$$

where φ contains no profile quantifiers (but possibly quantifiers over voters and candidates). Instead, we will view universal axioms semantically as follows.[15] Let $\mathscr{S}$ be the set of all pairs $\langle \mathbf{P}, Y \rangle$ where $\mathbf{P}$ is an anonymous profile and $Y \subseteq X(\mathbf{P})$.

Definition 6.1. An *n-ary universal axiom* is a subset $Ax \subseteq \mathscr{S}^n$. An anonymous voting method F *satisfies* Ax if $\{\langle \mathbf{P}, F(\mathbf{P}) \rangle \mid \mathbf{P} \in \mathrm{dom}(F)\}^n \subseteq Ax$.

We illustrate this concept by formalizing some of the axioms discussed in Section 2.

Example 6.2.

(1) Given an anonymous profile $\mathbf{P}$, a *Condorcet winner* is a candidate $a \in X(\mathbf{P})$ such that for all $b \in X(\mathbf{P}) \setminus \{a\}$, the margin of a over b in $\mathbf{P}$ is positive. A voting method F is *Condorcet consistent* if for all $\mathbf{P} \in \mathrm{dom}(F)$, if $\mathbf{P}$ has a Condorcet winner, then $F(\mathbf{P})$ contains only that candidate. Condorcet consistency corresponds to the unary axiom

$$\mathsf{CC} = \{\langle \mathbf{P}, X \rangle \mid \text{if } \mathbf{P} \text{ has a Condorcet winner, then } X \text{ contains only that candidate}\}.$$

(2) majoritarianism corresponds to the binary axiom

$$\mathsf{majoritarian} = \{\langle \langle \mathbf{P}, X \rangle, \langle \mathbf{P}', X' \rangle \rangle \mid \text{if } M(\mathbf{P}) = M(\mathbf{P}'), \text{ then } X = X'\}.$$

(3) neutrality corresponds to the binary axiom

$$\mathsf{N} = \{\langle \langle \mathbf{P}, X \rangle, \langle \mathbf{P}', X' \rangle \rangle \mid \text{if } \mathbf{P}' = \mathbf{P}_{a \leftrightarrows b}, \text{ then } X' = X_{a \leftrightarrows b}\}.$$

(4) binary quasi-resoluteness corresponds to the axiom

$$\mathsf{BQR} = \{\langle \mathbf{P}, X \rangle \mid \text{if } M(\mathbf{P}) \text{ is a tournament with 2 nodes, then } |X| = 1\}.$$

(5) quasi-resoluteness corresponds to the axiom

$$\mathsf{QR} = \{\langle \mathbf{P}, X \rangle \mid \text{if } \mathcal{M}(\mathbf{P}) \text{ is uniquely weighted, then } |X| = 1\}.$$

[15] For those familiar with mathematical logic (see, e.g., [**16**]), consider a two-sorted model $\mathfrak{A}$ whose domains contains all profiles and all nonempty finite sets of candidates. Where P_i is a variable for a profile and X_i is a variable for a set of candidates, a formula $\varphi(\mathrm{P}_1, \ldots, \mathrm{P}_n, \mathrm{X}_1, \ldots, \mathrm{X}_n)$ defines the set of all tuples $\langle \mathbf{P}_1, \ldots, \mathbf{P}_n, X_1, \ldots, X_n \rangle$ such that

$$\mathfrak{A} \models \varphi(\mathrm{P}_1, \ldots, \mathrm{P}_n, \mathrm{X}_1, \ldots, \mathrm{X}_n)[\mathbf{P}_1, \ldots, \mathbf{P}_n, X_1, \ldots, X_n].$$

We repackage $\langle \mathbf{P}_1, \ldots, \mathbf{P}_n, X_1, \ldots, X_n \rangle$ as $\langle \langle \mathbf{P}_1, X_1 \rangle, \ldots, \langle \mathbf{P}_n, X_n \rangle \rangle$ in Definition 6.1.

6.2. Majoritarian and homogeneously pairwise axioms. Inspired by the notions of majoritarian/pairwise voting methods, we will define the notions of majoritarian/pairwise axioms. In fact, we are interested in what we will call *homogeneously* pairwise axioms, inspired by the notion of a voting method F being homogeneous (recall Definition 3.1).

DEFINITION 6.3. Let Ax be an n-ary universal axiom. We say that Ax is *majoritarian* (resp. *homogeneously pairwise*) if for all anonymous profiles $\mathbf{P}_1, \ldots,$ $\mathbf{P}_n$ and $\mathbf{Q}_1, \ldots, \mathbf{Q}_n$ such that for $1 \leq i \leq n$, $M(\mathbf{P}_i) = M(\mathbf{Q}_i)$ (resp. for which there is a $k \in \mathbb{Z}^+$ such that for $1 \leq i \leq n$, $k\mathcal{M}(\mathbf{P}_i) = \mathcal{M}(\mathbf{Q}_i)$), we have that for all $X_1 \subseteq X(\mathbf{P}_1), \ldots, X_n \subseteq X(\mathbf{P}_n)$,

$$\langle\langle \mathbf{P}_1, X_1\rangle, \ldots, \langle \mathbf{P}_n, X_n\rangle\rangle \in Ax \text{ if and only if } \langle\langle \mathbf{Q}_1, X_1\rangle, \ldots, \langle \mathbf{Q}_n, X_n\rangle\rangle \in Ax.$$

Note that being majoritarian implies being homogeneously pairwise.

The axioms from Example 6.2 can be classified using Definition 6.3 as follows.

EXAMPLE 6.4.

(1) Condorcet consistency is majoritarian because if $\langle \mathbf{P}, X\rangle \in \mathsf{CC}$ and $M(\mathbf{P}) = M(\mathbf{Q})$, then clearly $\langle \mathbf{Q}, X\rangle \in \mathsf{CC}$.[16]
(2) The majoritarian axiom is obviously majoritarian.
(3) Neutrality is not homogeneously pairwise, because there are profiles $\mathbf{P}$, $\mathbf{P}'$, $\mathbf{Q}$, $\mathbf{Q}'$ such that $\mathbf{P}' = \mathbf{P}_{a\leftrightarrows b}$, $\mathcal{M}(\mathbf{P}) = \mathcal{M}(\mathbf{Q})$, and $\mathcal{M}(\mathbf{P}') = \mathcal{M}(\mathbf{Q}')$, but $\mathbf{Q}' \neq \mathbf{Q}_{a\leftrightarrows b}$. Then for $X, X' \subseteq X(\mathbf{P})$ such that $X' \neq X_{a\leftrightarrows b}$, we have $\langle\langle \mathbf{P}, X\rangle, \langle \mathbf{P}', X'\rangle\rangle \notin \mathsf{N}$ but trivially $\langle\langle \mathbf{Q}, X\rangle, \langle \mathbf{Q}', X'\rangle\rangle \in \mathsf{N}$.
(4) BQR is obviously majoritarian.
(5) QR is homogeneously pairwise (for $\mathcal{M}(\mathbf{P})$ is uniquely weighted iff $k\mathcal{M}(\mathbf{P})$ is), but it is not majoritarian, because there are profiles $\mathbf{P}, \mathbf{P}'$ such that $M(\mathbf{P}) = M(\mathbf{P}')$, $\mathcal{M}(\mathbf{P})$ is uniquely weighted, but $\mathcal{M}(\mathbf{P}')$ is not. Then for $X \subseteq X(\mathbf{P})$ with $|X| > 1$, $\langle \mathbf{P}, X\rangle \notin \mathsf{QR}$ but trivially $\langle \mathbf{P}', X\rangle \in \mathsf{QR}$.

We now verify that all majoritarian (resp. homogeneously pairwise) universal axioms are preserved by the operation of majoritarian (resp. pairwise) projection from Section 5.

LEMMA 6.5. *Let F be an anonymous voting method. For any majoritarian (resp. homogeneously pairwise) universal axiom Ax, if F satisfies Ax, then $F_{maj,Y}$ (resp. $F_{pair,Y,m}$) satisfies Ax.*

PROOF. Suppose Ax is a majoritarian n-ary universal axiom. To show that $F_{maj,Y}$ satisfies Ax, consider any $\mathbf{P}_1, \ldots, \mathbf{P}_n \in \mathrm{dom}(F_{maj,Y})$. We must show that

$$\langle\langle \mathbf{P}_1, F_{maj,Y}(\mathbf{P}_1)\rangle, \ldots, \langle \mathbf{P}_n, F_{maj,Y}(\mathbf{P}_n)\rangle\rangle \in Ax.$$

By definition of $F_{maj,Y}$ (Definition 5.2), this is equivalent to

$$\langle\langle \mathbf{P}_1, F(\mathfrak{P}_Y(M(\mathbf{P}_1)))\rangle, \ldots, \langle \mathbf{P}_n, F(\mathfrak{P}_Y(M(\mathbf{P}_n)))\rangle\rangle \in Ax.$$

By Lemma 4.8, for all i, $M(\mathbf{P}_i) = M(\mathfrak{P}_Y(M(\mathbf{P}_i)))$. Then since Ax is majoritarian, setting $\mathbf{Q}_i = \mathfrak{P}_Y(M(\mathbf{P}_i))$ in Definition 6.3, the previous membership statement is equivalent to

$$\langle\langle \mathfrak{P}_Y(M(\mathbf{P}_1)), F(\mathfrak{P}_Y(M(\mathbf{P}_1)))\rangle, \ldots, \langle \mathfrak{P}_Y(M(\mathbf{P}_n)), F(\mathfrak{P}_Y(M(\mathbf{P}_n)))\rangle\rangle \in Ax,$$

[16]Other examples of majoritarian axioms include the Condorcet loser criterion [**42**] and Smith criterion [**52**].

which holds by our assumption that F satisfies Ax.

Now suppose Ax is a homogeneously pairwise n-ary universal axiom. To show that $F_{pair,Y,m}$ satisfies Ax, consider any $\mathbf{P}_1, \ldots, \mathbf{P}_n \in \mathrm{dom}(F_{pair,Y,m})$. We must show that

$$\langle\langle \mathbf{P}_1, F_{pair,Y,m}(\mathbf{P}_1)\rangle, \ldots, \langle \mathbf{P}_n, F_{pair,Y,m}(\mathbf{P}_n)\rangle\rangle \in Ax.$$

By definition of $F_{pair,Y,m}$ (Definition 5.3), this is equivalent to

$$\langle\langle \mathbf{P}_1, F(\mathfrak{P}_{Y,m}(\psi\mathcal{M}(\mathbf{P}_1)))\rangle, \ldots, \langle \mathbf{P}_n, F(\mathfrak{P}_{Y,m}(\psi\mathcal{M}(\mathbf{P}_n)))\rangle\rangle \in Ax.$$

By Lemma 4.12, for all i, $\psi\mathcal{M}(\mathbf{P}_i) = \mathcal{M}(\mathfrak{P}_{Y,m}(\psi\mathcal{M}(\mathbf{P}_i)))$. Then since Ax is homogeneously pairwise, setting $\mathbf{Q}_i = \mathfrak{P}_{Y,m}(\psi\mathcal{M}(\mathbf{P}_i))$ in Definition 6.3, the previous membership statement is equivalent to

$$\Big\langle \langle \mathfrak{P}_{Y,m}(\psi\mathcal{M}(\mathbf{P}_1)), F(\mathfrak{P}_{Y,m}(\psi\mathcal{M}(\mathbf{P}_1)))\rangle, \ldots,$$
$$\langle \mathfrak{P}_{Y,m}(\psi\mathcal{M}(\mathbf{P}_n)), F(\mathfrak{P}_{Y,m}(\psi\mathcal{M}(\mathbf{P}_n)))\rangle \Big\rangle \in Ax,$$

which holds by our assumption that F satisfies Ax. □

Lemma 6.5 provides a method of extending impossibility theorems proved for majoritarian or pairwise methods to all voting methods. However, it does not go far enough for us, because the variable candidate axioms of binary γ and α-resoluteness are not homogeneously pairwise axioms (see Lemma 6.8).

6.3. Variable candidate axioms. Next we extend Lemma 6.5 to handle variable candidate axioms—axioms that concern adding or removing a candidate from each voter's ballot.

DEFINITION 6.6. Let Ax be an n-ary universal axiom. We say that Ax is a *variable candidate axiom* if for all profiles $\mathbf{P}_1, \ldots, \mathbf{P}_n$ and $\mathbf{Q}_1, \ldots, \mathbf{Q}_n$ such that

(1) for $1 \leq i \leq n$, $X(\mathbf{P}_i) = X(\mathbf{Q}_i)$, and
(2) for $1 \leq i, j \leq n$, if $\mathbf{P}_i = (\mathbf{P}_j)|_{X(\mathbf{P}_i)}$, then $\mathbf{Q}_i = (\mathbf{Q}_j)|_{X(\mathbf{Q}_i)}$,

we have that for all $X_1 \subseteq X(\mathbf{P}_1), \ldots, X_n \subseteq X(\mathbf{P}_n)$,

(3) if $\langle\langle \mathbf{Q}_1, X_1\rangle, \ldots, \langle \mathbf{Q}_n, X_n\rangle\rangle \in Ax$, then $\langle\langle \mathbf{P}_1, X_1\rangle, \ldots, \langle \mathbf{P}_n, X_n\rangle\rangle \in Ax$.

The intuition is that if the $\mathbf{Q}_i$'s are related to each other in terms of restriction in all of the ways that the $\mathbf{P}_i$'s are, and $X_1, \ldots, X_n$ are acceptable choices for the $\mathbf{Q}_i$'s according to the relevant variable candidate axiom (e.g., binary γ or α-resoluteness), then they are also acceptable choices for the $\mathbf{P}_i$'s.

EXAMPLE 6.7. Binary γ and α-resoluteness correspond to the following axioms:

$$\mathsf{BG} = \left\{ \langle\langle \mathbf{P}, X\rangle, \langle \mathbf{P}', X'\rangle, \langle \mathbf{P}'', X''\rangle\rangle \;\middle|\; \begin{array}{l} \text{if } \mathbf{P}' = \mathbf{P}_{-b} \text{ for some } b \in X(\mathbf{P}), \\ \mathbf{P}'' = \mathbf{P}|_{\{a,b\}} \text{ for } a \in X(\mathbf{P}'),\ a \in X', \\ \text{and } X'' = \{a\}, \text{ then } a \in X \end{array} \right\};$$

$$\mathsf{AR} = \left\{ \langle\langle \mathbf{P}, X\rangle, \langle \mathbf{P}', X'\rangle, \langle \mathbf{P}'', X''\rangle\rangle \;\middle|\; \begin{array}{l} \text{if } \mathbf{P}' = \mathbf{P}_{-b} \text{ for some } b \in X(\mathbf{P}), \\ \mathbf{P}'' = \mathbf{P}|_{\{a,b\}} \text{ for } a \in X(\mathbf{P}'),\ a \in X', \\ \text{and } X'' = \{a\}, \text{ then } |X| \leq |X'| \end{array} \right\}.$$

LEMMA 6.8. BG *and* AR *are variable candidate axioms but neither majoritarian nor homogeneously pairwise axioms.*

PROOF. We give only the proof for BG, as the proof for AR is almost the same. Suppose (i) that $\mathbf{P}, \mathbf{P}', \mathbf{P}''$ and $\mathbf{Q}, \mathbf{Q}', \mathbf{Q}''$ are related as in Definition 6.6.1-2. Further suppose (ii) that $\langle\langle \mathbf{Q}, X\rangle, \langle \mathbf{Q}', X'\rangle, \langle \mathbf{Q}'', X''\rangle\rangle \in \mathsf{BG}$. Finally, suppose (iii) that $\mathbf{P}' = \mathbf{P}_{-b}$ for some $b \in X(\mathbf{P})$, $\mathbf{P}'' = \mathbf{P}|_{\{a,b\}}$ for $a \in X(\mathbf{P}')$, $a \in X'$, and $X'' = \{a\}$. By (i) and (iii), we have that $\mathbf{Q}' = \mathbf{Q}_{-b}$ for $b \in X(\mathbf{Q}) = X(\mathbf{P})$, $\mathbf{Q}'' = \mathbf{Q}|_{\{a,b\}}$ for $a \in X(\mathbf{Q}') = X(\mathbf{P}')$, $a \in X'$, and $X'' = \{a\}$. Then by (ii), we have $a \in X$. Thus, we conclude that $\langle\langle \mathbf{P}, X\rangle, \langle \mathbf{P}', X'\rangle, \langle \mathbf{P}'', X''\rangle\rangle \in \mathsf{BG}$.

To see that BG is not homogeneously pairwise, it is easy to find $\langle\langle \mathbf{P}_1, X_1\rangle, \langle \mathbf{P}_2, X_2\rangle, \langle \mathbf{P}_3, X_3\rangle\rangle$ and $\langle\langle \mathbf{Q}_1, X_1\rangle, \langle \mathbf{Q}_2, X_2\rangle, \langle \mathbf{Q}_3, X_3\rangle\rangle$ such that while $\mathcal{M}(\mathbf{P}_i) = \mathcal{M}(\mathbf{Q}_i)$ and $\mathbf{P}_2 = (\mathbf{P}_1)_{-b}$, we have $\mathbf{Q}_2 \neq (\mathbf{Q}_1)|_{X(\mathbf{Q}_2)}$; then even if $\langle\langle \mathbf{P}_1, X_1\rangle, \langle \mathbf{P}_2, X_2\rangle, \langle \mathbf{P}_3, X_3\rangle\rangle \notin \mathsf{BG}$, from $\mathbf{Q}_2 \neq (\mathbf{Q}_1)|_{X(\mathbf{Q}_2)}$ we trivially have $\langle\langle \mathbf{Q}_1, X_1\rangle, \langle \mathbf{Q}_2, X_2\rangle, \langle \mathbf{Q}_3, X_3\rangle\rangle \in \mathsf{BG}$. □

Fortunately, not only majoritarian (resp. homogeneously pairwise) axioms but also variable candidate axioms are preserved by our operation of majoritarian (resp. pairwise) projection, thanks to the fact that our representation of weak (resp. weighted weak) tournaments commutes with restriction, in contrast to McGarvey and Debord's constructions.

PROPOSITION 6.9. *Let F be an anonymous voting method satisfying a variable candidate universal axiom Ax. Then for any finite $Y \subseteq \mathcal{X}$, $F_{maj,Y}$ (resp. $F_{pair,Y,m}$) satisfies Ax.*

PROOF. Suppose Ax is a majoritarian n-ary universal axiom. To show that $F_{maj,Y}$ satisfies Ax, consider any $\mathbf{P}_1, \ldots, \mathbf{P}_n \in \mathrm{dom}(F_{maj,Y})$. We must show that

$$\langle\langle \mathbf{P}_1, F_{maj,Y}(\mathbf{P}_1)\rangle, \ldots, \langle \mathbf{P}_n, F_{maj,Y}(\mathbf{P}_n)\rangle\rangle \in Ax.$$

By definition of $F_{maj,Y}$, this is equivalent to

$$\langle\langle \mathbf{P}_1, F(\mathfrak{P}_Y(M(\mathbf{P}_1)))\rangle, \ldots, \langle \mathbf{P}_n, F(\mathfrak{P}_Y(M(\mathbf{P}_n)))\rangle\rangle \in Ax.$$

By Proposition 4.14 and Lemma 4.13, we have

$$\begin{aligned}\mathfrak{P}_Y(M(\mathbf{P}_j))|_{X(\mathfrak{P}_Y(M(\mathbf{P}_i)))} &= \mathfrak{P}_Y(M(\mathbf{P}_j)|_{X(\mathfrak{P}_Y(M(\mathbf{P}_i)))})\\ &= \mathfrak{P}_Y(M((\mathbf{P}_j)|_{X(\mathfrak{P}_Y(M(\mathbf{P}_i)))}))\\ &= \mathfrak{P}_Y(M((\mathbf{P}_j)|_{X(\mathbf{P}_i)})).\end{aligned}$$

Thus, for $1 \leq i, j \leq n$, we have that

$$\text{if } \mathbf{P}_i = (\mathbf{P}_j)|_{X(\mathbf{P}_i)}, \text{ then } \mathfrak{P}_Y(M(\mathbf{P}_i)) = \mathfrak{P}_Y(M(\mathbf{P}_j))|_{X(\mathfrak{P}_Y(M(\mathbf{P}_i)))}.$$

Then since Ax is a variable candidate axiom, setting $\mathbf{Q}_i = \mathfrak{P}_Y(M(\mathbf{P}_i))$ in Definition 6.6, the membership statement displayed above is a consequence of

$$\langle\langle \mathfrak{P}_Y(M(\mathbf{P}_1)), F(\mathfrak{P}_Y(M(\mathbf{P}_1)))\rangle, \ldots, \langle \mathfrak{P}_Y(M(\mathbf{P}_n)), F(\mathfrak{P}_Y(M(\mathbf{P}_n)))\rangle\rangle \in Ax,$$

which holds by our assumption that F satisfies Ax. The proof for the pairwise case is analogous. □

Proposition 6.9 is the key to extending impossibility theorems involving variable candidate axioms proved for majoritarian or pairwise methods to all voting methods, as we show next.

6.4. Proof of the transfer lemmas. We are now in a position to prove the transfer lemmas introduced in Section 3.

LEMMA 3.5. *For any finite* $Y \subseteq \mathcal{X}$*, if there is a voting method with domain* $\{\mathbf{P} \in \mathsf{prof} \mid X(\mathbf{P}) \subseteq Y\}$ *satisfying anonymity, neutrality, binary quasi-resoluteness, binary* γ*, and* α*-resoluteness, then there is a* majoritarian *voting method with domain* $\{\mathbf{P} \in \mathsf{prof} \mid X(\mathbf{P}) \subseteq Y\}$ *satisfying anonymity, neutrality, binary quasi-resoluteness, binary* γ*, and* α*-resoluteness.*

PROOF. Fix a finite $Y \subseteq \mathcal{X}$, and let F be a voting method satisfying the assumption. For any profile $\mathbf{P}$, let $a(\mathbf{P})$ be the anonymous profile such that $a(\mathbf{P})(L) = |\{i \in V(\mathbf{P}) \mid \mathbf{P}_i = L\}|$ for $L \in \mathcal{L}(X(\mathbf{P}))$. Note that if F is anonymous, then for any $\mathbf{P}, \mathbf{P}' \in \mathrm{dom}(F)$, if $a(\mathbf{P}) = a(\mathbf{P}')$, then $F(\mathbf{P}) = F(\mathbf{P}')$. Also note that for any $a, b \in \mathcal{X}$, $a(\mathbf{P})_{a \leftrightarrows b} = a(\mathbf{P}_{a \leftrightarrows b})$.

Fix a set $V_0 \subseteq \mathcal{V}$ of voters with $|V| = |Y|!$, and consider the following restricted domain of profiles:

$$D = \{\mathbf{P} \in \mathsf{prof} \mid X(\mathbf{P}) \subseteq Y, V(\mathbf{P}) = V_0\}.$$

Now define the following domain of anonymous profiles:

$$D_A = \{\mathbf{P} \in \mathsf{anon\text{-}prof} \mid X(\mathbf{P}) \subseteq Y, \sum_{L \in \mathcal{L}(X(\mathbf{P}))} \mathbf{P}(L) = |Y|!\}.$$

For each anonymous profile $\mathbf{P} \in D_A$, fix a de-anonymized profile $d(\mathbf{P}) \in D$ such that $a(d(\mathbf{P})) = \mathbf{P}$. Now define an anonymous voting method G on D_A by

$$G(\mathbf{P}) = F(d(\mathbf{P})).$$

Then since F satisfies anonymity, neutrality, binary quasi-resoluteness, binary γ, and α-resoluteness, it is easy to see that the anonymous method G satisfies neutrality, binary quasi-resoluteness, binary γ, and α-resoluteness. For instance, for neutrality, since $\mathbf{P} = a(d(\mathbf{P}))$, we have $\mathbf{P}_{a \leftrightarrows b} = a(d(\mathbf{P}))_{a \leftrightarrows b} = a(d(\mathbf{P})_{a \leftrightarrows b})$, which with $\mathbf{P}_{a \leftrightarrows b} = a(d(\mathbf{P}_{a \leftrightarrows b}))$ implies $a(d(\mathbf{P}_{a \leftrightarrows b})) = a(d(\mathbf{P})_{a \leftrightarrows b})$. Then

$$\begin{aligned} G(\mathbf{P}_{a \leftrightarrows b}) &= F(d(\mathbf{P}_{a \leftrightarrows b})) \\ &= F(d(\mathbf{P})_{a \leftrightarrows b}) \text{ since } a(d(\mathbf{P}_{a \leftrightarrows b})) = a(d(\mathbf{P})_{a \leftrightarrows b}) \text{ and } F \text{ is anonymous} \\ &= F(d(\mathbf{P}))_{a \leftrightarrows b} \text{ by the neutrality of } F \\ &= G(\mathbf{P})_{a \leftrightarrows b}. \end{aligned}$$

The proofs that G satisfies the other axioms involve similar reasoning. Now recall from Definition 5.2 that

$$\mathrm{dom}(G_{maj,Y}) = \{\mathbf{P} \in \mathsf{anon\text{-}prof} \mid X(\mathbf{P}) \subseteq Y \text{ and } \mathfrak{P}_Y(M(\mathbf{P})) \in \mathrm{dom}(G)\}.$$

Since $\mathrm{dom}(G) = D_A$ and by Definition 4.7, $\mathfrak{P}_Y(M(\mathbf{P})) \in D_A$ for all $\mathbf{P}$ with $X(\mathbf{P}) \subseteq Y$, it follows that $\mathrm{dom}(G_{maj,Y}) = \{\mathbf{P} \in \mathsf{anon\text{-}prof} \mid X(\mathbf{P}) \subseteq Y\}$. Moreover, $G_{maj,Y}$ is an anonymous voting method that is majoritarian and neutral by Lemma 5.4, satisfies binary quasi-resoluteness by Lemma 6.5 (for majoritarian axioms), and satisfies binary γ and α-resoluteness by Lemma 6.8 and Proposition 6.9. Finally, define a voting method H such that for any profile $\mathbf{P}$, we have $H(\mathbf{P}) = G_{maj,Y}(a(\mathbf{P}))$. Then it is easy to see that H is a majoritarian voting method with domain $\{\mathbf{P} \in \mathsf{prof} \mid X(\mathbf{P}) \subseteq Y\}$ satisfying anonymity, neutrality, binary quasi-resoluteness, binary γ, and α-resoluteness. □

LEMMA 3.6. *For any finite $Y \subseteq \mathcal{X}$, if there is a voting method with domain $\{\mathbf{P} \in \mathsf{prof} \mid X(\mathbf{P}) \subseteq Y\}$ satisfying anonymity, neutrality, binary γ, and quasi-resoluteness, then for any $m \in \mathbb{Z}^+$, there is a* pairwise *voting method with domain*

$$\{\mathbf{P} \in \mathsf{prof} \mid X(\mathbf{P}) \subseteq Y,\ m \text{ is the maximum weight in } \mathcal{M}(\mathbf{P})\}$$

satisfying anonymity, neutrality, binary γ, and quasi-resoluteness.

PROOF. The proof is analogous to that of Lemma 3.5 above, only now we fix a set $V_0 \subseteq \mathcal{V}$ of voters with $|V| = m|Y|!$, since the profile $\mathfrak{P}_{Y,m}(\mathcal{M}(\mathbf{P}))$ from Definition 4.9 has $m|Y|!$ voters, and we move from the anonymous method G satisfying neutrality, quasi-resoluteness, and binary γ to $G_{pair,Y,m}$ being an anonymous voting method on the domain

$$\{\mathbf{P} \in \mathsf{anon\text{-}prof} \mid X(\mathbf{P}) \subseteq Y,\ m \text{ is the maximum weight in } \mathcal{M}(\mathbf{P})\}$$

that is pairwise and neutral by Lemma 5.4, quasi-resolute by Lemma 6.5 (for pairwise axioms), and satisfies binary γ by Lemma 6.8 and Proposition 6.9. □

This completes the proofs of Theorems 2.8.1 and 2.11 as in Section 3.

7. Conclusion

In this paper, we have proved impossibility theorems illustrating the tradeoff between binary expansion consistency and the power of voting methods to narrow down the set of winners. For further research, a natural refinement of these theorems would optimize not only the number of candidates but also the number of *voters* needed to obtain the impossibility results, in the spirit of [**11**, **14**].[17] As for the practical consequences of these theorems for the design of voting methods, a natural response is to seek variants of binary expansion consistency, still aimed at mitigating spoiler effects, that are consistent with the quasi-resoluteness property studied in this paper (see, e.g., [**32**]).

Another avenue for further research concerns applying the proof strategy of this paper to other impossibility theorems in voting theory. Though the proof strategy may appear tailored to the specific axioms appearing in our impossibility theorems, in fact it applies more broadly. Our majoritarian and pairwise projections preserve neutrality (Lemma 5.4.3), all majoritarian axioms and homogeneously pairwise axioms (Lemma 6.5), and all variable candidate universal axioms (Proposition 6.9). Furthermore, they preserve other well-known axioms such as monotonicity and Maskin monotonicity.[18] Thus, impossibility theorems involving monotonicity (resp. Maskin monotonicity) together with members of the other

[17]Inspection of our proofs shows that we use $6! = 760$ voters to generate the profiles in our proof of Theorem 2.8.1 and $12 \times 4! = 288$ voters to generate the profiles in our proof of Theorem 2.11.

[18]Monotonicity states that if $x \in F(\mathbf{P})$ and $\mathbf{P}'$ is obtained from $\mathbf{P}$ by moving x up in one voter's ballot, keeping the relative order of all other candidates the same, then $x \in F(\mathbf{P}')$. Maskin monotonicity (for non-anonymous profiles) states that if $x \in F(\mathbf{P})$ and $\mathbf{P}'$ is a profile with the same voters such that if a voter ranks x above y in $\mathbf{P}$, then the voter still ranks x above y in $\mathbf{P}'$, then $x \in F(\mathbf{P}')$ (which can be suitably adapted to anonymous profiles). To see that these axioms are preserved, one can check that if profiles $\mathbf{P}$ and $\mathbf{P}'$ are related by a monotonic (resp. Maskin monotonic) improvement of a candidate x, then $\mathfrak{P}_Y(M(\mathbf{P}))$ and $\mathfrak{P}_Y(M(\mathbf{P}'))$ are related in the same way, as are $\mathfrak{P}_{Y,m}(\mathcal{M}(\mathbf{P}))$ and $\mathfrak{P}_{Y,m}(\mathcal{M}(\mathbf{P}'))$, so the monotonicity (resp. Maskin monotonicity) of F implies the monotonicity (resp. Maskin monotonicity) of $F_{maj,Y}$ and $F_{pair,Y,m}$.

classes of axioms above fall under the scope of our proof strategy. We leave it to future work to explore impossibility theorems of such forms.

Appendix A. SAT solving for majoritarian methods

In this Appendix and the next, we explain how the SAT solving approach of Brandt and Geist [**10**] can be applied to prove our Propositions 3.2 and 3.4 for majoritarian and pairwise voting methods, respectively.

Jupyter notebooks with our SAT formalization are available at `https://github.com/szahedian/binary-gamma`.

A.1. Results. Instead of working with profiles and majoritarian methods, our first SAT approach works with tournaments and tournament solutions.

DEFINITION A.1. A *weak tournament solution* (resp. *tournament solution*) is a function F on a set of weak tournaments (resp. tournaments) such that for all $T \in \text{dom}(F)$, we have $\varnothing \neq F(T) \subseteq X(T)$, where $X(T)$ is the set of nodes of T.

The axioms of neutrality, binary quasi-resoluteness, binary γ, and α-resoluteness from Section 2.2 apply in the obvious way to weak tournament solutions.

DEFINITION A.2. Let F be a weak tournament solution.

(1) F is *neutral* if for all $T, T' \in \text{dom}(F)$, $F(T') = h[F(T)]$ for any isomorphism $h : T \to T'$.
(2) F satisfies *binary quasi-resoluteness* if for any tournament $T \in \text{dom}(F)$ with $|X(T)| = 2$, we have $|F(T)| = 1$.
(3) F satisfies *binary* γ if for every $T \in \text{dom}(F)$ and $a, b \in X(T)$, if $a \in F(T_{-b})$ and $F(T|_{\{a,b\}}) = \{a\}$, then $a \in F(T)$.
(4) F satisfies *α-resoluteness* if for any $T \in \text{dom}(F)$ and $a, b \in X(T)$, if $a \in F(T_{-b})$ and $F(T|_{\{a,b\}}) = \{a\}$, then $|F(T)| \leq |F(T_{-b})|$.

We used SAT to find the dividing line in terms of number of candidates between the consistency and inconsistency of these axioms.

PROPOSITION A.3.

(1) *For any finite $Y \subseteq \mathcal{X}$ with $|Y| \geq 6$, there is no neutral tournament solution satisfying binary quasi-resoluteness, binary γ, and α-resoluteness on the domain $\{T \in \mathbb{T} \mid X(T) \subseteq Y\}$.*
(2) *There is a neutral weak tournament solution satisfying binary quasi-resoluteness, binary γ, and α-resoluteness on the domain $\{T \in \mathbb{WT} \mid |X(T)| \leq 5\}$.*

This proposition is a consequence of Proposition A.12 and Lemma A.10 below.

A.2. SAT encoding: Canonical tournaments and solutions. For computational feasibility, following Brandt and Geist [**10**] we work with tournament solutions on a smaller domain of *canonical* tournaments up to a certain number of candidates.

DEFINITION A.4. Given $T, T' \in \mathbb{WT}$, let $T \cong T'$ if T and T' are isomorphic. For each $T \in \mathbb{WT}$, define the equivalence class $[T] = \{T' \in \mathbb{WT} \mid T \cong T'\}$ and pick a *canonical representative* $T_C \in [T]$. Let $\mathbb{WT}_C = \{T_C \mid T \in \mathbb{WT}\}$ and $\mathbb{T}_C = \{T_C \mid T \in \mathbb{T}\}$ be the sets of *canonical weak tournaments* and *canonical tournaments*, respectively. A *canonical weak tournament solution* (resp. *canonical tournament solution*) is a weak tournament solution (resp. tournament solution) whose domain is a subset of $\mathbb{WT}_C$ (resp. $\mathbb{T}_C$).

Next we consider analogues for canonical weak tournament solutions of the axioms on weak tournament solutions from Definition A.2. Binary quasi-resoluteness is the most straightforward.

DEFINITION A.5. A canonical weak tournament solution F satisfies *binary quasi-resoluteness* if for every $T \in \mathbb{T}_C$ with $|X(T)| = 2$, we have $|F(T)| = 1$.

As for neutrality, there is a one-to-one correspondence between neutral weak tournament solutions and canonical weak tournament solutions satisfying what Brandt and Geist call the *orbit condition.* Given $T \in \mathbb{WT}_C$ and $a, b \in X(T)$, let $a \sim_T b$ if there exists an automorphism h of T such that $h(a) = b$. The *orbit of a in T* is $\{b \in T \mid a \sim_T b\}$. Let $\mathcal{O}_T$ be the set of all orbits of elements of T.

DEFINITION A.6. Given a canonical weak tournament T and $Y \subseteq X(T)$, we say that (T, Y) satisfies the *orbit condition* if for all $O \in \mathcal{O}_T$, we have $O \subseteq Y$ or $O \cap Y = \varnothing$. A canonical weak tournament solution F satisfies the *orbit condition* if for all $T \in \mathrm{dom}(F)$, $(T, F(T))$ satisfies the orbit condition. Let

$$\mathscr{O}(T) = \{Y \subseteq X(T) \mid Y \neq \varnothing \text{ and } (T, Y) \text{ satisfies the orbit condition}\}.$$

To state the one-to-one correspondence, we need the following preliminary lemma.

LEMMA A.7. *Let F be a canonical weak tournament solution satisfying the orbit condition. Then for any $T \in \mathbb{WT}$ and isomorphisms $g : T_C \to T$ and $h : T_C \to T$, we have $g[F(T_C)] = h[F(T_C)]$.*

PROOF. Suppose $a \in g[F(T_C)]$, so $g^{-1}(a) \in F(T_C)$. Since $h^{-1} \circ g$ is an automorphism of T_C such that $h^{-1}(g(g^{-1}(a))) = h^{-1}(a)$, we have $g^{-1}(a) \sim_{T_C} h^{-1}(a)$. Then since F satisfies the orbit condition, $g^{-1}(a) \in F(T_C)$ implies $h^{-1}(a) \in F(T_C)$, so $a \in h[F(T_C)]$. The other direction is analogous. □

Lemma A.7 ensures that the function F^* in the following is well defined.

LEMMA A.8.

(1) *Given a canonical weak tournament solution F satisfying the orbit condition, the function F^* on $\{T \in \mathbb{WT} \mid T_C \in \mathrm{dom}(F)\}$ defined by $F^*(T) = h_T[F(T_C)]$ for an isomorphism $h_T : T_C \to T$ is a neutral weak tournament solution.*

(2) *Given a neutral weak tournament solution F, the function F_* on $\{T_C \mid T \in \mathrm{dom}(F)\}$ defined by $F_*(T_C) = F(T_C)$ is a canonical weak tournament solution satisfying the orbit condition.*

(3) *For any canonical weak tournament solution F satisfying the orbit condition, $(F^*)_* = F$; and for any neutral weak tournament solution F, $(F_*)^* = F$.*

Next we state the analogues of binary γ and α-resoluteness for canonical weak tournament solutions. Given weak tournaments T and T', an *embedding of T' into T* is an injective function $e : X(T') \to X(T)$ such that for all $a, b \in X(T')$, we have $a \to b$ in T' if and only if $e(a) \to e(b)$ in T, in which case we write $e : T' \hookrightarrow T$.

DEFINITION A.9. Let F be a canonical weak tournament solution.

(1) F satisfies *canonical binary* γ if for any $T, T', T'' \in \text{dom}(F)$, if $|X(T)| = |X(T')| + 1$, $|X(T'')| = 2$, and there are embeddings $e : T' \hookrightarrow T$ and $f : T'' \hookrightarrow T$ such that $X(T) \setminus e[X(T')] \subseteq f[X(T'')]$, then for $a \in e[X(T')] \cap f[X(T'')]$, if $e^{-1}(a) \in F(T')$ and $F(T'') = \{f^{-1}(a)\}$, then $a \in F(T)$.
(2) F satisfies *canonical α-resoluteness* if for any $T, T', T'' \in \text{dom}(F)$, if $|X(T)| = |X(T')| + 1$, $|X(T'')| = 2$, and there are embeddings $e : T' \hookrightarrow T$ and $f : T'' \hookrightarrow T$ such that $X(T) \setminus e[X(T')] \subseteq f[X(T'')]$, then for $a \in e[X(T')] \cap f[X(T'')]$, if $e^{-1}(a) \in F(T')$ and $F(T'') = \{f^{-1}(a)\}$, then $|F(T)| \leq |F(T')|$.

The correspondence in Lemma A.8 preserves the relevant axioms as follows.

LEMMA A.10.

(1) *If F is a canonical weak tournament solution satisfying the orbit condition and canonical binary γ (resp. canonical α-resoluteness, binary quasi-resoluteness), then F^* defined as in Lemma* A.8.1 *is a neutral weak tournament solution satisfying binary γ (resp. α-resoluteness, binary quasi-resoluteness).*
(2) *If F is a neutral weak tournament solution satisfying binary γ (resp. α-resoluteness, binary quasi-resoluteness), then F_* defined as in Lemma* A.8.2 *is a canonical weak tournament solution satisfying the orbit condition and canonical binary γ (resp. canonical α-resoluteness, binary quasi-resoluteness).*

A.3. SAT encoding: formulas. We assume familiarity with the basics of propositional logic [**24**, § 1.1], using $\neg$ for 'not', $\wedge$ for 'and', and $\vee$ for 'or'. In this section, we explain how the axioms in Section A.2 can be encoded as formulas of propositional logic for the purposes of SAT solving. Following [**10**], our propositional variables are of the form $A_{T,Y}$, where T is a canonical weak tournament, $\varnothing \neq Y \subseteq X(T)$, and $Y \in \mathscr{O}(T)$ (Definition A.6). The intended interpretation is that $A_{T_C,Y}$ evaluates to true if and only if Y is assigned as the set of winners of T. Next we introduce propositional formulas, written in conjunctive normal form for processing by the SAT solver, which formalize the relevant axioms.

Fix a finite set $\mathbb{D}$ of weak tournaments. The following formula $\mathsf{func}_{\mathbb{D}}$ ensures that the assignment of sets of winners to canonical weak tournaments in $\mathbb{D}$ yields a function:

$$\bigwedge_{T\in\mathbb{D}} \Big(\bigvee_{Y\in\mathscr{O}(T)} A_{T,Y} \Big) \wedge \bigwedge_{T\in\mathbb{D}} \bigwedge_{Y,Z\in\mathscr{O}(T):Y\neq Z} (\neg A_{T,Y} \vee \neg A_{T,Z}).$$

The first part of the formula (before the small '$\wedge$') says that for each weak tournament $T \in \mathbb{D}$, there is some subset Y of T such that (T, Y) satisfies the orbit condition and Y is assigned as the set of winners of T. Since $\mathbb{D}$ is finite, this is a finite formula. Similarly, the second part says that for each weak tournament $T \in \mathbb{D}$ and distinct subsets Y, Z of T such that (T, Y) and (T, Z) both satisfy the orbit condition, either Y is not assigned as the set of winners of T or Z is not assigned as the set of winners of T, i.e., it is not the case that both Y and Z are assigned as the set of winners. The formulas to follow can be read similarly.

To formalize binary quasi-resoluteness, where T_2 is the canonical tournament with two nodes, say 0 and 1, we define $\mathsf{bqr}_{\mathbb{D}}$ by:

$$A_{T_2,\{0\}} \vee A_{T_2,\{1\}}.$$

To formalize binary γ and α-resoluteness, for any $T \in \mathbb{D}$, define:

$$E(T) = \{(T', e) \mid e : T' \hookrightarrow T \text{ and } |X(T)| = |X(T')| + 1\}$$
$$F(T) = \{(T'', f) \mid f : T'' \hookrightarrow T \text{ and } |X(T'')| = 2\}.$$

Then let $\mathsf{bg}_{\mathbb{D}}$ be the formula:

$$\bigwedge_{T \in \mathbb{D}} \bigwedge_{(T',e)\in E(T)} \bigwedge_{(T'',f)\in F(T):\, X(T)\setminus e[X(T')]\subseteq f[X(T'')]} \varphi$$

with φ defined as follows, where $e[X(T')] \cap f[X(T'')] = \{a\}$:[19]

$$\varphi = \bigwedge_{\substack{Y' \in \mathscr{O}(T'):\, e^{-1}(a)\in Y' \\ Y''\in\mathscr{O}(T''):\, Y''=\{f^{-1}(a)\}}} (\neg A_{T',Y'} \vee \neg A_{T'',Y''} \vee \bigvee_{Y\in\mathscr{O}(T):\, a\in Y} A_{T,Y}).$$

The formula $\mathsf{ar}_{\mathbb{D}}$ is defined in the same way but with the last disjunction replaced by

$$\bigvee_{Y\in\mathscr{O}(T):\, |Y|\le|Y'|} A_{T,Y}.$$

Propositional valuations $v : \{A_{T,Y} \mid T \in \mathbb{D}, Y \in \mathscr{O}(T)\} \to \{0,1\}$ satisfying the formula $\mathsf{func}_{\mathbb{D}} \wedge \mathsf{bqr}_{\mathbb{D}} \wedge \mathsf{bg}_{\mathbb{D}} \wedge \mathsf{ar}_{\mathbb{D}}$ correspond to canonical weak tournament solutions on the domain $\mathbb{D}$ satisfying the relevant axioms.

LEMMA A.11.

(1) *If a valuation v satisfies $\mathsf{func}_{\mathbb{D}} \wedge \mathsf{bqr}_{\mathbb{D}} \wedge \mathsf{bg}_{\mathbb{D}} \wedge \mathsf{ar}_{\mathbb{D}}$, then*

$$\{(T,Y) \mid T \in \mathbb{D}, Y \in \mathscr{O}(T), v(A_{T,Y}) = 1\}$$

is a canonical weak tournament solution on $\mathbb{D}$ satisfying the orbit condition, binary quasi-resoluteness, canonical binary γ, and canonical α-resoluteness.

(2) *Given a canonical weak tournament solution F on $\mathbb{D}$ satisfying the orbit condition, binary quasi-resoluteness, canonical binary γ, and canonical α-resoluteness, the valuation v defined on $\{A_{T,Y} \mid T \in \mathbb{D}, Y \in \mathscr{O}(T)\}$ by $v(A_{T,Y}) = 1$ if and only if $F(T) = Y$ satisfies $\mathsf{func}_{\mathbb{D}} \wedge \mathsf{bqr}_{\mathbb{D}} \wedge \mathsf{bg}_{\mathbb{D}} \wedge \mathsf{ar}_{\mathbb{D}}$.*

Using a SAT solver and Lemma A.11, we obtain the following.

PROPOSITION A.12.

(1) *For any finite $Y \subseteq \mathcal{X}$ with $|Y| \geq 6$, is no canonical tournament solution satisfying the orbit condition, binary quasi-resoluteness, canonical binary γ, and canonical α-resoluteness on the domain $\{T \in \mathbb{T}_C \mid X(T) \subseteq Y\}$.*

(2) *There is a canonical weak tournament solution satisfying the orbit condition, binary quasi-resoluteness, canonical binary γ, and canonical α-resoluteness on the domain $\{T \in \mathbb{WT}_C \mid |T| \leq 5\}$.*

Finally, combining Proposition A.12 and Lemma A.10 establishes the results for weak tournament solutions in Proposition A.3 from Section A.1.

---

[19]Since $|X(T'')| = 2$, we have $|f[X(T'')]| = 2$. Then since $|X(T)| = |X(T')| + 1$, we have $|X(T) \setminus e[X(T')]| = 1$, which with $X(T) \setminus e[X(T')] \subseteq f[X(T'')]$ implies $|e[X(T')] \cap f[X(T'')]| = 1$.

A.4. From tournament solutions to majoritarian voting methods. Having obtained Proposition A.3 for (weak) tournament solutions, it is straightforward using McGarvey's theorem (Theorem 4.5) to obtain an analogous statement for majoritarian voting methods.

PROPOSITION 3.2. *For any finite $Y \subseteq \mathcal{X}$ with $|Y| \geq 6$, there is no* majoritarian *voting method with domain $\{\mathbf{P} \mid X(\mathbf{P}) \subseteq Y\}$ satisfying anonymity, neutrality, binary quasi-resoluteness, binary γ, and α-resoluteness.*

PROOF. Suppose for contradiction there is such a method F. We define a canonical tournament solution G on the domain $\{T \in \mathbb{T}_C \mid X(T) \subseteq Y\}$ as follows: for $T \in \text{dom}(G)$, let $G(T) = F(\mathsf{Mc}(T))$. Since F is majoritarian and satisfies the relevant axioms, it is not difficult to show that G satisfies the tournament solution versions of the relevant axioms, contradicting Proposition A.3.1. □

PROPOSITION A.13. *There is an anonymous, neutral, and majoritarian voting method satisfying binary quasi-resoluteness, binary γ, and α-resoluteness on the domain $\{\mathbf{P} \mid |X(\mathbf{P})| \leq 5\}$.*

PROOF. Define the method F as follows: where G is the weak tournament solution from Proposition A.3.2, given a profile $\mathbf{P}$, let $F(\mathbf{P}) = G(M(\mathbf{P}))$. □

A.5. Prospects for a human-readable proof. After obtaining Proposition A.12.1 using SAT solving, we investigated the prospects of a human-readable proof. First, a canonical tournament solution satisfying canonical α-resoluteness either (i) picks only the majority winner in the two-candidate canonical tournament T_2 or (ii) picks only the minority winner in T_2. Thus, we can simply prove the impossibility assuming (i), as the proof in case (ii) then follows by reversing all edges in tournaments. With this simplification, the CNF formula used for Proposition A.12.1 contains 132,626 conjuncts. To obtain an MUS (recall Section 3), following Brandt and Geist [**10**], we used the PicoMUS tool from PicoSAT [**4**], which produced an MUS with 266 conjuncts. By analyzing the MUS, we identified *nine* canonical tournaments ranging from size two to six—down from 75 canonical tournaments total from size two to six, as shown in Table 3—such that no canonical tournament solution on the domain of those nine canonical tournaments satisfies the relevant axioms. Nine canonical tournaments initially sounds promising for a human readable proof, but the conjuncts in the MUS contain 20 bg conjuncts and 30 ar conjuncts, involving a total of 19 embeddings between canonical tournaments. Naively, this means a human-readable proof might contain 20 appeals to binary γ and 30 appeals to α-resoluteness, involving the consideration of 19 ways a smaller tournament can embed into a larger one. Of course, it is possible that several such steps could be handled at once. It is also possible that there is a smaller MUS more amenable to use for a human-readable proof.[20]

Appendix B. SAT solving for pairwise methods

B.1. Results. Using SAT we also obtained an impossibility theorem for the conjunction of binary γ, quasi-resoluteness, and pairwiseness, assuming anonymity and neutrality. However, the relevant SAT approach works not with profiles and

[20]Following Brandt and Geist [**10**], we tried applying the CAMUS tool [**36**] to find a smallest MUS, but it did not terminate in a reasonable amount of time.

TABLE 3. Number of canonical tournaments (i) from size two to six (second column) and (ii) sufficient for an impossibility theorem obtained from an MUS (third column).

candidates	canonical tournaments from initial set	canonical tournaments from MUS
2	1	1
3	2	1
4	4	2
5	12	3
6	56	2
2 to 6	75	9

pairwise voting methods but with weighted tournaments and weighted tournament solutions.

DEFINITION B.1. A *weighted tournament solution* is a function F on a set of weighted tournaments such that for all $\mathcal{T} \in \text{dom}(F)$, we have $\varnothing \neq F(\mathcal{T}) \subseteq X(\mathcal{T})$, where $X(\mathcal{T})$ is the set of nodes in the underlying tournament of $\mathcal{T}$.

The binary γ and quasi-resoluteness axioms from Section 2.2 apply in the obvious ways to weighted tournament solutions.

DEFINITION B.2. Let F be a weighted tournament solution.

(1) F satisfies *binary* γ if for every $\mathcal{T} \in \text{dom}(F)$ and $a, b \in X(\mathcal{T})$, if $a \in F(\mathcal{T}_{-b})$ and $F(\mathcal{T}|_{\{a,b\}}) = \{a\}$, then $a \in F(\mathcal{T})$.
(2) F satisfies *quasi-resoluteness* if for every uniquely weighted $\mathcal{T} \in \text{dom}(F)$, we have $|F(\mathcal{T})| = 1$.

The following result explains the phenomenon noted in Section 2.2 that no known (anonymous and neutral) pairwise voting method satisfies binary γ and quasi-resoluteness. Note that since the SAT formalization requires a finite domain, we fix a finite set of weights for our weighted tournaments.

PROPOSITION B.3. *There is no neutral weighted tournament solution satisfying quasi-resoluteness and binary* γ *on the domain*

$$\{\mathcal{T} \in \mathbb{T}^w \mid |X(\mathcal{T})| \leq 4,\ \mathcal{T} \textit{ uniquely weighted,} \\ \textit{and all positive weights belong to } \{2, 4, 6, 8, 10, 12\}\}.$$

B.2. SAT encoding. The SAT encoding strategy for weighted tournament solutions follows the strategy for tournament solutions in Sections A.2 and A.3: for computational feasibility, we work with weighted tournament solutions defined on a smaller domain of *canonical* weighted tournaments. In fact, we restrict our domain $\mathbb{D}$ to only those canonical weighted tournaments in which all weights are distinct. Moreover, we build quasi-resoluteness into our encoding, by taking only propositional variables of the form $A_{\mathcal{T},Y}$ where $\mathcal{T} \in \mathbb{D}$ and Y is a singleton. Note that for $\mathcal{T} \in \mathbb{D}$, there are no non-trivial automorphisms of $\mathcal{T}$, since all weights are distinct, so we no longer need to check the orbit condition. Functionality and binary γ are formalized as formulas $\mathsf{func}_{\mathbb{D}}$ and $\mathsf{bg}_{\mathbb{D}}$ as in Section A.3 but using canonical weighted tournaments $\mathcal{T}$ instead of canonical tournaments T, setting $\mathscr{O}(T) = \{\{x\} \mid x \in X(\mathcal{T})\}$, and using weighted tournament embeddings[21] instead of tournament embeddings for binary γ. The unsatisfiability of $\mathsf{func}_{\mathbb{D}} \wedge \mathsf{bg}_{\mathbb{D}}$ according to the SAT solver yields Proposition B.3 by reasoning analogous to that in Section A.2.

B.3. From weighted tournament solutions to pairwise voting methods. It is a short step to transfer the impossibility result in Proposition B.3 from weighted tournament solutions to pairwise voting methods in Proposition 3.4. The proof of Proposition 3.4 is analogous to that of Proposition 3.2, except that we use Debord's representation of weighted tournaments instead of McGarvey's representation of tournaments.

PROPOSITION 3.4. *For any $Y \subseteq \mathcal{X}$ with $|Y| \geq 4$, there is no* pairwise *voting method with domain*

$$\{\mathbf{P} \in \mathsf{prof} \mid X(\mathbf{P}) \subseteq Y, \mathcal{M}(\mathbf{P}) \textit{ is uniquely weighted, and all positive weights belong to } \{2,4,6,8,10,12\}\}$$

satisfying anonymity, neutrality, binary γ, and quasi-resoluteness.

Recall that Proposition 3.4 together with the transfer lemma, Lemma 3.6, yields the desired impossibility result in Theorem 2.11.

B.4. Prospects for a human-readable proof. After obtaining Proposition B.3 using the SAT solving strategy, we investigated the prospects of a human-readable proof. There are 2,086 canonical uniquely weighted tournaments from size two to four with edge weights from $\{2,4,6,8,10,12\}$ (see the second column of Table 4), yielding 8,172 propositional variables. Then our CNF formula used to prove Proposition B.3 contains 38,092 conjuncts. To obtain an MUS (recall Sections 3 and A.5), we again used PicoMUS, which produced an MUS with 9,294 conjuncts.[22] By analyzing the MUS, we identified 1,650 canonical weighted tournaments (see the third column of Table 4) from the original 2,086 such that no canonical tournament solution on the domain of those 1,650 satisfies the relevant axioms. In fact, we were able to do better by a naive approach. In the same spirit as MUS extraction, we tried to find a smallest subset of the 2,086 canonical weighted tournaments for which SAT still produces an impossibility result, simply by iterating over these weighted tournaments (ordered by descending number of candidates and in an arbitrary order for weighted tournaments with the same number of candidates) and retaining a weighted tournament if and only if removing it leads to a satisfiability result.[23] The smallest set found so far contains 716 canonical weighted tournaments (see the fourth column of Table 4), which is unfortunately still too many for human inspection. Recall, by contrast, that in our proof assuming homogeneity and cancellativity for Proposition 2.14, we only had to consider three 4-candidate margin graphs, along with some of their subgraphs.

[21] I.e., tournament embeddings $f : T \to T'$ for which $w(x,y) = w'(f(x), f(y))$, where $\mathcal{T} = (T, w)$ and $\mathcal{T}' = (T', w')$.

[22] By contrast, in Brandt and Geist's [**10**] application of SAT to an impossibility theorem in social choice, PicoMUS returned an MUS with 55 conjuncts from an initial formula with over three million.

[23] We also tried this approach for Proposition A.12.1, but in that case the naive approach yielded more canonical tournaments than the approach based on PicoMUS.

TABLE 4. Number of canonical uniquely weighted tournaments (i) from size two to four with edges weights from $\{2, 4, 6, 8, 10, 12\}$ (second column), (ii) sufficient for an impossibility theorem obtained from an MUS (third column), and (iii) sufficient for an impossibility theorem obtained from the naive approach described in the main text (fourth column).

candidates	canonical weighted tournaments from initial set	canonical weighted tournaments from MUS	canonical weighted tournaments from naive approach
2	6	6	6
3	160	160	140
4	1,920	1,484	570
2 to 4	2,086	1,650	716

Appendix C. Lean verification of SAT encoding

To increase confidence in the SAT-based Proposition B.3, we produced a formally verified SAT encoding in the Lean Theorem Prover (Lean 3), building on recent work of Codel [**17**]. The Lean verification is available at `https://github.com/chasenorman/verified-encodings-social-choice`. In this appendix, we explain what exactly we verified. A similar approach could be used for Proposition A.12.1, but we concentrated our efforts on Proposition B.3.

Let $\mathbb{G}$ be the set of uniquely weighted tournaments $T = (X(T), w)$ such that $X(T) = \{0, \ldots, n\}$ for $n \in \{0, \ldots, 3\}$ and for each $a, b \in X(T)$ with $a \neq b$, we have $|w(a, b)| \in \{2, 4, 6, 8, 10, 12\}$. Given $T \in \mathbb{G}$ and $a, b \in X(T)$, let $T_{a,b}$ be the weighted tournament with $X(T_{a,b}) = \{0, 1\}$ and $w(0, 1) = |w(a, b)|$.

To prove Proposition B.3, it suffices to show that for some set $\mathcal{E}$ of weighted tournament embeddings between elements of $\mathbb{G}$, there is no resolute weighted tournament solution satisfying the following condition of *binary γ with respect to $\mathcal{E}$*.

DEFINITION C.1. Let F be a function on $\mathbb{G}$ such that for each $T \in \mathbb{G}$, $F(T) \in X(T)$. Let $\mathcal{E}$ be a family of maps $e : T \hookrightarrow T'$ where $T, T' \in \mathbb{G}$. Then F satisfies *binary γ with respect to $\mathcal{E}$* if for any $T, T' \in \mathbb{G}$ and map $e \in \mathcal{E}$ from T' to T, if

(1) $|X(T)| = |X(T')| + 1$,
(2) $a \in e[X(T')]$,
(3) $b \in X(T) \setminus e[X(T')]$,
(4) $F(T') = e^{-1}(a)$, and
(5) $F(T_{a,b}) = 0$ if $w(a, b) > 0$ and $F(T_{a,b}) = 1$ otherwise,

then $F(T) = a$.

Modulo the switching between images and pre-images of the embedding, the items of the definition capture the assumptions of binary γ, namely that we are considering removing from a situation T one candidate to obtain T', as in (1), where a is one of the candidates not removed, as in (2), whereas b is the candidate removed, as in (3), such that a wins without b, as in (4), and a wins against b head-to-head, as in (5). Note that to capture the condition that a wins head-to-head against b, we check in (5) whether the counterpart of a in $T_{a,b}$ wins.

The repository linked above contains Lean code that for a given $\mathbb{D} \subseteq \mathbb{G}$ and set $\mathcal{E}$ of maps generates a propositional formula $\mathsf{func}_{\mathbb{D}} \wedge \mathsf{bg}_{\mathcal{E}}$ in CNF. We then give a formal proof in Lean of the following result, showing that $\mathsf{func}_{\mathbb{D}} \wedge \mathsf{bg}_{\mathcal{E}}$ correctly encodes the existence of a function on $\mathbb{D}$ satisfying binary γ with respect to $\mathcal{E}$.

LEMMA C.2. *Let F be a function on $\mathbb{G}$ such that for each $T \in \mathbb{G}$, $F(T) \in X(T)$. Let $\mathbb{D} \subseteq \mathbb{G}$, and let $\mathcal{E}$ be a family of maps $e : T \hookrightarrow T'$ where $T, T' \in \mathbb{G}$. If F satisfies binary γ with respect to $\mathcal{E}$, then the valuation v defined on*

$$\{A_{T,c} \mid T \in \mathbb{D}, c \in X(T)\}$$

by

$$v(A_{T,c}) = 1 \text{ if and only if } F(T) = c$$

satisfies $\mathsf{func}_{\mathbb{D}} \wedge \mathsf{bg}_{\mathcal{E}}$.

We then defined in Lean a set $\mathbb{D}$ and a set $\mathcal{E}$ of maps for which we formally verified the following:

(i) $\mathbb{D}$ is indeed a subset of $\mathbb{G}$;
(ii) each map in $\mathcal{E}$ is indeed a weighted tournament embedding between elements of $\mathbb{G}$.

In fact, $\mathbb{D}$ contains the 716 canonical weighted tournaments obtained as described in Section B.4, and $\mathcal{E}$ was obtained using Python code to find a minimal set of embeddings sufficient for the impossibility theorem. However, the provenance of $\mathbb{D}$ and $\mathcal{E}$ are irrelevant, thanks to the Lean verification of (i) and (ii). There is no need to trust our Python code.

The final step was to test the satisfiability of the formula $\mathsf{func}_{\mathbb{D}} \wedge \mathsf{bg}_{\mathcal{E}}$ produced by Lean using a SAT solver. As Lean does not include a SAT solver (as of this writing), we had to leave the Lean environment for this last step. Standard SAT solvers returned unsatisfiability for the formula $\mathsf{func}_{\mathbb{D}} \wedge \mathsf{bg}_{\mathcal{E}}$. Thus, in order to trust our SAT-based proof of Proposition B.3, one need only trust that it is not the case that multiple standard SAT solvers contain bugs producing an incorrect unsatisfiability result for our formula.

Acknowledgments

For helpful comments, we thank the anonymous referee and Jennifer Wilson. We are also grateful for the opportunity to present versions of this work at the Logic@IITK seminar in February 2021, the AMS Special Session on The Mathematics of Decisions, Elections and Games at the Joint Mathematics Meeting in April 2022, and the UMD Logic Seminar in October 2022.

References

[1] Kenneth J. Arrow, *Social Choice and Individual Values*, Cowles Commission Monograph No. 12, John Wiley & Sons, Inc., New York, N. Y.; Chapman & Hall, Ltd., London, 1951. MR0039976

[2] J. M. Baldwin, *A technique of the Nanson preferential majority system of election*, Transactions and Proceedings of the Royal Society of Victoria **39** (1926), 45–52.

[3] J. S. Banks, *Sophisticated voting outcomes and agenda control*, Social Choice and Welfare **1** (1985), no. 4, 295–306.

[4] Armin Biere, *PicoSAT essentials*, Journal on Satisfiability, Boolean Modeling and Computation **4** (2008), no. 2-4, 75–79.

[5] Georges Bordes, *On the possibility of reasonable consistent majoritarian choice: Some positive results*, Journal of Economic Theory **31** (1983), no. 1, 122–132.

[6] Florian Brandl, Felix Brandt, Manuel Eberl, and Christian Geist, *Proving the incompatibility of efficiency and strategyproofness via SMT solving*, J. ACM **65** (2018), no. 2, Art. 6, 28, DOI 10.1145/3125642. MR3761673

[7] Florian Brandl, Felix Brandt, Christian Geist, and Johannes Hofbauer, *Strategic abstention based on preference extensions: positive results and computer-generated impossibilities*, J. Artificial Intelligence Res. **66** (2019), 1031–1056. MR4052122

[8] Felix Brandt and Chris Dong, *Local rationalizability and choice consistency*, `arXiv:2204.05062 [econ.TH]`, 2022.

[9] Felix Brandt and Felix Fischer, *Pagerank as a weak tournament solution*, Internet and Network Economics: Third International Workshop, WINE 2007 (Berlin) (Xiaotie Deng and Fan Chung Graham, eds.), Springer-Verlag, 2007, pp. 300–305.

[10] Felix Brandt and Christian Geist, *Finding strategyproof social choice functions via SAT solving*, J. Artificial Intelligence Res. **55** (2016), 565–602, DOI 10.1613/jair.4959. MR3478031

[11] Felix Brandt, Christian Geist, and Dominik Peters, *Optimal bounds for the no-show paradox via SAT solving*, Math. Social Sci. **90** (2017), 18–27, DOI 10.1016/j.mathsocsci.2016.09.003. MR3716254

[12] Felix Brandt, Christian Geist, and Hans Georg Seedig, *Identifying k-majority digraphs via SAT solving*, Proceedings of the 1st AAMAS Workshop on Exploring Beyond the Worst Case in Computational Social Choice, 2014.

[13] Felix Brandt and Paul Harrenstein, *Set-rationalizable choice and self-stability*, J. Econom. Theory **146** (2011), no. 4, 1721–1731, DOI 10.1016/j.jet.2011.03.006. MR2887760

[14] Felix Brandt, Marie Matthäus, and Christian Saile, *Minimal voting paradoxes*, Working paper, 2021.

[15] Felix Brandt, Christian Saile, and Christian Stricker, *Voting with ties: Strong impossibilities via SAT solving*, Proceedings of the 17th International Conference on Autonomous Agents and Multiagent Systems (AAMAS 2018) (M. Dastani, G. Sukthankar, E. André, and S. Koenig, eds.), International Foundation for Autonomous Agents and Multiagent Systems, 2018, pp. 1285–1293.

[16] C. C. Chang and H. J. Keisler, *Model theory*, 3rd ed., Studies in Logic and the Foundations of Mathematics, vol. 73, North-Holland Publishing Co., Amsterdam, 1990. MR1059055

[17] Cayden Codel, *Verifying SAT encodings in Lean*, 2022.

[18] Clyde Hamilton Coombs, *A theory of data*, John Wiley and Sons, New York, 1964.

[19] A. H. Copeland, *A 'reasonable' social welfare function*, Notes from a seminar on applications of mathematics to the social sciences, University of Michigan, 1951.

[20] Bernard Debord, *Caractérisation des matrices des préférences nettes et méthodes d'agrégation associées* (French, with English summary), Math. Sci. Humaines **97** (1987), 5–17, 83. MR890684

[21] John Duggan, *Uncovered sets*, Soc. Choice Welf. **41** (2013), no. 3, 489–535, DOI 10.1007/s00355-012-0696-9. MR3095869

[22] John Duggan, *Weak rationalizability and arrovian impossibility theorems for responsive social choice*, Public Choice **179** (2019), 7–40.

[23] Bhaskar Dutta and Jean-Francois Laslier, *Comparison functions and choice correspondences*, Soc. Choice Welf. **16** (1999), no. 4, 513–532, DOI 10.1007/s003550050158. MR1713186

[24] Herbert B. Enderton, *A mathematical introduction to logic*, 2nd ed., Harcourt/Academic Press, Burlington, MA, 2001. MR1801397

[25] Christian Geist and Ulle Endriss, *Automated search for impossibility theorems in social choice theory: ranking sets of objects*, J. Artificial Intelligence Res. **40** (2011), 143–174, DOI 10.1613/jair.3126. MR2805240

[26] Christian Geist and Dominik Peters, *Computer-aided methods for social choice theory*, Trends in computational social choice, AI Access, [2017] ©2017, pp. 249–267. MR4481034

[27] Bernard Grofman and Scott L. Feld, *If you like the alternative vote (a.k.a. the instant runoff), then you ought to know about the coombs rule*, Electoral Studies **23** (2004), no. 4, 641–659.

[28] Michael C. Herron and Jeffrey B. Lewis, *Did Ralph Nader spoil a Gore presidency? a ballot-level study of Green and Reform Party voters in the 2000 presidential election*, Quarterly Journal of Political Science **2** (2007), no. 3, 205–226.

[29] Clarence Hoag and George Hallett, *Proportional representation*, Macmillan, New York, 1926.

[30] Wesley H. Holliday and Eric Pacuit, *Split Cycle: A new Condorcet consistent voting method independent of clones and immune to spoilers*, Public Choice, **197** (2023), 1–62.

[31] Wesley H. Holliday and Eric Pacuit, *Axioms for defeat in democratic elections*, Journal of Theoretical Politics **33** (2021), no. 4, 475–524, arXiv:2008.08451.

[32] Wesley H. Holliday and Eric Pacuit, *Stable voting*, Constitutional Political Economy, **34** (2023), 421–433.

[33] Sean Horan and Yves Sprumont, *Two-stage majoritarian choice*, Theor. Econ. **17** (2022), no. 2, 521–537. MR4431703

[34] Boas Kluiving, Adriaan de Vries, Pepijn Vrijbergen, Arthur Boixel, and Ulle Endriss, *Analysing irresolute multiwinner voting rules with approval ballots via sat solving*, Proceedings of the 24th European Conference on Artificial Intelligence (ECAI-2020), August 2020, pp. 131–138.

[35] Gerald H. Kramer, *A dynamical model of political equilibrium*, J. Econom. Theory **16** (1977), no. 2, 310–334, DOI 10.1016/0022-0531(77)90011-4. MR524414

[36] Mark H. Liffiton and Karem A. Sakallah, *Algorithms for computing minimal unsatisfiable subsets of constraints*, J. Automat. Reason. **40** (2008), no. 1, 1–33, DOI 10.1007/s10817-007-9084-z. MR2387924

[37] Christopher S. P. Magee, *Third-party candidates and the 2000 presidential election*, Social Science Quarterly **84** (2003), no. 3, 29–35.

[38] David C. McGarvey, *A theorem on the construction of voting paradoxes*, Econometrica **21** (1953), 608–610, DOI 10.2307/1907926. MR62415

[39] Nicholas R. Miller, *A new solution set for tournaments and majority voting: Further graph-theoretical approaches to the theory of voting*, American Journal of Political Science **24** (1980), no. 1, 68–96.

[40] E. J. Nanson, *Methods of election*, Transactions and Proceedings of the Royal Society of Victoria **19** (1882), 197–240.

[41] Emerson M. S. Niou, *A note on Nanson's rule*, Public Choice **54** (1987), no. 2, 191–193.

[42] Hannu Nurmi, *Comparing voting systems*, D. Reidel, Dordrecht, 1987.

[43] Ömer Eğecioğlu, *Uniform generation of anonymous and neutral preference profiles for social choice rules*, Monte Carlo Methods Appl. **15** (2009), no. 3, 241–255, DOI 10.1515/MCMA.2009.014. MR2573215

[44] Ömer Eğecioğlu and Ayça E. Giritligil, *The impartial, anonymous, and neutral culture model: a probability model for sampling public preference structures*, J. Math. Sociol. **37** (2013), no. 4, 203–222, DOI 10.1080/0022250X.2011.597012. MR3173646

[45] Raúl Pérez-Fernández and Bernard De Baets, *The supercovering relation, the pairwise winner, and more missing links between Borda and Condorcet*, Soc. Choice Welf. **50** (2018), no. 2, 329–352, DOI 10.1007/s00355-017-1086-0. MR3749646

[46] Donald G. Saari, *Disposing dictators, demystifying voting paradoxes*, Cambridge University Press, Cambridge, 2008. Social choice analysis, DOI 10.1017/CBO9780511754265. MR2449532

[47] Markus Schulze, *A new monotonic, clone-independent, reversal symmetric, and Condorcet-consistent single-winner election method*, Soc. Choice Welf. **36** (2011), no. 2, 267–303, DOI 10.1007/s00355-010-0475-4. MR2745981

[48] Markus Schulze, *The Schulze method of voting*, arXiv:1804.02973, 2018.

[49] Thomas Schwartz, *The logic of collective choice*, Columbia University Press, New York, 1986.

[50] Amartya Sen, *Choice functions and revealed preference*, The Review of Economic Studies **38** (1971), no. 3, 307–317.

[51] Paul B. Simpson, *On defining areas of voter choice: Professor Tullock on stable voting*, The Quarterly Journal of Economics **83** (1969), no. 3, 478–490.

[52] John H. Smith, *Aggregation of preferences with variable electorate*, Econometrica **41** (1973), 1027–1041, DOI 10.2307/1914033. MR441227

[53] Richard Stearns, *The voting problem*, Amer. Math. Monthly **66** (1959), 761–763, DOI 10.2307/2310461. MR109087

[54] Kotaro Suzumura, *Rational choice, collective decisions, and social welfare*, Cambridge University Press, Cambridge, 1983.

[55] T. N. Tideman, *Independence of clones as a criterion for voting rules*, Soc. Choice Welf. **4** (1987), no. 3, 185–206, DOI 10.1007/BF00433944. MR908272

[56] Christopher J. Tyson, *Cognitive constraints, contraction consistency, and the satisficing criterion*, J. Econom. Theory **138** (2008), no. 1, 51–70, DOI 10.1016/j.jet.2007.01.019. MR2888494

[57] Robert Wilson, *Social choice theory without the Pareto principle*, J. Econom. Theory **5** (1972), no. 3, 478–486, DOI 10.1016/0022-0531(72)90051-8. MR449494

[58] T. M. Zavist and T. N. Tideman, *Complete independence of clones in the ranked pairs rule*, Soc. Choice Welf. **6** (1989), no. 2, 167–173, DOI 10.1007/BF00303170. MR995431

Department of Philosophy and Group in Logic and the Methodology of Science, University of California, Berkeley, Berkeley, California 94720
Email address: wesholliday@berkeley.edu

Computer Science Department, Carnegie Mellon University, Pittsburgh, Pennsylvania 15213
Email address: chasen@cmu.edu

Department of Philosophy, University of Maryland, College Park, Maryland 20742
Email address: epacuit@umd.edu

Department of Economics, Northwestern University, Evanston, Illinois 60208
Email address: saamzahedian2029@u.northwestern.edu

Contemporary Mathematics
Volume **795**, 2024
https://doi.org/10.1090/conm/795/15969

Connecting Arrow's Theorem, voting theory, and the Traveling Salesperson Problem

Donald G. Saari

Abstract. Problems with majority voting over pairs as represented by Arrow's Theorem and those of finding the lengths of closed paths as captured by the Traveling Salesperson Problem (TSP) appear to have nothing in common. In fact, they are connected. As shown, pairwise voting and a version of the TSP share the same domain where each system is simplified by restricting it to complementary regions that eliminate extraneous terms. Central for the analysis is the Borda Count, which, as demonstrated, most accurately reflects the voter preferences.

1. Introduction

Among the many challenges posed by discrete mathematics and the social sciences are aspects of voting theory as characterized by Arrow's Theorem and the properties of closed paths on a graph as typified by the Traveling Salesperson Problem (TSP). Surprisingly, both topics can be analyzed in essentially the same way. After introducing the commonality with examples, each situation is then simplified by emphasizing different regions of the associated geometry.

1.1. Voting theory and Arrow's Theorem. Central to Arrow's Theorem [**1**] is the "Independence of Irrelevant Alternatives" (IIA) condition, which requires a profile's conclusion to be completely determined by the rankings of its associated paired comparisons.[1] A $N = 70$ person majority vote example over the $n = 3$ alternatives $\{A_j\}_{j=1}^3$ is where

(1.1) 25 prefer $A_1 \succ A_2 \succ A_3$, 23 prefer $A_2 \succ A_3 \succ A_1$, 22 prefer $A_3 \succ A_1 \succ A_2$.

This leads to the cyclic pairwise majority vote outcomes of

(1.2) $A_1 \succ A_2$ by 47:23, $A_2 \succ A_3$ by 48:22, and $A_3 \succ A_1$ by 45:25,

Key words and phrases. Voting theory, Arrow's Theorem, Traveling Salesperson Problem.
This work is part of a National Science Foundation project under NSF Award Number CMMI-1923164.

[1]Even stronger, a decision method satisfies "IIA" iff it can be expressed in terms of independent paired comparison methods [**10,11**]. This means it is not a negative feature should a decision method, such as the plurality vote, fails to satisfy IIA; it only means that the method cannot be so expressed. Conversely, any method that cannot be expressed in this manner does not satisfy IIA. Consequently, Arrow's Theorem is a result describing the problems that accompany paired comparisons.

where "A_1 is better than A_2," "A_2 is better than A_3," but, to complete the cycle, "A_3 is better than A_1." Not only does this three-cycle violate Arrow's objective of obtaining a transitive conclusion, but it reflects a long-standing goal of replacing cyclic voting outcomes with appropriate transitive rankings, or at least with a "best choice." Prominent approaches were developed by the mathematicians Dodgson [**2**] in 1876 and Kemeny [**3**] in 1959.

A way to compare how an alternative fares in a paired comparison is to compute how its tally differs from the average score of $\frac{\text{number of voters}}{2} = \frac{N}{2}$. (In what follows, N is the number of voters, n is the number of alternatives.) So

$$d_{i,j} = \frac{1}{2}[A_i\text{'s tally} - A_j\text{'s tally}] = A_i\text{'s tally} - \frac{N}{2} = -d_{j,i}. \tag{1.3}$$

For the Eq. 1.2 values, where $N = 70$ and $n = 3$,

$$d_{1,2} = 47 - \frac{70}{2} = 12 = -d_{2,1},\ d_{2,3} = 13 = -d_{3,2},\ \text{and } d_{1,3} = -10 = -d_{3,1}. \tag{1.4}$$

1.2. TSP. To introduce the particular TSP system discussed here, suppose a company reimburses a salesperson for the average cost of traveling between cities. This policy creates an incentive for the salesperson to make extra money by selecting routes with below average travel costs. For instance, if city H is on top of a hill while city V is in a valley, the cost of going uphill from $V \to H$ exceeds that of going downhill from $H \to V$. If, for instance, traveling from H to V costs \$30 while the uphill trek of $V \to H$ costs \$40, the company will reimburse the average cost of \$35. The salesperson is more interested in how costs differ from the average, which are $\pm$\$5 depending on the route. Denote this relevant information as $V \xrightarrow{5} H$, which also means that $H \xrightarrow{-5} V$. More generally, if $d_{i,j}$ represents the "difference from the average" cost of going from A_i to A_j, then

$$A_i \xrightarrow{d_{i,j}} A_j \text{ and } A_j \xrightarrow{-d_{i,j}} A_i \text{ are equivalent,} \tag{1.5}$$

or, as true with Eq. 1.3,

$$d_{i,j} = -d_{j,i}. \tag{1.6}$$

Figure 1 is a typical TSP example, where "differences from averages" information about the alternatives $\{A_j\}_{j=1}^{6}$ and its $\binom{6}{2} = 15$ pairs are geometrically represented with a graph. (Subscripts "A" and "S" refer, respectively, to whether the graph has an asymmetric or a symmetric structure.) The weight on an arrow between vertices specifies the positive cost (the amount above average) of traveling between them; traveling counter to an arrow is a negative, or the amount below the average cost (Eqs. 1.5, 1.6). The zero on the dashed line connecting A_2 and A_5 is where the travel costs in both directions agree, so there is a zero difference from the average.

A standard TSP objective is to discover the longest and shortest connected path that starts and ends at the same vertex (this is a closed path; presumably the vertex is the salesperson's home city) and passes once through each of the other vertices—this is called a "Hamiltonian circuit" or a "Hamiltonian path." To appreciate the way in which these concerns will be addressed, before reading more, let me ask the reader to find the longest such path in Fig. 1. To illustrate the computations with a Hamiltonian path that is *not* the longest, by summing the

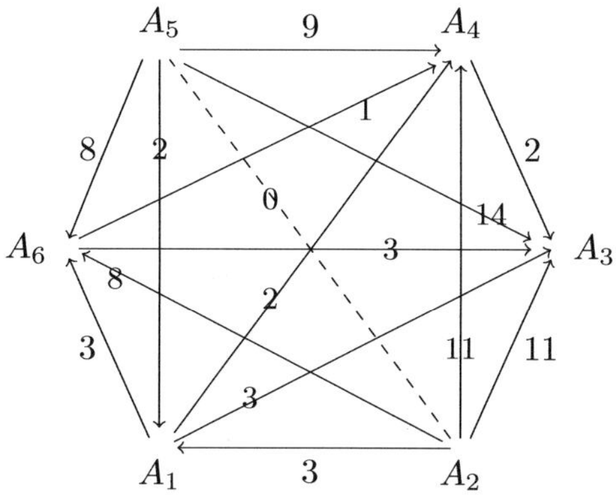

FIGURE 1. $\mathcal{G}^6_A$.

costs required to travel between each pair of vertices, path

$$A_1 \xrightarrow{-3} A_2 \xrightarrow{11} A_4 \xrightarrow{-1} A_6 \xrightarrow{3} A_3 \xrightarrow{-14} A_5 \xrightarrow{2} A_1$$

has length -2, and its reversal has length 2. In fact, thanks to Eqs. 1.5, 1.6, the tasks of finding the longest and shortest Hamiltonian paths coincide; the reversal of one is the other.

1.3. A common space. The voting and TSP questions involve the $\binom{n}{2}$ pairs defined by n alternatives where costs/differences between vertices of a $\{i, j\}$ pair are measured by $d_{i,j} = -d_{j,i}$ values.[2] As this equality requires $d_{i,i} = 0$, the $d_{i,i}$ terms are ignored. Thus, the voting and TSP settings share the same $\mathbb{R}_A^{\binom{n}{2}}$ domain, where $\mathbf{d}^n_A \in \mathbb{R}_A^{\binom{n}{2}}$ has the form

$$\mathbf{d}^n_A = (d_{1,2}, d_{1,3}, \dots, d_{1,n}; d_{2,3}, \dots, d_{2,n}; d_{3,4} \dots; d_{n-1,n}); \; d_{i,j} = -d_{j,i}. \tag{1.7}$$

Semicolons indicate changes in the first subindex. The $\mathbb{R}_A^{\binom{n}{2}}$ Euclidean space structure is developed next.

To start, if a triplet $\{A_i, A_j, A_k\}$ defines a transitive ranking, it can be expressed with these $d_{s,u}$ values; e.g., should $d_{i,j} > 0$ and $d_{j,k} > 0$, then transitivity requires that $d_{i,k} > 0$. These inequalities are borrowed from the structure of points on a line where $p_i > p_j$ and $p_j > p_k$ require $p_i > p_k$. But points on a line satisfy the stronger algebraic relationship $(p_i - p_j) + (p_j - p_k) = (p_i - p_k)$. The following mimics this equality.

DEFINITION 1 ([**12**]). Vector $\mathbf{d}^n_A \in \mathbb{R}_A^{\binom{n}{2}}$ is strongly transitive if and only if each triplet $\{i, j, k\}$ satisfies

$$d_{i,j} + d_{j,k} = d_{i,k}. \tag{1.8}$$

This condition was introduced for voter profiles in [**9**]; the decision theory version used here comes from [**12**]. The subspace of strongly transitive vectors, $\mathbb{ST}^n_A$, is described next.

---

[2]Rather than "difference from average," $d_{i,j}$ can be anything where $d_{i,j} = -d_{j,i}$.

THEOREM 1 ([**12**]). *The set of strongly transitive* $\mathbf{d}^n_{A,st} \in \mathbb{R}_A^{\binom{n}{2}}$*, denoted by* $\mathbb{ST}^n_A$*, is a* $(n-1)$*-dimensional linear subspace of* $\mathbb{R}_A^{\binom{n}{2}}$.

That $\mathbb{ST}^n_A$ is a linear subspace is a standard exercise.[3] The assertion about its dimension follows from the fact that any $d_{j,k} \in \mathbb{ST}^n_A$ can be expressed as $d_{j,k} = d_{j,1} + d_{1,k}$ (Eq. 1.8), where the path from A_j to A_k is diverted to pass through A_1. As such, all $d_{j,k}$ values for $\mathbf{d}^n_{A,st} \in \mathbb{ST}^n_A$ follow from the $(n-1)$ terms $\{d_{1,s}\}_{s=2}^n$.

The dimensions of $\mathbb{R}_A^{\binom{n}{2}}$ and $\mathbb{ST}^n_A$ dictate that $\mathbb{C}^n_A$, the orthogonal complement of $\mathbb{ST}^n_A$, has dimension $\binom{n-1}{2}$. To motivate the form of these orthogonal vectors, expressing Eq. 1.8 as $x + y = z$, or as the plane $x + y - z = 0$, identifies $(1, 1, -1)$ as a normal vector. Thus the $x = y = -z = 1$ values define a normal vector $d_{i,j} = d_{j,k} = d_{k,i} = 1$, which define a cycle.

THEOREM 2 ([**12**]). *The* $\binom{n-1}{2}$ *dimensional linear subspace* $\mathbb{C}^n_A$ *consists of cycles where one choice of a basis, which consists of three-cycles, is* $\{\mathbf{d}^n_{1,j,k}\}_{1<j<k\leq n}$. *The only non-zero terms of* $\mathbf{d}^n_{1,j,k}$ *are* $d_{1,j} = d_{j,k} = d_{k,1} = 1$.

More generally, any set $\mathcal{SET}^n_A$ is a basis if it consists of $\binom{n-1}{2}$ three-cycles where the weight (cost) between vertices on each cycle is unity and each cycle has an arc not in any other cycle of $\mathcal{SET}^n_A$. The proof is immediate: the cycles are linearly independent because each has an arc not in any other $\mathcal{SET}^n_A$ cycle. As the number of cycles equals the dimension of $\mathbb{C}^n_A$, $\mathcal{SET}^n_A$ is a basis.

According to these theorems, $\mathbb{R}_A^{\binom{n}{2}}$ nicely separates into linear subspaces of strongly transitive vectors, $\mathbb{ST}^n_A$, and cyclic vectors, $\mathbb{C}^n_A$, where a basis for $\mathbb{C}^n_A$ consists of a special type of three-cycles. As such, $\mathbf{d}^n_A \in \mathbb{R}_A^{\binom{n}{2}}$ has a unique decomposition

$$\mathbf{d}^n_A = \mathbf{d}^n_{A,st} + \mathbf{d}^n_{A,cyclic} \tag{1.9}$$

where $\mathbf{d}^n_{A,st}$ and $\mathbf{d}^n_{A,cyclic}$ are, respectively, the orthogonal projection of $\mathbf{d}^n_A$ to $\mathbb{ST}^n$ and $\mathbb{C}^n_A$.

An immediate consequence of Eq. 1.9 is that if a transitive vector is not strongly transitive, it is a hybrid.

COROLLARY 1. *For* $n \geq 3$, *if* $\mathbf{d}^n_A$ *is transitive but not strongly transitive, then there are unique non-zero vectors* $\mathbf{d}^n_{A,st} \in \mathbb{ST}^n_A$ *and* $\mathbf{d}^n_{A,cyclic} \in \mathbb{C}^n_A$ *so that* $\mathbf{d}^n_A = \mathbf{d}^n_{A,st} + \mathbf{d}^n_{A,cyclic}$.

What limits the use of transitive vectors is that they fail to define a linear subspace. For instance, $d_{1,2} = d_{2,3} = 2$, $d_{1,3} = 1$ and $d^*_{1,2} = d^*_{2,3} = -1$, $d^*_{1,3} = -2$ are transitive triplets, but their sums (e.g., $d_{i,j} + d^*_{i,j} = d^+_{i,j}$) define the cycle $d^+_{1,2} = d^+_{2,3} = 1$, $d^+_{1,3} = -1$ (where, $A_1 \succ A_2$, $A_2 \succ A_3$, but $A_3 \succ A_1$). According to Cor. 1, a transitive vector is a strongly transitive choice that is lightly contaminated with cyclic terms. The following theorem partly explains the "lightly contaminated" modifier by showing that rankings of transitive $\mathbf{d}^n_A$ and its strongly transitive component $\mathbf{d}^n_{A,st}$ can differ, but not radically. This holds even if $\mathbf{d}^n_A$ is not transitive, but at least it has a Condorcet winner and loser. (A Condorcet winner is a candidate who beats each of the other candidates in majority vote comparisons.

[3] In [**9, 14**], $\mathbb{ST}^n_A$ is called the "transitivity plane."

A Condorcet loser loses all pairwise majority votes. Both can exist even without transitivity; e.g., for $n = 5$, A_1 could be the Condorcet winner, A_5 the Condorcet loser, and A_2, A_3, A_4 define a cycle.) The simple proof of Thm. 3 is in Sect. 5.

THEOREM 3. *If $\mathbf{d}_A^n$ has a unique Condorcet winner A_1 and a unique Condorcet loser A_n, then A_1 is strictly ranked above A_n in $\mathbf{d}_{A,st}^n$.*

For the converse of this assertion, it must be determined what happens by adding cyclic terms to a $\mathbf{d}_{A,st}^n$. If the resulting $\mathbf{d}_A^n$ is "wildly cyclic" (that is, vertices are connected by many different cycles), not much can be stated. But if A_1 and A_n are, respectively, the top and bottom ranked candidates of the $\mathbf{d}_{A,st}^n$ and if $\mathbf{d}_A^n$ remains transitive, the question is whether A_1 must be ranked above A_n in $\mathbf{d}_A^n = \mathbf{d}_{A,st}^n + \mathbf{d}_{A,cyclic}^n$. Proving this kind of result can be messy. But as a way to illustrate Thm. 2, the basic idea is developed in Sect. 5 for $n = 3, 4$. The details of these proofs are similar to earlier relationships that were developed about the Kemeny and paired voting rankings relative to the Borda ranking [**9**, **14**].

2. Voting theory

A central objective in voting and decision theory is to obtain transitive outcomes, where the cyclic components of $\mathbf{d}_A^n$ introduce problems. A resolution is obvious; drop the troubling $\mathbf{d}_{A,cyclic}^n$ cyclic term and retain only $\mathbf{d}_{A,st}^n$. Dodgson's method partly does so by replacing $\mathbf{d}_A^n$ with a vector that might have cyclic terms, but at least it has a Condorcet winner. (A_j is a Condorcet winner if and only if $d_{j,k} > 0$ for all $k \neq j$. That is, A_j is "better than" all other alternatives in paired comparisons.) Dodgson's method, then, projects $\mathbf{d}_A^n \in \mathbb{R}_A^{\binom{n}{2}}$ to the nearest $\mathbb{R}_A^{\binom{n}{2}}$ subset where all vectors have a Condorcet winner. For $n \geq 4$, these regions can have cyclic components; e.g., for $n = 4$, the dimension of $\mathbb{ST}_A^4$ is only three, while the Condorcet subset has the full $\binom{4}{2} = 6$ dimension. For instance, A_1 is the Condorcet winner in the eight $\mathbb{R}_A^6$ orthants where $d_{1,2} > 0, d_{1,3} > 0, d_{1,4} > 0$. Without imposing restrictions on the signs of $d_{2,3}, d_{2,4}, d_{3,4}$, cyclic behavior is allowed among $\{A_2, A_3, A_4\}$.

Kemeny adopted the more ambitious goal [**3**] of replacing $\mathbf{d}_A^n$ with a transitive outcome. Similar to Dodgson, he designed a projection mapping that sends $\mathbf{d}_A^n$ into the nearest $\mathbb{R}_A^{\binom{n}{2}}$ subset consisting of transitive rankings.

Both methods provide electoral relief at a first level, but a deeper investigation reveals a host of other subtle difficulties that cast doubt on the reliability of these approaches. Important results in this direction are described in a series of papers by Ratliff [**4**–**6**]. For instance, the above "projection" descriptions makes it reasonable to expect that the Dodgson and Kemeny outcomes are related; perhaps the Dodgson winner always is the top-ranked Kemeny candidate. But Ratliff proved that, in general, such conjectures are false. A small sample of his findings follows.

THEOREM 4 ([**4**–**6**]). *For $n \geq 4$ candidates, select an integer k, where $1 \leq k \leq n$. There exist paired comparison examples where the Dodgson winner is in k^{th} place in the Kemeny ranking.*

Generalize Dodgson's method to select a committee of $k \geq 2$ candidates by using Dodgson's projection method to a subset of $\mathbb{R}_A^{\binom{n}{2}}$ where all vectors have k candidates where each is ranked above the $(n - k)$ remaining candidates. For integers k and *s satisfying* $1 \leq s < n-k$, $k \neq s$, *there exist examples where, with the Dodgson's generalized approach, no candidate in the Dodgson committee of s is in the Dodgson committee of k candidates.*

As Ratliff proved, rather than being top-ranked, the Dodgson winner can be anywhere in a Kemeny outcome, even bottom-ranked. Moreover, the Dodgson winner need not be in a Dodgson-Ratliff committee of two or three. These unexpected conclusions are consequences of the cyclic terms that remain even after using the Dodgson and Kemeny projections. Indeed, unless Kemeny's outcome is strongly transitive it contains cyclic components (Cor. 1) that can cause other difficulties.

Stated differently, the Dodgson and Kemeny procedures remove only as many of the $\mathbf{d}^n_A$ cyclic components as needed to attain their objectives. Without a thorough cleansing, it is reasonable to expect other mysterious behavior: this happens. This also holds for Arrow's Theorem; its negative conclusion is strictly a consequence of the $\mathbf{d}^n_{A,cyclic}$ component. Completely removing $\mathbf{d}^n_{A,cyclic}$ converts Arrow's assertion into a positive result [**10, 11**]. Indeed, the ultimate goal for decision and voting problems should be to eliminate *all cyclic components of a given* $\mathbf{d}^n_A$. Doing so is a common mathematical computation.

2.1. Orthogonal projection. The standard way to eliminate the unwanted cyclic terms from a given $\mathbf{d}^n_A$ is with the orthogonal projection

$$P : \mathbb{R}^{\binom{n}{2}}_A \to \mathbb{ST}^n_A. \tag{2.1}$$

The computations require finding a basis for $\mathbb{ST}^n_A$ (using Eq. 1.8) and carrying out the associated vector analysis. The resulting approach follows; see [**12**] for details and examples.

DEFINITION 2. For alternative A_j, the weighted sum is

$$\mathcal{S}_A(A_j) = \sum_{k=1}^{n} d_{j,k};\ j = 1, \dots, n. \tag{2.2}$$

With Eq. 1.4 values,

$$S_A(A_1) = 12 - 10 = 2,\ S_A(A_2) = -12 + 13 = 1,\ S_A(A_3) = -13 + 10 = -3. \tag{2.3}$$

THEOREM 5 ([**12**]). *For* $\mathbf{d}^n_A \in \mathbb{R}^{\binom{n}{2}}_A$, *the* $d_{i,j}$ *value in* $\mathbf{d}^n_{A,st}$ *is*

$$d_{i,j} = \frac{1}{n}[\mathcal{S}_A(A_i) - \mathcal{S}_A(A_j)]. \tag{2.4}$$

Of surprise, this projection of $\mathbf{d}^n_A$ to $\mathbf{d}^n_{A,st}$ is equivalent to the well known Borda Count. Recall, the Borda Count tallies an n-candidate ballot by assigning $n-j$ points to the jth ranked candidate. A candidate's Borda tally, $\mathcal{B}(A_j)$, is the sum of points assigned to A_j over all N ballots. With Eq. 1.1, $\mathcal{B}(A_1) = 25(3-1)+23(3-3)+22(3-2) = 72$, $\mathcal{B}(A_2) = 25(3-2)+23(3-1)+22(3-3) = 71$, and $\mathcal{B}(A_3) = 25(3-3)+23(3-2)+22(3-1) = 67$.

THEOREM 6 ([**12**]). *For* $n \geq 3$, *the ranking of the alternatives in the orthogonal projection of* $\mathbf{d}^n_A \in \mathbb{R}^{\binom{n}{2}}_A$ *to* $\mathbb{ST}^n_A$ *from Eq. 2.1, which is* $\mathbf{d}^n_{A,st}$, *agrees with the Borda Count ranking. More precisely,*

$$\mathcal{B}(A_j) = S_A(A_j) + (n-1)\frac{N}{2},\ j = 1, \dots, n. \tag{2.5}$$

and

$$\mathcal{B}(A_j) - \mathcal{B}(A_k) = S_A(A_j) - S_A(A_k) = nd_{j,k}. \tag{2.6}$$

PROOF. As known (e.g., [**8**–**10**]), a way to compute $\mathcal{B}(A_j)$ is to sum A_j's tallies over each of its $(n-1)$ majority vote paired comparisons. (As an example, using the Eq. 1.2 tallies, in the $\{A_1, A_2\}$ and $\{A_1, A_3\}$ elections, A_1 receives, respectively, 47 and 25 votes where the $47 + 25 = 72$ value agrees with A_1's above computed $\mathcal{B}(A_1) = 72$.) According to Eq. 1.3,

$$\mathcal{B}(A_j) = \sum_{k=1,k\neq j}^{n} [d_{j,k} + \frac{N}{2}] = S_A(A_j) + (n-1)\frac{N}{2}. \tag{2.7}$$

Thus, for voting theory, $S_A(A_j) - S_A(A_k) = \mathcal{B}(A_j) - \mathcal{B}(A_k) = nd_{j,k}$. □

For an interesting consequence of Thm. 6, recall that for many voting methods, its outcome is determined by scores it assigns to each candidate; e.g., this includes almost all standard methods such as the plurality vote, all positional methods, cumulative voting, Approval Voting, etc. These scores satisfy strong transitivity, which means that the outcome is in $\mathbb{ST}^n_A$. This leads to a natural objective, which is to select the $\mathbb{ST}^n_A$ outcome that most accurately reflects the views (i.e., preferences) of the voters. That is, find the $\mathbb{ST}^n_A$ outcome that is closest (as $\mathbb{R}^n_A$ is an Euclidean space, "closest" is with the Euclidean distance) to the data $\mathbf{d}^n_A$. This is the orthogonal projection of $\mathbf{d}^n_A$ to $\mathbb{ST}^n_A$, which is $\mathbf{d}^n_{A,st}$ and the Borda Count (Thm. 6).

Combining comments from the above provide new explanations for earlier results. For instance, the following, which has the spirit of Thm. 3, was proved by using these tools.

THEOREM 7 ([**9**, **14**]). *For $n \geq 3$, Kemeny's method ranks a Borda winner over the Borda loser. Conversely, the Borda Count ranks a Kemeny winner over the Kemeny bottom ranked candidate.*

For paired comparison majority votes, the Borda outcome ranks the Condorcet winner over the Condorcet bottom ranked candidate. Conversely, if the paired comparisons define a transitive ranking, the Borda winner is ranked over the Borda bottom ranked candidate.

The Kemeny outcome is a transitive vector, while the strongly transitive Borda ranking is the orthogonal projection of $\mathbf{d}^n_A$ to $\mathbb{ST}^n_A$. As Cor. 1 asserts, a transitive ranking is a strongly transitive ranking clouded by cyclic terms, which captures the flavor, and potential disadvantages, of the Kemeny outcome.

Theorem 6 provides an explanation why the Borda Count has so many positive properties; examples with decision methods are in [**12**]. The following completes the earlier introductory comments about Arrow's Theorem.

THEOREM 8 ([**8**, **9**]). *If $\mathbf{d}^n_A \in \mathbb{ST}^n_A$, then the Borda Count and majority vote paired comparisons both satisfy Arrow's Theorem. In this space, the Borda Count is the only positional voting method that satisfies Arrow's conditions.*

The last assertion holds because the Borda Count is the only positional method where its outcomes are determined by the outcomes of majority votes over pairs. The outcome for all other positional methods need not be related, in any manner, to the paired comparison outcomes [**7**].

3. Turning to the TSP

According to Eq. 1.9, $\mathbf{d}^n_A \in \mathbb{R}^{\binom{n}{2}}$ can be uniquely expressed as $\mathbf{d}^n_A = \mathbf{d}^n_{A,st} + \mathbf{d}^n_{A,cyclic}$, where $\mathbf{d}^n_{A,st} \in \mathbb{ST}^n_A$ and $\mathbf{d}^n_{A,cyclic} \in \mathbb{C}^n_A$. As described, voting and decision theories seek transitive outcomes, which means that the cyclic $\mathbf{d}^n_{A,cyclic}$ component imposes obstacles. The natural resolution is to eliminate $\mathbf{d}^n_{A,cyclic}$ and retain $\mathbf{d}^n_{A,st}$ by projecting $\mathbf{d}^n_A$ into $\mathbb{ST}^n_A$.

In contrast, TSP and other closed path concerns involve *cycles*, so linear behavior imposes barriers; that is, the strongly transitive $\mathbf{d}^n_{A,st}$ term switches from being the desired component into becoming a troublemaker. To analyze TSP issues, mimic what was done for voting by orthogonally projecting $\mathbf{d}^n_A$ into $\mathbb{C}^n_A$ to eliminate $\mathbf{d}^n_{A,st}$ and retain $\mathbf{d}^n_{A,cyclic}$.

What simplifies finding $\mathbf{d}^n_{A,cyclic}$ is that $\mathbf{d}^n_{A,cyclic} = \mathbf{d}^n_A - \mathbf{d}^n_{A,st}$, and $\mathbf{d}^n_{A,st}$, which is equivalent to the Borda Count (Thm. 6), is easily computed (Thm. 5). The decomposition of the Fig. 1 graph is in Fig. 2 where the $\mathcal{G}^6_{A,cpi}$ and $\mathcal{G}^6_{A,cyclic}$ entries represent, respectively, $\mathbf{d}^6_{A,st}$ and $\mathbf{d}^6_{A,cyclic}$.

The computation of $\mathcal{G}^6_{A,cpi}$, which is the graph version of $\mathbf{d}^6_{A,st}$ and, as explained later, "cpi" stands for "closed path independent," follows from Def. 2 and Thm. 5. For instance, $S_A(A_1) = [-3+3+2-2+3] = 3$ while $S_A(A_2) = [3+8+0+11+11] = 33$, so $d_{1,2} = \frac{1}{6}[S_A(A_1) - S_A(A_2)] = -5$ is the $\mathbf{d}^6_{A,st}$ entry; it is the length of the $A_1 \to A_2$ leg in $\mathcal{G}^6_{cpi}$. (To agree with the literature, the "length" and "cost" of a path are the same.)

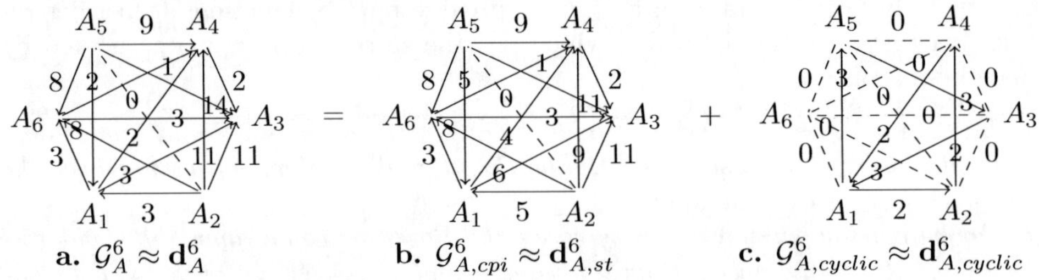

FIGURE 2. Decomposition of a $\mathcal{G}^6_A$.

Thanks to the linearity of the decomposition, the length of any closed path (not just Hamiltonian circuits) in $\mathcal{G}^6_A$ equals the sum of the path's lengths in $\mathcal{G}^6_{A,cpi}$ and $\mathcal{G}^6_{A,cyclic}$. For instance, the -4 length of the $\mathcal{G}^6_A$ closed path $A_1 \xrightarrow{-3} A_2 \xrightarrow{8} A_6 \xrightarrow{3} A_3 \xrightarrow{-14} A_5 \xrightarrow{2} A_1$ equals the sum of its path's lengths of $A_1 \xrightarrow{-5} A_2 \xrightarrow{8} A_6 \xrightarrow{3} A_3 \xrightarrow{-11} A_5 \xrightarrow{5} A_1$ in $\mathcal{G}^6_{A,cpi}$ and $A_1 \xrightarrow{2} A_2 \xrightarrow{0} A_6 \xrightarrow{0} A_3 \xrightarrow{-3} A_5 \xrightarrow{-3} A_1$ in $\mathcal{G}^6_{A,cyclic}$. But this path's length in $\mathcal{G}^6_{A,cpi}$ is zero, so its -4 length in $\mathcal{G}^6_{A,cyclic}$ equals the $\mathcal{G}^6_A$ path length. That this $\mathcal{G}^6_{A,cpi}$ closed path has length zero is not a lucky coincidence. Instead, *all closed paths in $\mathcal{G}^n_{A,cpi}$ have length zero,* which leads to the terminology that $\mathcal{G}^n_{A,cpi}$ closed paths are "closed path independent."

THEOREM 9. *For $n \geq 3$, a closed path in $\mathcal{G}^n_{A,cpi}$ has length zero. The length of a closed path in $\mathcal{G}^n_A$ equals its length in $\mathcal{G}^n_{A,cyclic}$.*

The length of any $\mathcal{G}^n_A$ path starting at A_i and ending at A_j equals its length in $\mathcal{G}^n_{A,cyclic}$ plus the $A_i \to A_j$ length in $\mathcal{G}^n_{A,cpi}$.

The proof of the first assertion follows immediately from the strong transitivity of the $\mathcal{G}^n_{A,cpi}$ components. To check that $\mathcal{G}^6_{A,cpi}$ has this property, select any triplet from Fig. 2b—perhaps $\{A_1, A_5, A_3\}$; the goal is to show that $d_{1,5} + d_{5,3} = d_{1,3}$. From $\mathcal{G}^6_{A,cpi}$, this requires $d_{1,5} + d_{5,3} = -5 + 11$ to equal $d_{1,3} = 6$, which it does.

A standard induction exercise shows that if $\mathbf{d}^n_{A,st}$ is strongly transitive, then *any connected path* from A_j to A_k, no matter how it may meander, has the same length as the direct $A_j \to A_k$ path. Reversing this last arc creates a closed path with length zero. The rest of the theorem follows immediately.

The last assertion of Thm 9 merely uses that a path's length in $\mathcal{G}^n_A$ is the sum of its lengths in $\mathcal{G}^n_{A,cpi}$ and $\mathcal{G}^n_{A,cyclic}$. But the length of the curve in $\mathcal{G}^n_{A,cpi}$ equals the length of the arc connecting the first and last vertices. For instance, the "-1" length of $\mathcal{G}^6_A$ path,

$$A_1 \xrightarrow{-3} A_2 \xrightarrow{11} A_4 \xrightarrow{2} A_3 \xrightarrow{-11} A_2 \xrightarrow{0} A_5,$$

which skips A_6 but visits A_2 twice, equals this path's length in $\mathcal{G}^6_{A,cyclic}$ plus -5. The -5 comes from the $A_1 \xrightarrow{-5} A_5$ arc in $\mathcal{G}^6_{A,cpi}$ (Fig. 2b). This path's length in $\mathcal{G}^6_{A,cyclic}$ is

$$A_1 \xrightarrow{2} A_2 \xrightarrow{2} A_4 \xrightarrow{0} A_3 \xrightarrow{0} A_2 \xrightarrow{0} A_5$$

or 4, so, as Thm. 9 promises, $-1 = 4 - 5$.

3.1. Finding the longest and shortest Hamiltonian circuits. According to Theorem 9, the longest/shortest Hamiltonian path in $\mathcal{G}^6_A$ (Fig. 2a) can be found by ignoring $\mathcal{G}^6_A$ and, instead, analyzing the reduced $\mathcal{G}^6_{A,cyclic}$ with its two three-cycles (Fig. 3) $A_1 \xrightarrow{3} A_5 \xrightarrow{3} A_3 \xrightarrow{3} A_1$ and $A_1 \xrightarrow{2} A_2 \xrightarrow{2} A_4 \xrightarrow{2} A_1$. The obvious strategy is to use these cycles as fully as possible. To avoid returning to a previously visited vertex, use at most two arcs from each three-cycle.

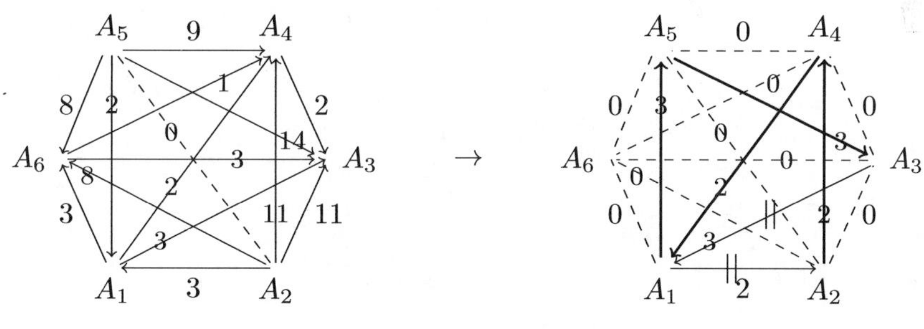

FIGURE 3. Finding $\mathcal{G}^6_A$'s longest Hamiltonian circuit via $\mathcal{G}^6_{A,cyclic}$.

Consequently, starting at A_1, use only the edges $A_1 \xrightarrow{3} A_5 \xrightarrow{3} A_3$ from the first three-cycle; this is why $A_3 \xrightarrow{3} A_1$ arc is crossed out in Fig. 3. Only one arc can leave A_1, which eliminates the crossed-out $A_1 \xrightarrow{2} A_2$ arc. Thus, the Hamiltonian circuit depends upon the four bold $\mathcal{G}^6_{A,cyclic}$ arcs in Fig. 3. Finding the longest Hamiltonian circuit now is immediate—use zero-length arcs to ensure all vertices are visited and the four basic ones are connected. An answer is the Hamiltonian path

$$(A_1 \xrightarrow{3} A_5 \xrightarrow{3} A_3) \xrightarrow{0} A_6 \xrightarrow{0} (A_2 \xrightarrow{2} A_4 \xrightarrow{2} A_1)$$

of length 10. The reversal of this closed path has length -10. It is clear from the graph's structure that these are the longest and shortest $\mathcal{G}^6_{A,cyclic}$ Hamiltonian circuits, so (by Thm. 3) they define the longest and shortest $\mathcal{G}^6_A$ Hamiltonian circuits with the same lengths.

To appreciate what is going on, what complicates computing the Fig. 2a path length $A_5 \xrightarrow{14} A_3 \xrightarrow{-3} A_1 \xrightarrow{-2} A_5$ of $14-3-2=9$ are the subtractions/cancellations. To see what they are, let u, v, and w be the unknown terms to be canceled, respectively, for arcs $\widehat{A_5A_3}, \widehat{A_3A_1}$, and $\widehat{A_1A_5}$. That is, after cancellations, the path length computation would be $(14-u)+(-3-v)+(-2-w)=9$ where $u+v+w=0$, which satisfies cpi from Eq. 1.8. That is, the cancellations in computing closed path lengths are linear expressions that define a $\mathbb{ST}^n_A$ graph. The optimal choice of cancellations for a $\mathcal{G}^n_A$ is the $\mathbb{ST}^n_A$ graph that most closely approximates $\mathcal{G}^n_A$; this is its orthogonal projection into $\mathbb{ST}^n_A$, or $\mathcal{G}^n_{A,cpi}$. Indeed, with Fig. 2b, the optimal cancellation is from $\mathcal{G}^6_{A,cpi}$ where $u+v+w=11-6-5=0$. Removing these cancellations leaves $\mathcal{G}^6_{A,cyclic}$ with the path length $A_5 \xrightarrow{3} A_3 \xrightarrow{3} A_1 \xrightarrow{3} A_5$ of 9 where no modifications are needed for the computations.

3.2. The symmetric case. For the standard symmetric TSP, the distance between vertices is the same in each direction; for instance, this may be the mileage between two cities, which agrees in both directions. Thus the $d_{i,j}=-d_{j,i}$ asymmetric condition is replaced with $d_{i,j}=d_{j,i}$, which creates a different $\mathbb{R}^{\binom{n}{2}}_S$ decomposition along with different inherent symmetries. Of importance, the decomposition provides a best possible simplification for closed paths. Details for the general symmetric and asymmetric cases are given elsewhere [**13**], but a flavor of what happens is described next.

An $n=6$ example is given in Fig. 4; recall, the S subscript in $\mathcal{G}^6_S$ refers to "symmetric." While a typical goal is to find the shortest Hamiltonian circuit, everything extends to the analysis of any connected path, whether closed or not.

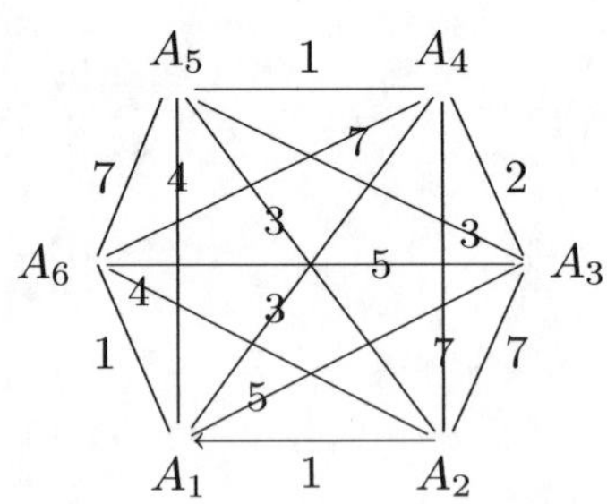

FIGURE 4. A symmetric $\mathcal{G}^6_S \in \mathbb{G}^6_S$.

Central to the discussion is the average Hamiltonian path length in $\mathcal{G}^n_S$ denoted by $T(\mathcal{G}^n_S)$.

DEFINITION 3. For a graph $\mathcal{G}^n_S$, let $S_S(A_j)$ be the sum of arc lengths attached to vertex A_j, $j=1, \dots, n$. Let $T(\mathcal{G}^n_S)=\frac{1}{n-1}\sum_{j=1}^n S_S(A_j)$.

The definitions of $S_S(A_j)$ (Def. 3) and $S_A(A_j)$ (Eq. 2) agree. That $T(\mathcal{G}^n_S)$ is the average Hamiltonian path length follows from the fact that $\frac{1}{n-1}S_S(A_j)$ is the average length of the $(n-1)$ arcs attached to A_j. Illustrating with Fig. 4, $S_S(A_1) = 1+5+3+4+1 = 14, S_S(A_2) = 22, S_S(A_3) = 22, S_S(A_4) = 20, S_S(A_5) = 18, S_S(A_6) = 24$, so $T(\mathcal{G}^6_S) = \frac{1}{5}(120) = 24$. When considering non-Hamiltonian closed paths, restrict the $S_S(A_j)$ values and the summation defining T to the relevant arcs and vertices.

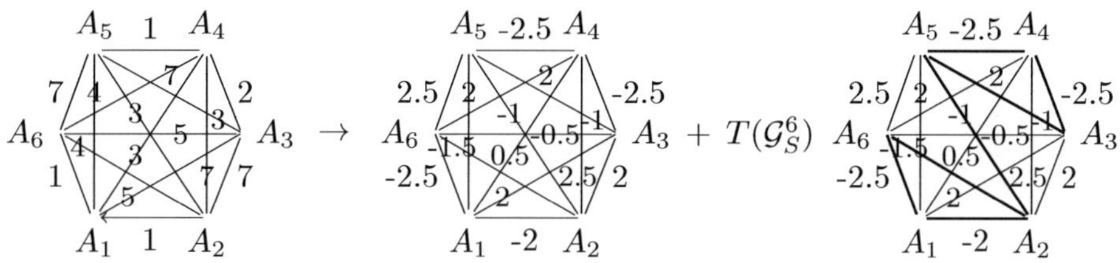

a. $\mathcal{G}^6_{S,cyclic}$ b. $\mathcal{G}^6_{S,cyclic} \in \mathbb{C}^6_S$ c. Below average links

FIGURE 5. Decomposition of a $\mathcal{G}^6_S \in \mathbb{G}^6_S$.

The decomposition for the symmetric setting replaces $\mathcal{G}^n_S$ with $\mathcal{G}^n_{S,cyclic}$, which turns out to be its sharpest possible reduction. It does so by replacing weights in $\mathcal{G}^n_S$ with values that are the differences from $\mathcal{G}^n_S$'s average arc lengths; an example is Fig. 5b. This reduction means that $\mathcal{G}^n_{S,cyclic}$'s negative values are "smaller than average" costs. The theorem is that *the length of a Hamiltonian circuit in* $\mathcal{G}^n_S$ *is the length of its path in* $\mathcal{G}^n_{S,cycle}$ *plus the average* $T(\mathcal{G}^n_S)$. Standard modifications handle paths that are not closed and/or incomplete graphs.

Each $\mathcal{G}^6_{S,cyclic}$ vertex has an arc with negative (i.e., below average) length; this holds for all vertices of any $\mathcal{G}^n_{S,cyclic}$. By observation and using arcs with negative values as often as possible, an Hamiltonian path in $\mathcal{G}^6_{S,cyclic}$ with shortest length of -11 is

$$A_1 \xrightarrow{-2.5} A_6 \xrightarrow{-0.5} A_3 \xrightarrow{-2.5} A_4 \xrightarrow{-2.5} A_5 \xrightarrow{-1} A_2 \xrightarrow{-2} A_1, \tag{3.1}$$

which defines the shortest Hamiltonian path in $\mathcal{G}^6_S$ with length $24 - 11 = 13$.

In general, the shortest $\mathcal{G}^n_S$ Hamiltonian path is bounded below by $T(\mathcal{G}^n_S)$ plus the sum of the n smallest $\mathcal{G}^n_{S,cyclic}$ arc lengths. With $\mathcal{G}^6_S$, this is $24 + (-2.5 - 2.5 - 2.5 - 2 - 1.5 - 1) = 12$. The Eq. 3.1 shortest Hamiltonian path length is larger than, but close to, this lower bound.

Although the $\mathcal{G}^6_S$ entries can be random numbers, an advantage of $\mathcal{G}^6_{S,cyclic}$ is that each entry has a meaning with respect to a path's length. Thus, a way to find the shortest Hamiltonian path is to rank the arcs according to length, where "smaller is better." Start with the first n arcs to determine whether they form a path; by doing so geometrically, these are the bold arcs in Fig. 5c. If a Hamiltonian path is not defined (as true in Fig. 5c), then iteratively add the smaller remaining arcs until a Hamiltonian path emerges. The Fig. 5c bold arcs emphasize using $\widehat{A_1A_6}, \widehat{A_3A_4}, \widehat{A_4A_5}$, where $\widehat{A_3A_6}$ forms a transition between the first and the second arc, and $\widehat{A_5A_2}$ includes the missing A_2, while $\widehat{A_2A_1}$ completes the journey. This is the Eq. 3.1 tour.

4. Conclusion

Figure 6 summarizes much of what was discussed. Namely, both the voting problem and TSP can be described in terms of a point $\mathbf{d}^n_A \in \mathbb{R}^{\binom{n}{2}}_A$. This space can be orthogonally decomposed into the strongly transitive linear subspace $\mathbb{ST}^n_A$ and the linear subspace $\mathbb{C}^n_A$ that consists of three-cycles. The $\mathbf{d}^n_{A,st}$ component of $\mathbf{d}^n_A$ in the $\mathbb{ST}^n_A$ direction resolves several concerns from voting theory, but it frustrates the analysis of TSP paths. Conversely, $\mathbf{d}^n_A$'s component in the $\mathbb{C}^n_A$ direction, $\mathbf{d}^n_{A,cyclic}$, determines all closed path properties for the TSP, but it erects barriers and introduces paradoxes for voting theory. A resolution for the difficulties of both topics is to project $\mathbf{d}^n_A$ into the appropriate subspace for the problem being considered.

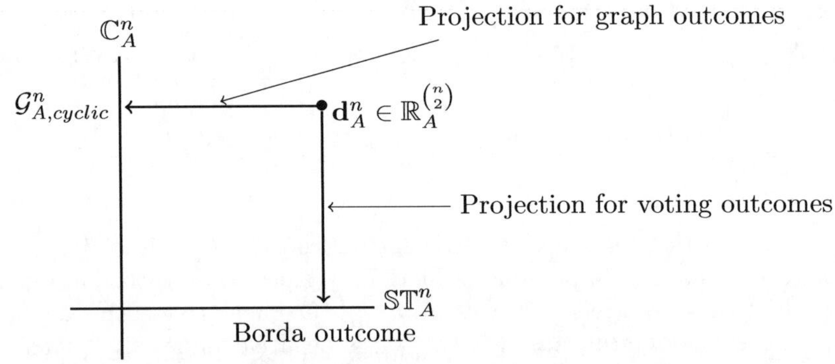

FIGURE 6. Resolutions via projections.

To introduce the rest of the material in this concluding section, let me brag about my two granddaughters. Heili has a deep interest in neurobiology and gymnastics, while Tatjana is heavily involved in the social sciences and ballet. Who is older? It is impossible to tell from this information; comparing ages requires having relevant inputs. More generally, to realize any specified objective, for the data to be useful, they must, in some manner, be directed toward the stated purpose. This truism extends to voting theory and the TSP.

To see this with Arrow's result, return to the introductory voting example where the alternatives are three cities. Suppose the 25 voters with the $A_1 \succ A_2 \succ A_3$ ranking judge cities according the ease of finding parking, the 23 who prefer $A_2 \succ A_3 \succ A_1$ evaluate cities based on the popularity of local professional basketball teams, and the 22 preferring $A_3 \succ A_1 \succ A_2$ assess them in terms of available types of craft beer. Thus the $A_1 \succ A_2$ outcome by 47:23 is a conglomeration of attitudes about parking, basketball, and beer. Of importance, *nothing* is directed toward Arrow's explicitly stated objective of having a transitive ranking. Consequently (with these "apple and orange" comparisons), rather than a surprise, non-transitive outcomes should be anticipated. Indeed, expect a transitive outcome *only if* the inputs contribute toward this transitivity target. This comment leads to the decomposition (Fig. 6); the $\mathbf{d}^n_{A,cyclic}$ component of $\mathbf{d}^n_A$, which runs counter to the stated objective of transitivity, is identified and dismissed. This leads to the approach described in the first paragraph of the proof of Thm. 5, which, in reference [**8**], is called the "summation method."

Another way to handle this problem is to express the inputs, the preferences of voters, in terms of the transitivity objective. To do so, describe the $A_1 \succ A_2 \succ A_3$ ranking as $(A_1 \succ A_2, 1), (A_1 \succ A_3, 2)$ and $(A_2 \succ A_3, 1)$, which restates this ranking in terms of pairs that now are in a strongly transitive format. Here $d_{1,2} = 1$ captures that A_1 is ranked one spot above A_2, while $d_{1,3} = 2$ means that A_1 is ranked two spots above A_3. Choosing the triplet (A_1, A_2, A_3) requires checking whether $d_{1,2} + d_{2,3} = 1 + 1$ equals $d_{1,3} = 2$, which it does. By using this strongly transitive approach, which is called "Intensity of IIA" or IIIA (introduced in [**8**]), the profile information for $\{A_1, A_3\}$ has 25 voters with $(A_1 \succ A_3, 2)$, 23 with $(A_1 \succ A_3, -1)$ and 22 with $(A_1 \succ A_3, -1)$, The IIIA tally for $\{A_1, A_3\}$ sums the products of the number of voters with each ranking times its intensity. Here the tally is $25(2) + 23(-1) + 22(-1) = 5$ leading to $A_1 \succ A_3$, rather than the above $A_3 \succ A_1$ (Eq. 1.2) that forced a cycle. Even stronger, the IIIA tallies over the three pairs not only are transitive, they are strongly transitive.

With any number of alternatives, the IIIA outcome for the majority vote always is strongly transitive. It must be because $\mathbb{ST}^n_A$ is a linear subspace, and the outcome is a summation of strongly transitive profiles. This IIIA method is equivalent to the Borda Count [**8**]; a conclusion that should be anticipated (particularly with Thm. 6). As shown in [**8**], by replacing IIA with IIIA, Arrow's Theorem now has a positive conclusion. That is, by using data that is consistent with the objective of transitivity causes the problems of Arrow's Theorem to disappear.

Similarly with TSP, each difference from the average cost between two vertices, as catalogued with $\mathcal{G}^6_A$ (Fig. 1), could be based on different attributes; e.g., the $d_{1,2}$ difference between A_1 and A_2 might reflect the topography while $d_{2,3}$ between A_2 and A_3 could be caused by traffic restrictions. As the objective concerns path lengths of closed curves, the goal must be to use inputs that contribute to the specified goal. Here the strongly transitive component $\mathbf{d}^n_{A,cpi}$ pushes the outcome in a linear fashion that conflicts with the goal of having circular paths. Dismissing this $\mathbf{d}^n_{A,cpi}$ term (as indicated in Fig. 6) leaves $\mathbf{d}^n_{A,cyclic}$ data, which are consistent with the global objective of finding closed path properties.

5. Proofs

Beyond proving Thm. 3, an intent of this section is to demonstrate the above tools.

Proof (*Thm.* 3). Assume that the paired rankings defined by $\mathbf{d}^n$ have A_1 and A_n, respectively, the unique Condorcet winner and loser. As A_1 is the unique Condorcet winner, the associated $\mathbf{d}^n_A$ must have $d_{1,k} > 0$ for $k = 2, \dots, n$. Similarly, as A_n is the unique Condorcet loser, it must be that $d_{n,k} < 0$ for all $k = 1, \dots, n-1$. From Def. 2, this means that $S_A(A_1) > 0$ and $S_A(A_n) < 0$. According to Eq. 2.6, A_1 is Borda ranked above A_n. As the Borda ranking is the $\mathbf{d}^n_{A,st}$, ranking, this proves the theorem. Notice, the Condorcet uniqueness conditions are unnecessary conveniences. □

It remains to show for $n = 3, 4$ that if A_1 and A_n are, respectively, the top and bottom ranked alternative for $\mathbf{d}^n_{A,st}$, and if $\mathbf{d}^n_A = \mathbf{d}^n_{A,st} + \mathbf{d}^n_{A,cyclic}$ is transitive, then $A_1 \succ A_n$. The approach uses the fact, which follows from strong transitivity, that $d_{1,n}$ for $\mathbf{d}^n_{A,st}$ is an upper bound for all other $d_{s,u}$ values. This size of $d_{1,n}$ requires the cyclic perturbations to affect and reverse smaller $d_{u,s}$ values, which violates transitivity before they can impact on the $d_{1,n}$ term to reverse $A_1 \succ A_n$. Assume the $\mathbf{d}^n_{A,st}$ ranking is $A_1 \succ A_2 \succ \cdots \succ A_{n-1} \succ A_n$.

The proof is immediate for $n = 3$. All cyclic $n = 3$ vectors are α multiples of $(1, -1; 1)$, so $\mathbf{d}^3_A = \mathbf{d}^3_{A,st} + \mathbf{d}^3_{A,cyclic} = (d_{1,2} + \alpha, d_{1,3} - \alpha; d_{2,3} + \alpha)$. If $\alpha \geq 0$, then the $A_1 \succ A_2$ and $A_2 \succ A_3$ rankings remain unchanged. The $A_1 \succ A_3$ ranking persists as long as $d_{1,3} - \alpha > 0$. As soon as $\alpha > d_{1,3}$, the $A_1 \succ A_3$ ranking reverses to become $A_3 \succ A_1$, which converts the set of pairwise rankings from transitive to cyclic.

For all $\alpha < 0$ values, $A_1 \succ A_3$. But the system becomes cyclic as soon as $-\alpha$ equals the second largest of $\{d_{1,2}, d_{2,3}\}$. For instance, suppose $d_{1,2} < d_{2,3}$. For $\alpha = -d_{1,2}$, the rankings are $A_1 \succ A_3, A_2 \succ A_3$, and $A_2 \sim A_1$ defining the transitive $A_1 \sim A_2 \succ A_3$. But once $\alpha = -d_{2,3}$, the rankings are the non-transitive $A_1 \succ A_3, A_3 \sim A_2, A_2 \succ A_1$. Stated differently, if $\mathbf{d}^3_A$ has a transitive ranking, then $A_1 \succ A_3$, which proves the assertion.

The theme of the proof for $n \geq 4$ is that, because of strong transitivity, $d_{1,n}$ from $\mathbf{d}^n_{A,st}$ is so large that before the cyclic components can reverse the $A_1 \succ A_n$ ranking, they change enough rankings of other pairs to define a cyclic outcome that violates the transitivity of $\mathbf{d}^n_A$. One approach follows:

(1) Assume that the cyclic terms force $A_n \succ A_1$. It must be shown that the pairs do not define a transitive outcome.
(2) The size of the perturbations of cyclic terms that accompany the $A_n \succ A_1$ assumption requires some other alternative, say A_k, to satisfy $A_k \succ A_n$. If $\mathbf{d}^n_A$ is transitive, as assumed, then $A_k \succ A_n \succ A_1$, and $A_k \succ A_1$.
(3) The size of the cyclic perturbations that cause $A_k \succ A_1$ forces some other alternative, A_j, to satisfy $A_j \succ A_k$. This requires $A_j \succ A_k \succ A_n \succ A_1$, or $A_j \succ A_1$.
(4) The next step is to show that this $A_j \succ A_1$ ranking drives the size of certain cyclic terms to be large enough so that $A_n \succ A_j$.
(5) Step 4 is the sought contradiction. The $A_j \succ A_n$ ranking follows from the assumption of the transitivity of $\mathbf{d}^n_A$, and the conflicting $A_n \succ A_j$ is a direct result of computations based on the impacts of the cyclic terms. Thus the assumption that $\mathbf{d}^n_A$ is transitive is false, which proves the conclusion.

To illustrate this program with $n = 4$, all of the following $d_{i,j}$ values come from $\mathbf{d}^4_{A,st}$. Assume that $\mathbf{d}^4_A$ is transitive. Using the Thm. 2 basis for $\mathbb{C}^4_A$, where $\alpha_{j,k}$ is the coefficient for $\mathbf{d}^4_{1,j,k}$, a representation for $\mathbf{d}^4_A = \mathbf{d}^4_{A,st} + \mathbf{d}^4_{A,cyclic}$ is

$$(d_{1,2} + \alpha_{2,3} + \alpha_{2,4}, d_{1,3} - \alpha_{2,3} + \alpha_{3,4}, d_{1,4} - \alpha_{24} - \alpha_{3,4}; \quad d_{2,3} + \alpha_{2,3}, d_{2,4} + \alpha_{2,4}, d_{3,4} + \alpha_{3,4}). \tag{5.1}$$

According to Eq. 5.1, if $A_4 \succ A_1$, then $d_{1,4} - \alpha_{2,4} - \alpha_{3,4} < 0$, or $0 < d_{1,4} < \alpha_{2,4} + \alpha_{3,4}$. This forces one or both of $\alpha_{3,4}$, $\alpha_{2,4}$ to be positive.

STEP 2. Start with the assumption that $\alpha_{3,4} > 0$, which requires from Eq. 5.1 that $A_3 \succ A_4$. From the transitivity of $\mathbf{d}^4_A$, this requires $A_3 \succ A_1$, or $d_{1,3} - \alpha_{2,3} + \alpha_{3,4} < 0$, which leads to the inequality

$$0 < d_{1,3} + \alpha_{3,4} < \alpha_{2,3}. \tag{5.2}$$

STEP 3. Combining $\alpha_{2,3} > 0$ (Eq. 5.2) with $\mathbf{d}^4_A$'s transitivity (and Eq. 5.1), requires $A_2 \succ A_3 \succ A_4 \succ A_1$. The $A_2 \succ A_1$ ranking forces (Eq. 5.1) $\alpha_{2,3} + \alpha_{2,4} <$

$-d_{1,2} < 0$, which, with Eq. 5.2, becomes

$$\alpha_{2.4} < -d_{1,2} - \alpha_{2,3} < -d_{1,2} - d_{1,3} - \alpha_{3,4} < 0. \tag{5.3}$$

STEP 4. As $\alpha_{2,4} < 0$ (Step 3), it follows from $A_4 \succ A_1$ and $d_{1,4} - \alpha_{2,4} - \alpha_{3,4} < 0$ that

$$\alpha_{3,4} > d_{1,4} \geq d_{j,k}. \tag{5.4}$$

The last inequality follows because $d_{1,4}$ is the largest $d_{j,k}$ value (strong transitivity).

STEP 5. By using Eq. 5.3, the computation for the $\{A_2, A_4\}$ ranking depends on the sign of $d_{2,4} + \alpha_{2,4} < -d_{1,2} - d_{1,3} + [-\alpha_{3,4} + d_{2,4}]$. Because $-\alpha_{3,4} + d_{2,4} < 0$ (Eq. 5.4), it follows that $d_{2,4} + \alpha_{2,4} < 0$, which is $A_4 \succ A_2$ and the desired contradiction.

The remaining case of $\alpha_{2,4} > 0, \alpha_{3,4} \leq 0$ is simpler because from $A_4 \succ A_1$ (Eq. 5.1)

$$\alpha_{2,4} > d_{1,4} \geq d_{j,k}. \tag{5.5}$$

STEP 2. According to Eq. 5.1, the $\alpha_{2,4} > 0$ inequality and the transitivity of $\mathbf{d}^4$ mandates $A_2 \succ A_4 \succ A_1$, or from $A_2 \succ A_1$ that $d_{1,2} + \alpha_{2,3} + \alpha_{2,4} < 0$. or that $\alpha_{2,3} < -d_{1,2} - \alpha_{2,4} < 0$.

STEP 3. Using this inequality and Eq. 5.5, the $\{A_2, A_3\}$ ranking equation is determined by $d_{2,3} + \alpha_{2,3} < -d_{1,2} + [d_{2,3} - \alpha_{2,4}] < 0$, or $A_3 \succ A_2 \succ A_4 \succ A_1$.

STEP 4. According to the $A_3 \succ A_1$ ranking, $d_{1,3} - \alpha_{2,3} + \alpha_{3,4} < 0$ or (Step 2) $d_{1,3} + \alpha_{3,4} < \alpha_{2,3} < -d_{1,2} - \alpha_{2,4} < 0$. What follows uses the inequality $\alpha_{3,4} < -d_{1,3} - d_{1,2} - \alpha_{2,4}$.

STEP 5. The $\{A_3, A_4\}$ ranking depends on the sign of $d_{3,4} + \alpha_{3,4} < -d_{1,2} - d_{1,3} + [-\alpha_{2,4} + d_{3,4}] < 0$. As the term in the brackets is negative, the value is negative, leading to the $A_4 \succ A_3$ contradiction. This completes the proof for $n = 4$.

Acknowledgments

This paper is a written version of my 4/8/2022 JMM presentation; these results provide applications of conclusions in [**12, 13**]. My thanks to George Hazelrigg for our several discussions of these topics.

References

[1] K. J. Arrow, *Social Choice and Individual Values*, Wiley, New York, 1951 (2nd ed, 1963).

[2] Charles Lutwidge Dodgson, *The political pamphlets and letters of Charles Lutwidge Dodgson and related pieces: A mathematical approach*, The Pamphlets of Lewis Carroll, vol. 3, Lewis Carroll Society of North America, New York; distributed by University Press of Virginia, Charlottesville, VA, 2001. Compiled, with introductory essays, notes, and annotations, by Francine F. Abeles. MR1887851

[3] J. Kemeny, *Mathematics without numbers*, Daedalus **88** (1959): 571-591.

[4] Thomas C. Ratliff, *A comparison of Dodgson's method and Kemeny's rule*, Soc. Choice Welf. **18** (2001), no. 1, 79–89, DOI 10.1007/s003550000060. MR1818578

[5] Thomas C. Ratliff, *A comparison of Dodgson's method and the Borda count*, Econom. Theory **20** (2002), no. 2, 357–372, DOI 10.1007/s001990100218. MR1925049

[6] Thomas C. Ratliff, *Some startling inconsistencies when electing committees*, Soc. Choice Welf. **21** (2003), no. 3, 433–454, DOI 10.1007/s00355-003-0209-y. MR2023718

[7] Donald G. Saari, *A dictionary for voting paradoxes*, J. Econom. Theory **48** (1989), no. 2, 443–475, DOI 10.1016/0022-0531(89)90037-9. MR1013282

[8] Donald G. Saari, *Basic Geometry of Voting*, Springer, New York, 1995.
[9] Donald G. Saari, *Mathematical structure of voting paradoxes. I. Pairwise votes*, Econom. Theory **15** (2000), no. 1, 1–53, DOI 10.1007/s001990050001. MR1731508
[10] Donald G. Saari, *Mathematics motivated by the social and behavioral sciences*, CBMS-NSF Regional Conference Series in Applied Mathematics, vol. 91, Society for Industrial and Applied Mathematics (SIAM), Philadelphia, PA, 2018, DOI 10.1137/1.9781611975185.ch1. MR3767233
[11] Donald G. Saari, *Arrow, and unexpected consequences of his theorem*, Public Choice, **179** (2019), 133-144.
[12] Donald G. Saari, *Seeking consistency with paired comparisons: A systems approach*, Theory and Decision **90** (2021), 1-26
[13] Donald G. Saari, *Inherent symmetries of graphs, paths, and Traveling Salesperson Problems*, UCI preprint. Jan. 2022. `arXiv:4275433` (25 April, 2022)
[14] Donald G. Saari and Vincent R. Merlin, *A geometric examination of Kemeny's rule*, Soc. Choice Welf. **17** (2000), no. 3, 403–438, DOI 10.1007/s003550050171. MR1762588

Departments of Mathematics and of Economics, University if California, Irvine, California 92697-5100

Current address: 20 Harvey Ct., Irvine, California 92617

Email address: `dsaari@uci.edu`

Contemporary Mathematics
Volume **795**, 2024
https://doi.org/10.1090/conm/795/15970

An iterative procedure for apportionment and its use in the 2016 Georgia Republican primary

Michael A. Jones, David McCune, and Jennifer Wilson

ABSTRACT. For the 2016 U.S. presidential primary, the Georgia Republican Party created a new method, which we call the *Iterated Lower Quota* (ILQ) method, to apportion delegates using a series of rounds that is reminiscent of a fair division procedure. We analyze the ILQ method and compare it to Hamilton's method (the most common method of delegate apportionment) using simplicial geometry. We use the geometry to estimate the likelihood that the two methods agree. To understand the bias exhibited by the ILQ method towards the strongest candidates, we calculate the threshold values for ILQ and estimate the expected deviation of the allocations from quota. We determine the maximum number of rounds required to execute the ILQ method, and investigate the apportionment paradoxes to which ILQ is susceptible. In practice, before using the ILQ method, the Georgia Republican Party uses cutoffs to eliminate candidates with low vote totals. We explain how these cutoffs, also used in the Democratic Party's primaries and in other Republican state primaries, affect the theoretical results on ILQ. We conclude by considering the advantages and disadvantages of ILQ in the context of delegate allocation.

1. Introduction

The Republican and Democratic Presidential Primaries are a series of state contests held every four years prior to the U.S. presidential election in November. The purpose of the primaries is to select each party's nominee to run in the general election. Each state holds a primary (or caucus) in which individual voters can select among the party's candidates. Each candidate receives a number of delegates based on the outcome of the contest. For both the Democratic and Republican primaries, the candidate that receives a majority of the delegates, by adding the number of delegates allocated in each state, wins the party's nomination and then represents the party in the general election. In what follows we provide a brief history of how delegate allocation has evolved and explain why the focus of this paper is on the allocation procedure used by the 2016 Georgia Republican party due to its unique mathematical properties.

The current processes by which the two parties select their candidates date from the late 1960's. Prior to this, decisions were largely made during national conventions, where compromises were brokered among party officials and leaders

2020 *Mathematics Subject Classification.* Primary 91B32; Secondary 91B12, 91F10.
Key words and phrases. Apportionment, 2016 Republican primary, Iterated Lower Quota, Hamilton's method.

who had influence over large blocs of state delegates [**9**]. The number of statewide primaries fluctuated, and although candidates often ran in select primaries, the results did not have a sizable influence on the final nomination. The Democratic Party was first to enact significant changes to this process, spurred to action in part by demonstrations at the 1968 Democratic National Convention. Among these changes was a switch from state caucuses to more state primaries and, after 1976, a ban on winner-take-all rules in favor of proportional allocation with a 10% cutoff—in which candidates receiving less than 10% of the popular vote were allowed no delegates. In 1990, the Democratic National Committee's Rules and Bylaws Committee instituted a uniform 15% cutoff for all states, and mandated how the proportional allocation should be implemented (using what is called Hamilton's method in the U.S.). These rules remain in place up to the current time.

The Republican Party was slower to amend its rules, often changing them in response to Democratic Party changes. Over time, the number of states holding primaries (rather than caucuses) increased in both parties, and the organization of delegates became more standardized. In 1976, the Republican Party gave each state a fixed number of delegates per congressional district as well as a number of at-large delegates and bonus delegates based on the results of recent elections. Besides the number of delegates, the national party let the state parties decide most of the details for how the delegates were to be allocated; the state parties largely continued to use winner-take-all rules to allocate the delegates.

After the 2008 presidential election, the Republican Temporary Delegate Selection Committee enacted more sweeping changes in time for the 2012 presidential primary. These included a condition that all states contests be held in March or later, with the exception of four states (Iowa, New Hampshire, Nevada and South Carolina). States holding primaries or caucuses before April were required to use a method of proportional allocation to award their state-wide delegates, but could still use winner-take-all for district-wide delegates. States holding primaries after April 1 could continue to use whatever method they desired. Any state not abiding by these rules was penalized by losing 50% of their delegates.

Unlike in the Democratic party, the new Republican rules did not specify a particular method for proportional allocation. Suggestions for "rounding rules" were given, but states were free to devise their own rules. In Georgia, the state Republican party which, prior to 2012, had used a winner-take-all approach, adopted a method for proportional allocation which they modified for the 2016 primary. In this paper, we analyze the delegate allocation process used by the Georgia Republican Party during the 2016 primary. (In 2020, the state party delayed the primary until June because of the Covid pandemic and reverted to a de facto winner-take-all election in which incumbent President Trump was the only candidate on the ballot; see Rule 7B of [**23**].)

The evolution of delegate selection rules in both parties has been the subject of extensive research with respect to its effect on elections [**1, 13, 25**]. Additionally, Meinke, Staton and Wuhs [**21**] examined an empirical model to explain why parties might be motivated to adopt more open processes. However, the mathematical analysis of delegate allocation rules has been more limited. Geist, Jones, and Wilson [**11**] examine the effect of using Hamilton's method in the Democratic Party Delegate Selection Rules, and we consider the effect of cutoffs on this selection process in [**14**]. While in [**15**], we survey the seven proportional delegate apportionment rules—including the method used by the 2016 Georgia Republican Party—used in Republican state primaries, focusing on bias toward the top and bottom vote-getting candidates. Several Republican state parties invented apportionment methods which had not been previously analyzed. In our view, the apportionment method invented by the 2016 Georgia Republican party is the most mathematically interesting of these new methods. The purpose of this article is to provide an in depth mathematical analysis of this new method. Such a detailed analysis was not considered in [**15**] and is not considered in the broader work on both Democratic and Republican delegate apportionment in [**16**].

In countries with proportional representation, apportionment methods are often used to determine the number of seats assigned to each political party based on the number of votes each party receives; Pukelsheim [**24**] provides a thorough introduction to the theory of apportionment in this context. In the U.S., apportionment methods have been used to determine the number of seats a state receives in the House of Representatives based on each state's population; Balinski and Young [**4**] provide the definitive treatment of apportionment in this context. For allocating house members, there has been an emphasis on using apportionment methods that advantage neither large nor small states; favoring states based on their population (large or small) has been referred to as *bias* in the apportionment literature.

Apportionment methods used in delegate selection arguably have a different objective than methods used to determine representation in governmental bodies. For primaries, a favorable bias toward the vote leader, corresponding to a state with a large population in the house seat context, and the inclusion of cutoffs to help consolidate support for more popular candidates may be desirable. As considered in [**15**], there is a balance needed between selecting a nominee quickly so that the nominee can focus on the general election and allowing the party to hone their platform positions to better attract and represent voters.

When used to allocate house seats, the apportionment problem is often described as a problem of rounding fractional quotas to whole numbers of seats to meet the fixed number of seats in the house. The method created by the 2016 Georgia Republican party to allocate delegates is fundamentally different: rather than rounding decimals to whole numbers to attain a fixed number of delegates, an initial rounding is determined, and then any remaining delegates are allocated in a series of rounds until no delegates are left; we call this the Iterated Lower Quota (ILQ) method. For this reason, the method used in the 2016 Georgia Republican presidential primary is similar to fair division procedures, such as cake-cutting procedures, in which an initial allocation is awarded and a remaining piece is then allocated (these include, for example, Selfridge and Conway's trimming procedure, as described by Brams and Taylor [**8**]). When implemented in the 2016 primary, the ILQ method was used in conjunction with vote cutoffs, which are minimal vote shares a candidate must surpass to be eligible to receive any delegates. We refer to the ILQ method paired with the cutoffs in the 2016 Georgia Republican primary as the GA method.

The paper is organized as follows. In Section 2 we provide our notation and foundational definitions, including formal definitions of the ILQ and GA methods. In Section 3 we adapt the geometric interpretation of apportionment for three candidates used by Bradberry [**5**], Lucas [**20**], and Balinski and Young [**4**] to visualize the set of possible apportionments in the two-dimensional simplex (2-simplex). This visualization will be used throughout the rest of the paper. In Section 4 we compare the ILQ method to Hamilton's method, which is the most commonly used method for apportioning delegates. To examine the bias of the ILQ method, in Section 5, we apply the standard definition of the thresholds of representation (the thresholds of inclusion and exclusion, as considered by Lijphart and Gibberd [**19**]). In Section 6, we provide an alternative avenue for examining bias by analyzing the tendency with which the ILQ method favors the top candidates, focusing on a measure of bias developed by Schuster et al. [**29**]. In Section 7 we consider the mechanics of the ILQ method to determine the maximum number of rounds that may occur, given a fixed number of candidates. In Section 8 we examine the susceptibility of the ILQ method to various paradoxes, including classical paradoxes such as the Alabama and population paradoxes, as discussed in [**4**], and also newer paradoxes such as the elimination paradox [**14**]. Section 9 concludes the paper with a discussion of the 2016 Georgia Republican presidential primary.

2. Preliminaries

We first define delegate apportionment methods and then describe the ILQ method, the GA method—the version of the ILQ method with cutoffs that was used in the 2016 Georgia Republican primary—and Hamilton's method. An apportionment method is a rule for allocating a number of objects among a set of constituents usually in a "proportional" way. The objects, constituents, and criterion for measuring proportionality all vary depending on the application, however the crux of the difficulty is that the objects to be distributed are discrete. In the context of the presidential primaries, discrete delegates must be allocated to the presidential candidates based on each candidate's rational proportion of the vote in the state primary.

Formally, let D be the number of delegates to be awarded among n candidates $C_1, \ldots, C_n$. Assume C_i receives $v_i > 0$ votes where the candidates are ordered so that $v_1 \geq v_2 \geq \cdots \geq v_n$. Let $V = \sum_{i \leq n} v_i$ be the total number of votes, $p_i = v_i/V$ be C_i's percentage of the vote and $q_i = Dp_i$ be C_i's *quota*: this is the fraction of the delegate pool C_i should receive if delegates were divisible. Candidate C_i's *lower quota* is denoted $\lfloor q_i \rfloor$ and is equal to the greatest integer less than or equal to q_i; $\lfloor \cdot \rfloor$ is called the floor function. Similarly, C_i's *upper quota* is denoted $\lceil q_i \rceil$ and is equal to the least integer greater than or equal to q_i; $\lceil \cdot \rceil$ is called the ceiling function.

DEFINITION 1. For fixed D and n, an *apportionment method* F is a mapping of $\mathbf{v} = (v_1, \ldots, v_n)$ to a non-empty set of n-tuples $\mathbf{d} = (d_1, \ldots, d_n)$ of non-negative integers such that $\sum_{i \leq n} d_i = D$, where $d_i \geq 0$ is the number of delegates that candidate i receives.

Because of the possibility of ties at some point in the apportionment process, F is a set-valued function. For instance, if $D = 20$ and $\mathbf{v} = (100, 100, 100)$, then most apportionment methods F would give $F(\mathbf{v}) = \{(7,7,6),(7,6,7),(6,7,7)\}$. We write $F(\mathbf{v}) = \mathbf{d}$, with the assumption that if F is multi-valued, $\mathbf{d}$ represents one of the possible apportionments. In practice, in the (unlikely) event of a tie occurring, we assume the final allocation is randomly selected among all possible $\mathbf{d}$ in $F(\mathbf{v})$.

Generally, apportionment methods are assumed to be *homogeneous*, that is $F(\lambda \mathbf{v}) = F(\mathbf{v})$ where $\lambda > 0$ is a positive constant [**4**]. Thus, they can be defined in terms of the vote distribution $\mathbf{v}$, the percentage distribution $\mathbf{p} = (p_1, \ldots, p_n)$, or the quota distribution $\mathbf{q} = (q_1, \ldots, q_n)$. All apportionment methods discussed in this paper are homogeneous, as are the classical methods used to allocate house seats and the other delegate allocation methods explored in [**15**]. For the remainder of the paper, we consider an apportionment method to be a mapping of a quota distribution $\mathbf{q}$ to a non-empty set of n-tuples $\mathbf{d}$.

As the name suggests, the ILQ method is based on the lower quota. Informally, ILQ can be defined through a series of rounds. In the first round, each candidate receives their lower quota, leaving some delegates unallocated. In the second round, new quotas for the candidates are calculated, based on the number of remaining delegates using the same vote distribution, and the candidates once again receive their lower quotas. This process repeats: in each round, new quotas are calculated based on the number of remaining delegates from the previous round and the candidates receive their lower quotas. The allocations continue until either all delegates are awarded or every candidate's quota is less than one (resulting in all lower quotas being zero), in which case the candidate(s) with the most votes gets the remaining delegates. Formally, we define the ILQ method as follows; recall that $v_1 \geq v_2 \geq \cdots \geq v_n$ and $q_i = Dv_i/V$.

DEFINITION 2. The *iterated lower quota (ILQ) method* is defined algorithmically by the following steps.

(1) Each C_i initially receives $\lfloor Dv_i/V \rfloor$ delegates. Let $R_1 = D - \sum_{i=1}^{n} \lfloor Dv_i/V \rfloor$ be the number of remaining delegates, and let $j = 1$.
(2) While $\lfloor R_j v_1/V \rfloor \geq 1$, do the following:
   (a) To each C_i, assign an additional $\lfloor R_j v_i/V \rfloor$ delegates.
   (b) Calculate the remaining delegates $R_{j+1} = D - \sum_{i=1}^{n} \lfloor R_j v_i/V \rfloor$.
   (c) Increase the value of j by 1.
(3) Assign the remaining R_j delegates to C_1.

Because C_1 is assumed to be the top vote-getter, the condition on the "while" loop in Step 2 of the algorithm ensures that when Step 3 is reached, all the candidates' lower quotas are 0. This follows because $\lfloor R_j v_i/V \rfloor \leq \lfloor R_j v_1/V \rfloor < 1$ for each i. The following example demonstrates the ILQ method.

EXAMPLE 1. Suppose there are 34 delegates and six candidates with vote totals as indicated in Table 1. Under the ILQ method, each candidate is allocated their lower quota, resulting in a Round 1 allocation of $(9, 9, 3, 3, 3, 3)$. New quotas are then calculated based on the remaining $R_1 = 4$ delegates, which are rounded down, leading to an additional distribution of $(1, 1, 0, 0, 0, 0)$ delegates. New quotas are calculated once more for the remaining $R_2 = 2$ delegates. Since each candidate's quota is now less than 1, C_1 receives the remaining delegates in Round 3. Adding the delegates apportioned in each round, the final delegate allocation is $(12, 10, 3, 3, 3, 3)$.

In Example 1, note that $q_1 = 9.88$ (in round 1) and the ILQ method apportions $d_1 = 12$ delegates to C_1. This is an example of an apportionment method *breaking quota*, where a candidate receives more than their upper quota or less than their lower quota. If an apportionment method always apportions $\lfloor q_i \rfloor$ or $\lceil q_i \rceil$ delegates to C_i for any i and any vote distribution, the method is said to *satisfy quota*. Given

TABLE 1. A six-candidate example of the ILQ method.

Cand.	Vote Total	Quota $D = 34$	Rd. 1	Quota $R_1 = 4$	Rd. 2	Quota $R_2 = 2$	Rd. 3
1	10,000	9.88	9	1.16	1	0.58	2
2	9,900	9.78	9	1.15	1	0.58	
3	3,700	3.66	3	0.43		0.22	
4	3,650	3.61	3	0.42		0.21	
5	3,600	3.56	3	0.42		0.21	
6	3,550	3.51	3	0.41		0.21	
Total	34,400	34	30	4	2	2	2

the clear favoritism that ILQ shows to the leading candidate, it is perhaps not surprising that the ILQ method does not satisfy quota.

In fact, the ILQ method can break quota, particularly for the leading candidate by a considerable amount. To see this, note that $0 \leq q_i - \lfloor q_i \rfloor < 1$, for each i. Summing over i gives $0 \leq D - \sum_i \lfloor q_i \rfloor < n$ which implies (since the $\lfloor q_i \rfloor$ are integers) that $R_1 = D - \sum_i \lfloor q_i \rfloor \leq n - 1$. Thus, there are up to $n-1$ delegates remaining to allocate after round 1. In many cases, C_1 receives at least 2 of them, ensuring that $d_1 > \lceil q_1 \rceil$. We explore this, and its implication for the number of rounds required by the ILQ algorithm, in Sections 6 and 7.

The Georgia Republican primary rules do not specify what happens in the event of a tie among two or more candidates under the ILQ method. Since lower-ranked candidates receive their lower quotas in each round, multiple apportionments are only possible among candidates tied for first place. For example, suppose that there are two candidates tied with the most votes. Further, if there are 3 delegates remaining and four candidates with quotas $(0.8, 0.8, 0.7, 0.7)$, since no candidate's lower quota is positive, then the 3 delegates should be awarded to the top vote-getting candidates, C_1 and C_2. Thus, the four possible allocations are $(3, 0, 0, 0)$, $(2, 1, 0, 0)$, $(1, 2, 0, 0)$ and $(0, 3, 0, 0)$.

The full method used by the 2016 Georgia Republican Party to apportion statewide delegates uses the ILQ method after applying a series of cutoffs, which are minimal vote shares a candidate must meet or surpass to be eligible to receive any delegates. We refer to the ILQ method under the series of cutoffs simply as the *GA method.* (The official rules use the term threshold rather than cutoff. However, we reserve the term threshold for the existing apportionment concept of thresholds of inclusion and exclusion discussed later in the paper.) Cutoffs are used in many state primaries. The Democratic state primaries employ a uniform 15% structure in which candidates receiving less that 15% of the vote are eliminated, along with their votes. The Republican state primaries use a variety of different cutoffs, but the cutoff structure used in Georgia is particularly elaborate. We define it below.

DEFINITION 3. Under the GA method, if a candidate receives more than 50% of the vote then that candidate is awarded all the delegates. If no candidate receives more than 50% of the vote then only candidates who receive more than 20% of the vote are eligible to receive delegates. If no candidate receives more than 20% of the vote, then the cutoff is reduced to 15%, and then, if necessary, to 10%. Eliminated candidates have their votes thrown out and then the ILQ method is used to apportion the delegates among the remaining candidates based on their vote totals.

EXAMPLE 2. Referring back to Example 1, notice that C_1 and C_2 each receive a percentage of the vote that is between 20% and 50%, whereas the other four candidates all receive less than 20% of the vote. Thus under the GA method, the bottom four candidates are eliminated, along with their vote totals and the quotas of C_1 and C_2 are adjusted, resulting in quota values of 17.085 and 16.915, respectively. Applying the ILQ method gives 18 delegates to C_1 and 16 delegates to C_2.

Of course, if none of the candidates are eliminated, (for example, if $n = 4$ and $0.2 \leq q_i < 0.5$ for each i), then the GA method reduces to the ILQ method.

Because the sequence of cutoffs used by the 2016 GA Republican party is arbitrary (from a mathematical point of view, at least) and can be changed from primary to primary, most of our results in this paper focus on the underlying apportionment method, the ILQ method.

The most commonly used method to apportion delegates in presidential primaries is Hamilton's method which, like the ILQ method, is based on lower quotas. We define it formally below. The adjective Hamilton is used in the U.S. because the method was originally proposed by Alexander Hamilton in 1792 to apportion the first U.S. House of Representatives. This method is used in all Democratic state primaries and some Republican state primaries. Because of the frequency with which Hamilton's method is used, we provide results in Sections 3 and 4, which compare the ILQ method to Hamilton's method.

DEFINITION 4. Under *Hamilton's method*, each candidate initially receives their lower quota. The remaining $R = D - \sum_{i=1}^{n} \lfloor q_i \rfloor$ delegates are given (one each) to the R candidates with the highest decimal remainders $q_i - \lfloor q_i \rfloor$.

Note that Hamilton's method satisfies quota by construction, in contrast to the ILQ method. Because candidate's decimal remainders may be equal, Hamilton's method is also set valued.

3. Geometry of the Hamilton, ILQ and GA methods

For $n = 3$ candidates, it is often useful to visualize an apportionment method using simplicial geometry. The geometry provides insight into the properties of a method and the insights may extend for more than three candidates. This section presents the geometry of the Hamilton, ILQ and GA methods.

Following Bradberry [**5**], we let $\Sigma_2^D = \{(q_1, q_2, q_3) \mid q_i \geq 0 \text{ and } \sum_{i=1}^{3} q_i = D\}$ be the two-dimensional simplex of size D. The 2-simplex Σ_2^D is the intersection of the plane $q_1 + q_2 + q_3 = D$ with the positive octant in $\mathbb{R}^3$; when placed in the plane, Σ_2^D is represented by an equilateral triangle (as in Figure 1) in which a candidate C_i has $q_i = 0$ along each side of the triangle. For a fixed apportionment method, we associate each point $\mathbf{q} = (q_1, q_2, q_3) \in \Sigma_2^D$ with the possible apportionments under the apportionment method F. (Note that in illustrating the simplex, we relax the assumptions that $v_1 \geq v_2 \geq v_3$ and $q_1 \geq q_2 \geq q_3$.)

Under Hamilton's method, this association of quotas $\mathbf{q}$ to the apportionment outcome $\mathbf{d}$ results in a partition of the simplex into hexagons each centered at points $\mathbf{d} = (d_1, d_2, d_3)$, where quotas $\mathbf{q}$ within the hexagons have $F(\mathbf{q}) = \mathbf{d}$; see

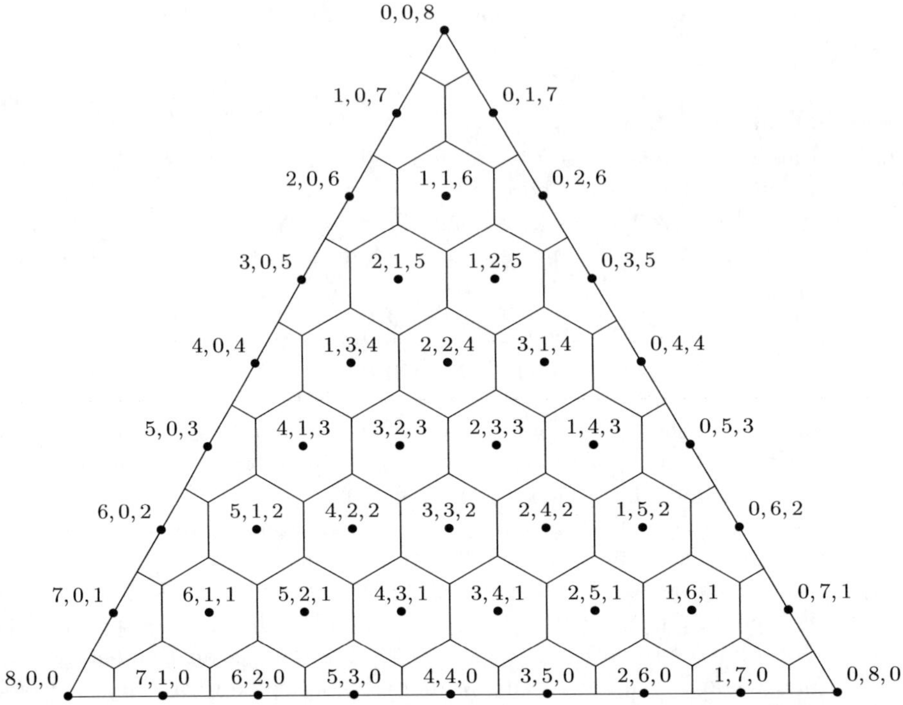

FIGURE 1. The simplex for Hamilton's method with $D = 8$ and $n = 3$.

Figure 1 for the case where $n = 3$ and $D = 8$. The geometry of the 2-simplex and Hamilton's method was described in depth by Bradberry [**5**] and Lucas [**20**]. The boundaries of the hexagons correspond to ties where more than one apportionment is possible.

Before we consider the geometry of the ILQ method for $n = 3$ candidates, we consider the two-candidate case. For two candidates under the ILQ method, the candidate with the largest (resp. smallest) percentage always receives its upper (resp. lower) quota. This result generalizes for $n > 2$ candidates in that the two candidates with the smallest percentages of votes always get their lower quota of delegates (assuming they are not in a tie for first-place), as proved in the following proposition.

PROPOSITION 1. *For $n > 2$ candidates, under the ILQ method, the two candidates with the least support receive their lower quotas, unless they are in a tie for first place. Hence, for $n = 3$, a unique top vote-getting candidate receives all remaining delegates after the first round.*

PROOF. As noted previously, $R_1 \leq n - 1$; hence, at most $n - 1$ delegates are allocated in the second round. Since $v_1 \geq v_2 \geq \cdots \geq v_n$, then $v_i/V \leq 1/i$ for all i, where the inequality is strict unless there is a tie among the top i players. Thus $0 < v_n/V \leq v_{n-1}/V < 1/(n-1)$, which implies that $R_1 \cdot (v_{n-1}/V) < 1$. Consequently, candidates $n - 1$ and n have their second round quotas rounded down to zero, thereby proving the proposition. □

From Proposition 1, it is easy to determine the simplex for the ILQ method when $n = 3$. As illustrated in Figure 2 for $D = 8$, the simplex is divided into rhombi, and portions of rhombi along the portions of the tie lines (which originate from the center of the triangle) where the top two candidates receive the same vote totals. (The geometry of the tie lines is explained in more detail below.)

To understand the labeling of each rhombus, consider the bottom left corner of the simplex where C_1 is the top vote-getting candidate. All quotas in this region get mapped to the delegate allocation (d_1, d_2, d_3), which lies on the bottom left vertex of the rhombi labeled (d_1, d_2, d_3). Symmetric behavior happens when either C_2 or C_3 is the top vote-getting candidate. Along the tie lines and at the center of the simplex, there are small differences based on the value of D modulo 3. Since $R_1 \leq 2$ when $n = 3$, the simplex can be divided into regions where $R_1 = 2$ (shaded regions) and where $R_1 = 1$ (white regions). Note that all remaining delegates in these areas are allocated to the leading vote-getter, as described in Proposition 1.

Unlike Hamilton's method, the boundaries of the regions in the ILQ simplex generally do not correspond to tie situations. For instance, the left and bottom edges of the rhombus labeled $(6, 1, 1)$ have the unique apportionment $(6, 1, 1)$. As

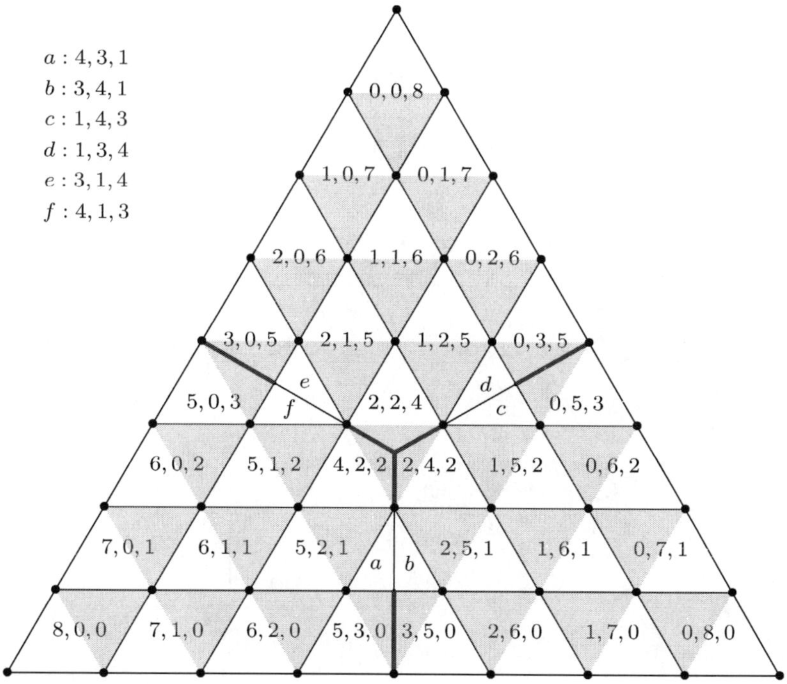

FIGURE 2. The simplex for the ILQ method with $D = 8$ and $n = 3$. Tie lines in the simplex indicate regions where two possible apportionments are possible (unbolded intervals between a and b, between c and d, and between e and f) and where unique apportionments such as $(4, 4, 0)$ occur (bolded intervals) are obtained by averaging the coordinates of the adjacent regions $(5, 3, 0)$ and $(3, 5, 0)$.

described in Section 2, the only points with more than one apportionment occur when there are ties among the top candidates. In Figure 2, there are three of these tie regions, one on each perpendicular bisector or tie line. For example, the region on the $q_1 = q_2$ line corresponding to the shared boundary between the regions labeled a and b has two possible apportionments: $(4,3,1)$ and $(3,4,1)$. The other intervals along the perpendicular bisectors (shown as bold in Figure 2) correspond to apportionments such as $(4,4,0)$, which is the average of the two adjacent regions.

The geometry of the GA method combines the geometry of the ILQ method with regions reflecting the effect of the cutoffs. When $n = 3$, either one candidate has a vote percentage greater than 50% and hence receives all the delegates, or the 20% cutoff criterion is applied. Figure 3 shows the resulting division of the simplex for $D = 8$. The white region near the center of the simplex corresponds to when no candidates are eliminated and matches the center of the ILQ simplex. The light gray regions correspond to apportionments when one candidate receives a majority of votes.

The dark gray regions represent points in which one candidate is eliminated. If C_3 is eliminated, for instance, the possible apportionments are $(5,3,-)$, $(4,4,-)$ or $(3,5,-)$, where we denote the elimination of a candidate by "$-$". Geometrically, determining specific apportionments when the third candidate is eliminated corresponds to projecting each point $\mathbf{q} = (q_1, q_2, q_3)$ to the bottom of the simplex by extending a line from $(0,0,D)$ through $\mathbf{q}$ to the point $\hat{\mathbf{q}} = (\frac{q_1 D}{q_1+q_2}, \frac{q_2 D}{q_1+q_2}, 0)$. When C_1 has the largest quota, this leads to the apportionment $(d_1, D-d_1, 0)$ where $d_1 - 1 < \hat{q}_1 \leq d_1$. For $D = 5$, Figure 4 shows this process applied to two points $\otimes$ and $\triangle$. Point $\otimes$ is projected to a point between $(4,1,-)$ and $(5,0,-)$, leading to

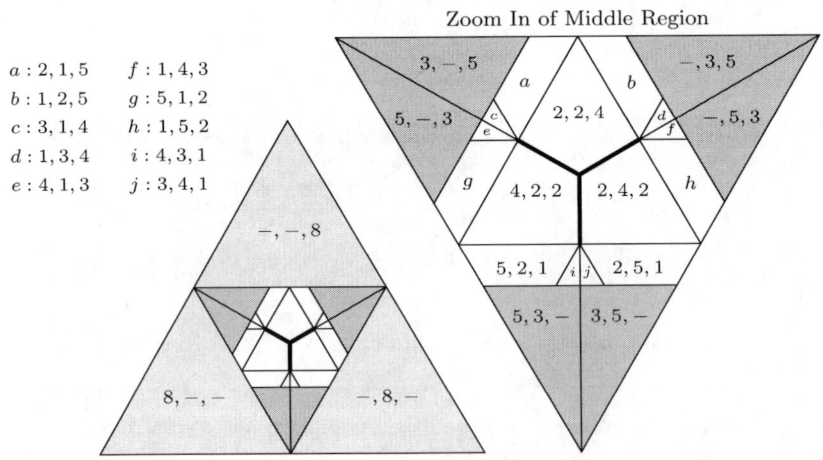

FIGURE 3. The simplex for the GA method with $D = 8$ and $n = 3$ is the smaller equilateral triangle. Tie lines in the simplex indicate regions where two possible apportionments are possible (unbolded intervals) and where unique apportionments such as $(3,3,2)$, $(3,2,3)$, and $(2,3,3)$ occur (bolded intervals). An apportionment with a "$-$" means that the candidate in that position was eliminated by a cutoff. The middle region is magnified on the right to show greater detail.

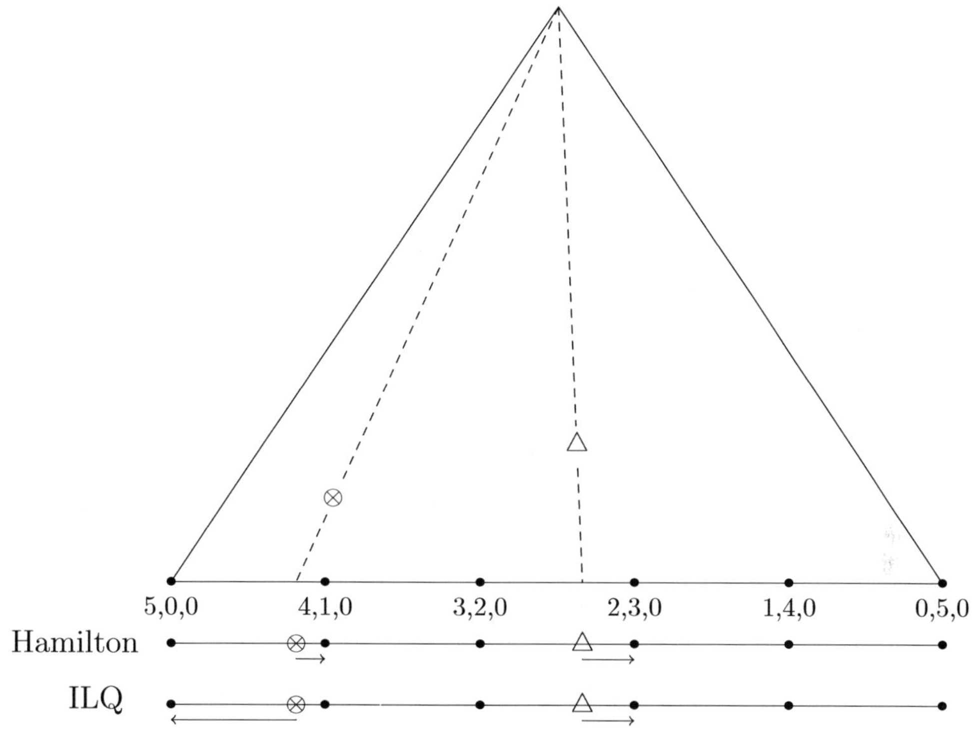

FIGURE 4. Visualizing the elimination of Candidate 3.

$(5, 0, -)$ because the remaining delegate goes to the top vote getter (C_1) when the least vote getter (C_3) is eliminated; for point $\triangle$, the remaining delegate goes to C_2.

The cutoffs used by the Georgia Republican party are unique to Georgia, and thus are not used in conjunction with any other apportionment methods. Nevertheless, the geometry associated with the elimination of a candidate is analogous for other methods. For instance, if C_3 is eliminated under Hamilton's method, the two-candidate Hamilton's method assigns the projection $\hat{\mathbf{q}}$ to the apportionment $(d_1, D - d_1, 0)$ where $d_1 - 0.5 < \hat{q}_1 \le d_1 + 0.5$. Figure 4 highlights the difference between the two-candidate ILQ and Hamilton's methods. Under Hamilton's method, the point $\otimes$ leads to the apportionment $(4, 1, 0)$, unlike the ILQ method; the point $\triangle$ leads to the apportionment $(2, 3, 0)$ for both methods.

Note that under the GA method, not all numbers of delegates are achievable. For instance, when $D = 8$, a candidate cannot receive either 6 or 7 delegates because the cutoffs restrict which apportionments are possible. More generally, for 3 candidates under the cutoffs of the GA method, geometry can be used to show that no candidate can receive k delegates for $0 < k < \lfloor D/5 \rfloor$ and $\lceil 5D/8 \rceil < k < D$. Cutoffs have a similar effect on other apportionment methods.

In general, the geometry of the simplices can be used to make comparisons between different apportionment methods. In the next section, we consider how often the ILQ and Hamilton's methods agree. In later sections, we examine how large the difference is and which candidates are most affected.

4. The likelihood that the Hamilton and ILQ methods agree

In this section we investigate the likelihood that the ILQ and Hamilton's methods agree. To get a sense of this probability, we assume that all quota distributions are equally probable. When $n = 3$, this means that the probability of agreement can be determined by calculating the area in which the ILQ and Hamilton's methods coincide in the simplex. We use the geometry of the simplices introduced in Section 3 to determine this probability when $n = 3$. Then we generalize this result to determine the asymptotic probability that the methods agree as D increases.

Recall under both the ILQ and Hamilton methods, candidates receive their lower quota in Round 1. Under Hamilton's method, the remaining delegates are given to the R_1 candidates with largest remainder; under the ILQ method, the top vote-getting candidate receives at least 1 delegate in each subsequent round. Therefore, the two methods agree if and only if $R_1 = 1$ and $d_1 - \lfloor Dp_1 \rfloor > \max_{i \neq 1} \{d_i - \lfloor Dp_i \rfloor\}$. We obtain the following theorem.

THEOREM 1. *Let $p(D)$ be the probability that the ILQ method agrees with Hamilton's method for D delegates under a uniform distribution over all possible quotas. If $n = 3$, then*

$$p(D) = \begin{cases} \frac{1}{6} + \frac{2D}{6D^2} & \text{if } D \equiv 0 \text{ or } 4 \bmod 6, \\ \frac{1}{6} + \frac{3D+2-2\lfloor D/2 \rfloor}{6D^2} & \text{if } D \equiv 1 \bmod 6, \\ \frac{1}{6} + \frac{2D-2}{6D^2} & \text{if } D \equiv 2 \bmod 6, \text{ and} \\ \frac{1}{6} + \frac{3D-2\lfloor D/2 \rfloor}{6D^2} & \text{if } D \equiv 3 \text{ or } 5 \bmod 6. \end{cases}$$

The proof appears in the Appendix and follows from the fact that exactly one-sixth of each hexagon in the interior of the Hamilton simplex corresponds to when $R_1 = 1$ and C_1 has the largest remainder. The additional terms are due to the hexagons along the edges and the hexagons whose apportionments correspond to regions of measure 0 in the ILQ simplex (where there are ties involving the leading candidate). The number of these depends on D mod 6, and hence so does the probability.

As the number of delegates increases, the number of interior hexagons grows faster than the number of regions around the edges and the number of regions with ties involving the top candidate. Thus for a large number of delegates, the probability that the two methods agree is approximately 1/6, and in the limit, $\lim_{D \to \infty} p(D) = 1/6$.

An analogous result holds for n candidates, as shown in the theorem below.

THEOREM 2. *Let D be the number of delegates, n be the number of candidates and $p(D, n)$ be the probability that the Hamilton and ILQ methods agree. Assuming a uniform distribution over all possible quotas, then $\lim_{D \to \infty} p(D, n) = 1/n!$.*

PROOF. As for the case $n = 3$, the $(n-1)$-dimensional simplex $\Sigma_{n-1}^D = \{(q_1, \dots, q_n) \mid \sum_{i=1}^n q_i = D\}$ can be divided up into regions $R(\mathbf{d})$ corresponding to specific apportionments $\mathbf{d} = (d_1, \dots, d_n)$ under Hamilton's method. Each interior region $R(\mathbf{d})$—that is, away from the boundary of the simplex—contains a subregion with volume $1/n!$ in which $R_1 = 1$ and the top vote-getting candidate receives the largest remainder. Since the number of interior regions grows faster than those at the edges of the simplex or those corresponding to ties involving the top candidate, the theorem holds. □

Theorems 1 and 2 indicate how often the apportionment methods agree assuming that every vote distribution is equally likely. In practice, this assumption is unlikely to hold. However, the results suggest that the two methods frequently disagree, and that the disagreement increases with the number of candidates. We explore the differences between the methods in the next two sections

5. Thresholds of representation

Thresholds of representation provide another tool for analyzing and comparing apportionment methods. Thresholds answer the question "What percentage of the vote leads to a candidate receiving a specific number of delegates?" For example, under Hamilton's method, if there are 3 candidates and 8 delegates, a candidate may receive 2 delegates if their quota is between 4/3 and 8/3. This is observable in Figure 1, where this pair of quotas corresponds to the minimum and maximum heights of the hexagons with $d_3 = 2$. (By height, we mean the z-coordinate in the planar representation of the simplex.) Whether the candidate receives exactly 2 delegates depends, of course, on the quotas of the other candidates. In this section, we determine the threshold values for the ILQ and GA methods when $n = 3$ and generalize some of the thresholds to an arbitrary number of candidates. Thresholds also provide a way to understand how an apportionment method deviates from proportionality: to receive k delegates, we would expect that candidates should have roughly a fraction k/D of the vote. Thus, large deviations of the thresholds from k/D can be indicative of where and how an apportionment method exhibits bias. We use this idea to compare the effect of the thresholds of the ILQ and Hamilton's methods.

Interest in thresholds, within the context of state representation, has focused on the *threshold of inclusion*, the minimum percentage required to possibly gain one seat [**27**], and the *threshold of exclusion*, the maximum percentage in which it is possible to be denied a seat [**26**]. Lijphart and Gibberd [**19**] generalize this work for thresholds for any number of seats, deriving values for Hamilton's, Jefferson's and Webster's methods and a modified version of the Webster/Sainte-Laguë method. Jones and Wilson [**17**] provide thresholds for divisor methods based on concave or convex rounding functions, including Hill-Huntington's method, currently used to apportion the U.S. House of Representatives. The thresholds of inclusion and exclusion for the ILQ method for candidates C_n and C_1 were given in [**15**] for $4 \leq n \leq D$.

In general, the lower (I_k) and upper (S_k) thresholds for a fixed number of k delegates are the minimum and maximum percentages of the vote respectively that a candidate may have to receive k delegate.

DEFINITION 5. The *lower threshold* for a candidate to receive k delegates, denoted I_k, is defined as

$$I_k = \inf\{\, p' \mid \text{ there exists } (\mathbf{p}_{-i}, p') \text{ with } (\mathbf{d}_{-i}, k) \in F(\mathbf{p})\}$$

and the *upper threshold* for a candidate to receive k delegates, denoted S_k, is defined as

$$S_k = \sup\{\, p' \mid \text{ there exists } (\mathbf{p}_{-i}, p') \text{ with } (\mathbf{d}_{-i}, k) \in F(\mathbf{p})\}$$

where $(\mathbf{p}_{-i}, p')$ and $(\mathbf{d}_{-i}, k)$ denote n-tuples in which the ith coordinates are equal to p' and k, respectively.

If $p_i \in [I_k, S_k]$, then candidate C_i may or may not receive k delegates depending on the vote distribution of the remaining candidates. However if $p_i \notin [I_k, S_k]$, then C_i is guaranteed not to receive k delegates. For most apportionment methods, including Hamilton's method, if $p_i < I_k$, then C_i receives fewer than k delegates; if $p_i > S_k$, then C_i receives more than k delegates. Thus, the threshold of inclusion is I_1 and the threshold of exclusion is S_0. In general, the minimum percentage of the vote required to be *guaranteed* to receive at least k delegates is S_{k-1}. We discuss this assumption for the ILQ and GA methods below.

When $n = 3$ and D is fixed, the threshold values for any apportionment method can be read off the appropriate simplex. Under Hamilton's method, I_k and S_k values correspond to the lower and upper heights of the hexagons corresponding to (d_1, d_2, k) (divided by D). These are $\frac{k-2/3}{D}$ and $\frac{k+2/3}{D}$ respectively (see Figure 1). For $n > 3$, these generalize to $\frac{k-(n-1)/n}{D}$ and $\frac{k+(n-1)/n}{D}$, as shown in [**19**].

The thresholds under ILQ when $n = 3$ can also be determined by analyzing the ILQ simplex. However, the I_k values vary depending on the behavior of D mod 3. They are also dependent on how delegates are allocated in the event of a tie among the first-place candidates. Note that since a candidate receiving $p \geq \frac{k}{D}$ percentage of the vote is guaranteed at least k delegates, $I_k \leq \frac{k}{D} \leq S_k \leq \frac{k+1}{D}$.

When $n = 3$ the S_k values follow easily from Proposition 1 which states that, in the absence of a tie, C_2 and C_3 always receive their quota rounded down while C_1 receives the remainder of the delegates.

PROPOSITION 2. *For $n = 3$, $D \geq 3$ and $1 \leq k \leq D$, the upper threshold values are*

$$S_k = \begin{cases} \frac{k+1}{D} & \text{if } k \leq \lfloor D/2 \rfloor - 1 \\ \frac{1}{2} & \text{if } k = \lfloor D/2 \rfloor \\ \frac{k}{D} & \text{if } k \geq \lfloor D/2 \rfloor + 1. \end{cases}$$

PROOF. Suppose $k \leq \lfloor D/2 \rfloor - 1$. Let $q_2 = k + 1 - \epsilon$, $q_3 = 2\epsilon$ and $q_1 = D - (k+1) - \epsilon$ for some $\epsilon > 0$. Then $q_1 \geq q_2 > q_3$, with $q_1 = q_2$ if $k = D/2 - 1$. Then the initial apportionment is $(D - k - 1, k, 0)$, with a final apportionment of $\mathbf{d} = (D - k, k, 0)$ (either as the single-valued apportionment, or as one of the multi-valued apportionments in the case of the tie.) Dividing by D, leads to $S_k = \frac{k+1}{D}$.

Suppose $k = \lfloor D/2 \rfloor$. Only C_1 can have a quota exceeding $D/2$, and if $q_1 > D/2$ then $d_1 \geq k + 1$. Hence $S_k \leq 1/2$. Let $q_1 = q_2 = D/2 - \epsilon$ and $q_3 = 2\epsilon$ for some $\epsilon > 0$. Then the initial apportionment is $(k - 1, k - 1, 0)$ if D is even and $(k, k, 0)$ if D is odd. In either case, since C_1 and C_2 are tied, the final apportionment is multi-valued, but at least one possible allocation corresponds to either C_1 or C_2 receiving k delegates. Hence $S_k = 1/2$.

Lastly, suppose $k \geq \lfloor D/2 \rfloor + 1$. Again only C_1 can have a quota exceeding k, and if $q_1 > k$ then $d_1 \geq k+1$ so $S_k \leq k/D$. Let $q_1 = k-\epsilon$, $q_2 = D-k-\epsilon$ and $q_3 = 2\epsilon$. Then the initial apportionment is $(k - 1, D - k - 1, 0)$, and since $2p_2 = 2q_2/D < 1$ for $k \geq \lfloor D/2 \rfloor + 1$, this implies that C_1 receives the remaining 2 delegates, leading to a final apportionment of $\mathbf{d} = (k, D - k, 0)$. Thus, $S_k \leq k/D$. □

Similar reasoning leads to the following.

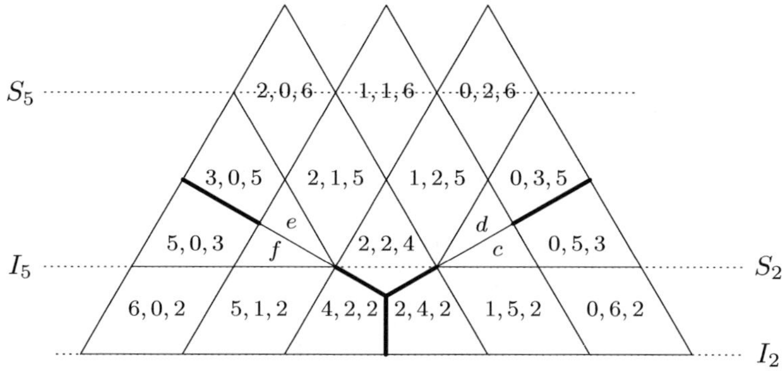

FIGURE 5. Under the ILQ method with $D = 8$, the thresholds I_2, I_5, S_2, and S_5 occur at the relative heights as shown where regions c, d, e, and f have apportionments $(1, 4, 3)$, $(1, 3, 4)$, $(3, 1, 4)$, and $(4, 1, 3)$, respectively. Refer to Figure 2 to see the entire ILQ simplex.

PROPOSITION 3. *For $n = 3$ and $1 \leq k \leq D$, the lower threshold values are*

$$I_k = \begin{cases} \frac{k}{D} & \text{if } k \leq \lfloor D/3 \rfloor \\ \frac{1}{3} & \text{if } k = \lfloor D/3 \rfloor + 1 \\ \frac{1}{3} & \text{if } k = \lfloor D/3 \rfloor + 2 \text{ and } D \equiv 0 \text{ or } 2 \bmod 3 \\ \frac{k-1.5}{D} & \text{if } k = \lfloor D/3 \rfloor + 2 \text{ and } D \equiv 1 \bmod 3 \\ \frac{k-2}{D} & \text{if } k \geq \lfloor D/3 \rfloor + 3. \end{cases}$$

PROOF. We use geometry to provide an informal proof. Consider the ILQ simplex for $n = 3$ and $D = 8$, a portion of which is illustrated in Figure 5. Focusing on C_3, the minimum percentage of the vote for which C_3 may receive k delegates corresponds to the minimum height of the apportionment regions (d_1, d_2, k) for some d_1 and d_2 (divided by $D = 8$). If $k = 2$, for instance, this is equal to $I_2 = 2/8$. (The value $S_2 = 3/8$ is also indicated.) If $k = 5$, $I_5 = 3/8$ (and $S_5 = 5/8$).

Applying this reasoning to other values of D and k, we see that $I_k = \frac{k}{D}$ for $k \leq \lfloor D/3 \rfloor$ and $I_k = \frac{k-2}{D}$ for $k \geq \lfloor D/3 \rfloor + 3$. For values of k near $D/3$, the formula varies with the geometry of the ILQ method near the center of simplex based on D modulo 3. $\square$

As mentioned previously, for most apportionment methods, if $p_i < I_k$, then C_i receives fewer than k delegates and if $p_i > S_k$, then C_i receives more than k delegates. Because of addition of apportionments over rounds under the ILQ method, this relationship between p_i and the lower and upper thresholds may not hold. To get an idea of why this occurs, assume that ties are not allowed. Then, notice that, in Figure 5, the region with the apportionment $(2, 2, 4)$ dips below the regions with apportionments $(5, 0, 3)$, $(0, 5, 3)$, $(4, 1, 3)$ (denoted by c), and $(1, 4, 3)$ (denoted by f). This indicates that I_4 is less than I_3. If ties are allowed, then the bold line segments between regions $(2, 2, 4)$ and $(4, 2, 2)$ and between regions $(2, 2, 4)$ and $(2, 4, 2)$ give apportionments $(3, 2, 3)$ and $(2, 3, 3)$, so I_3 is equal I_4.

The thresholds of most interest in the primaries, arguably, are those for k near D/n when the election is very close, and for k near $D/2$, indicating when a candidate may receive a majority of the delegates. From Propositions 2 and 3, we see that when $n = 3$, $S_k = (k+1)/D$ for $k \approx D/3$ but that I_k varies depending on the relationship between D and 3. In particular, if $D \equiv 0$ or 2 mod 3, then $I_k = 1/3$ for $k = \lfloor D/3 \rfloor + 1$ and $k = \lfloor D/3 \rfloor + 2$. Similar behavior occurs when there are n candidates with vote totals closely tied, as demonstrated in the following proposition.

PROPOSITION 4. *Suppose* $D = \lfloor D/n \rfloor n + r$ *for some* $r \leq n-1$. *Let* $k = \lfloor D/n \rfloor + r$ *if* $r \neq 0$ *and* $k = \lfloor D/n \rfloor + (n-1)$ *if* $r = 0$. *Then under the ILQ method,*

$$I_k = 1/n \text{ and } S_k = (k+1)/D.$$

PROOF. We prove the statement for I_k; the proof for S_k is analogous to that in Proposition 2. Suppose $r \neq 0$. If $p_i < 1/n$ then C_i cannot be the strongest candidate. Furthermore, since $R_1 \leq n-1$, the C_i's round 2 quota is less than $(n-1)/n < 1$; hence C_i receives $\lfloor D/n \rfloor < k$ delegates, which implies $I_k \geq 1/n$. Alternatively, if $p_1 = 1/n + \epsilon$, $p_i = 1/n$ for $i = 2, \ldots, n-r$ and $p_j = 1/n - \frac{\epsilon}{r}$ for $j = n-r+1, \ldots, n$, then the initial apportionment allocates $\lfloor D/n \rfloor$ delegates to the strongest $n-r$ candidates and $\lfloor D/n \rfloor - 1$ to the rest, resulting in $d_1 = \lfloor D/n \rfloor + r = k$ delegates. Thus, $I_i = 1/n$. The proof for $r = 0$ is similar. □

Proposition 4 shows that in a close election, the candidate with a slight edge may receive up to $n-1$ delegates in subsequent rounds. This advantage is also present for threshold values near $D/2$.

PROPOSITION 5. *Suppose* $D \geq 2n+1$. *Under the ILQ method if* $k = \lfloor D/2 \rfloor$ *or* $k = \lfloor D/2 \rfloor + 1$,

$$I_k = \frac{k-(n-1)}{D} \text{ and } S_k = \frac{k}{D}.$$

PROOF. Again we prove the statement for I_k as the proof for S_k is analogous to that in Proposition 2. First note that if $p_i < \frac{k-(n-1)}{D}$, then C_i initially receives fewer than $k-(n-1)$ delegates, and since $R_1 \leq n-1$, this implies $d_i < k$. Hence, $I_k \geq \frac{k-(n-1)}{D}$. Alternatively, suppose $p_1 = \frac{k-(n-1)}{D} + \epsilon$ and that $k-(n-1) = m(n-1)+r$ for some $m \geq 1$ and $0 \leq r \leq n-2$. If $p_i = \frac{m}{D} - \frac{\epsilon}{n-1}$ for $i = 2, \ldots, r+1$ and $p_j = \frac{m-1}{D} - \frac{\epsilon}{n-1}$ for the rest of the candidates then the initial apportionment allocates $k-(n-1)$ delegates to C_1, $m-1$ delegates to the next r candidates and $m-2$ delegates to the remaining candidates, leading to $d_1 = k$. □

Hence if $p_1 > I_{\lfloor D/2 \rfloor + 1} = \frac{\lfloor D/2 \rfloor + 1 - (n-1)}{D}$, then the leading candidate may receive a majority of the delegates. If $p_1 > S_{\lfloor D/2 \rfloor} = \frac{1}{2}$ the leading candidate is guaranteed to receive a majority of the delegates. The latter fact is not surprising: indeed we would consider such an allocation as fair. Less apparent is that, as the I_k value indicates, the leading candidate may receive a fraction of the vote that is significantly below the majority and yet receive a majority of the delegates.

When $n = 3$ the threshold values for the ILQ method and Hamilton's method are not that far apart: both remain close to k/D. However as n increases, the thresholds, especially the lower threshold, diverge. In fact, the proof of Proposition 5 can be adapted to show that the thresholds values $I_k = \frac{k-(n-1)}{D}$ and $S_k = \frac{k}{D}$ hold for all $k \geq D/2$ and that these thresholds values result from the leading candidate receiving an additional $n-1$ candidates after Round 1 of the ILQ algorithm. Thus, as n increases, this suggests that under some circumstances, the strongest candidate may be significantly advantaged in comparison to Hamilton's method. We illustrate this with the following example.

EXAMPLE 3. Suppose $D = 40$ and $n = 10$. Under ILQ, $I_{13} = 0.1$. However, under Hamilton's method, $I_5 = 0.1025$. Thus a candidate receiving 10% of the vote may receive 13 delegates under ILQ while receiving fewer than 5 delegates under Hamilton's method.

The upper and lower threshold values for the GA method differ significantly from those of the ILQ method because of the additional restrictions governing how candidates may receive delegates. We illustrate the effect of the cutoffs on the thresholds with Figure 3 where, since $n = 3$, either one candidate receives a majority of the vote (in which case this candidate receives all the delegates), or the 20% cutoff rule is implemented. Recall that there are some numbers of delegates that are not possible to obtain: in this case no candidate may receive 6 or 7 delegates. A candidate may get 0 delegates from being eliminated (either because they do not meet the 20% threshold or because another candidate has the majority of the vote); hence $I_0 = 0$ and $S_0 = 1/2$. A candidate receives 1 or 2 delegates only if they are not eliminated; otherwise their thresholds are the same as that for the ILQ method. Thus $I_1 = 0.2$, $S_1 = I_2 = 2/8$ and $S_2 = 3/8$. However, a candidate can receive 3 delegates either if no one is eliminated or if the candidate is the second-ranked candidate and the weakest candidate is eliminated. From Figure 3, we see that the latter can happen when the strongest candidate receives just under 50% of the vote leading to $I_3 = 0.3$ and $S_3 = 0.5$. Similar reasoning applies to the other values of k. The threshold values are summarized in Table 2, along with those of ILQ for comparison.

TABLE 2. I_k and S_k values—listed as (I_k, S_k)—under ILQ and GA methods for $n = 3$ and $D = 8$.

k	0	1	2	3	4	5	6	7	8
ILQ	$(0, \frac{1}{8})$	$(\frac{1}{8}, \frac{2}{8})$	$(\frac{2}{8}, \frac{3}{8})$	$(\frac{1}{3}, \frac{1}{2})$	$(\frac{1}{3}, \frac{1}{2})$	$(\frac{3}{8}, \frac{5}{8})$	$(\frac{4}{8}, \frac{6}{8})$	$(\frac{5}{8}, \frac{7}{8})$	$(\frac{6}{8}, 1)$
GA	$(0, \frac{1}{2})$	$(\frac{2}{10}, \frac{2}{8})$	$(\frac{2}{8}, \frac{3}{8})$	$(\frac{3}{10}, \frac{1}{2})$	$(\frac{1}{3}, \frac{1}{2})$	$(\frac{3}{8}, \frac{1}{2})$			$(\frac{1}{2}, 1)$

Note that with the exception of I_1 whose value is greater under GA than under ILQ in order to meet the cutoff, the threshold values generally decrease or remain the same when moving from the ILQ method to the GA method. This makes sense because imposing cutoffs should only help (or, at least, not hurt) the candidates who survive the cutoffs. However, paradoxically, it is possible for a leading candidate to lose delegates as a result of a cutoff. We explore such paradoxical behavior of the ILQ and GA methods in Section 8.

6. The ILQ method and the question of bias

The results from Section 5 indicate that the allocation of delegates under the ILQ method can depart significantly from that indicated by the candidates' quotas. And that the largest deviation occurs in favor of C_1. This suggests that the ILQ method exhibits a degree of bias towards the strongest or set of strongest candidates. Historically, the analysis of bias has played a large role in apportionment literature, particularly in the comparison of different apportionment methods' bias towards either larger or smaller constituencies in political representation [**4**]. In this section we analyze the bias that the ILQ method shows in favor of the strongest candidates—particularly C_1.

First, recall from Proposition 1 unless there is a tie among the strongest candidates, the weakest two candidates will always receive their lower quota. When $n = 3$, the ILQ method assigns the remaining delegates to C_1. Consequently, only the strongest candidate is favored; the other two candidates receive an allocation that is less than their quota (or receive their quota if it is an integer). This bias toward C_1 can be seen in the simplex for the ILQ method (see Figure 2), where each region "reaches" toward the corner of the corresponding leading candidate.

In general, any candidate other than the weakest two candidates may receive their upper quota or even more, as in Example 1 where the ILQ method breaks quota in favor of C_1. In fact, C_1 will always receive more than their upper quota whenever $R_1 \geq 2$. This is because C_1 receives an additional delegate in every round; in order for C_1 to get $\lceil q_1 \rceil$ delegates in total, the ILQ algorithm must end in Round 2 and there must be a tie among the top candidates so that C_1 receives only 1 of the remaining delegates. This leads to the following proposition.

PROPOSITION 6. *Assume* $q_1 > q_2 \geq q_3 \geq \cdots \geq q_n$. *If* $R_1 \geq 2$, *then* C_1 *will receive more than* $\lceil q_1 \rceil$ *delegates.*

To gain some understanding about how frequently this occurs, we use simulations to estimate the probability that quota is broken for a variety of n and D values. We again assume a uniform distribution of quotas across the simplex. These simulations were also done using Python 3 code, which is available upon request. The results, based on 1,000,000 runs, are shown in Table 3.

The first number in each entry denotes the probability that under ILQ, C_1 receives more than $\lceil q_1 \rceil$ delegates. Since ties happen with 0 probability, Proposition 6 implies that this is equal to the probability that $R_1 \geq 2$. Note that the probabilities increase with both D and n; however, the increases are more sensitive to changes in the number of candidates, an observation that corresponds to the sensitivity of the thresholds to n in Section 5. The last column provides an estimate for the asymptotic probability as $D \to \infty$. (Notice that when $n = 3$, $R_1 = 1$ and $R_1 = 2$ occur with equal probability, hence C_1 receives $\lceil q_1 \rceil + 1$ delegates half the time.) The second number, which indicates the probability that C_2 receives more than $\lceil q_2 \rceil$ delegates is virtually 0. This suggests that in practice, it is only the leading candidate whose delegate count will exceed upper quota.

TABLE 3. Probability of the ILQ method breaking quota for C_1 and C_2, assuming a uniform distribution across the simplex.

n/D	13	18	34	100	200000
3	$0.462, 0$	$0.473, 0$	$0.485, 0$	$0.495, 0$	$0.500, 0$
4	$0.793, 0$	$0.805, 0$	$0.819, 0$	$0.828, 0$	$0.833, 0$
5	$0.936, 0$	$0.943, 0$	$0.950, 0$	$0.956, 0$	$0.957, 0$
6	$0.983, 0$	$0.986, 0$	$0.989, 0$	$0.990, 0$	$0.992, 0$
7	$0.996, 0$	$0.997, 0$	$0.998, 0.001$	$0.998, 0.001$	$0.999, 0.001$
8	$0.999, 0$	$0.999, 0.001$	$1.000, 0.002$	$1.000, 0.003$	$1.000, 0.004$

In some instances, this bias in favor of C_1 can be extreme, as when C_1 receives $n-1$ additional delegates after round 1, leading to a total allocation of $\lceil q_1 \rceil + n - 2$ delegates. This happens when $p_2 < (1/R_1)$, ensuring that C_2 receives no delegates in round 2, as occurs in the following example.

EXAMPLE 4. Let $D = 45$ and $n = 10$. Suppose $q_1 = 16.9$, $q_2 = 4.9$, and $q_i = 2.9$ for $3 \le i \le 10$, leading to $R_1 = 9$. Since $p_2 = 4.9/45 < 1/9$, all the remaining delegates are assigned to C_1. Thus C_1 receives a total of 25 delegates, 8 delegates more than C_1's upper quota.

To measure the size of this bias on average, we turn to the notion of *seat bias*. This measure was first introduced in Schuster et al. [**29**] and is a standard way to quantify the bias of an apportionment method. For the context of this paper, we use the term *delegate bias* rather than seat bias. The delegate bias of a method measures how many extra delegates above its quota, on average, a candidate receives, assuming that all election outcomes are equally likely.

DEFINITION 6. For fixed n and D, the *delegate bias* for C_i is the expected deviation of assigned delegates from quota $(d_i - q_i)$ according to a uniform distribution over the simplex. For fixed n, the *asymptotic delegate bias* of C_i is the limit of C_i's delegate bias as $D \to \infty$.

In the literature, asymptotic delegate bias, or really asymptotic seat bias, is often referred to as bias. We will continue to use asymptotic delegate bias to distinguish it from delegate bias.

For Hamilton's method, the asymptotic delegate bias for all candidates is 0 for all n and all D (see Drton and Schwingenschlögl [**10**]). For $n = 3$ and any D, the asymptotic delegate bias under ILQ for C_1 is 1.0 and the delegate biases of the other two candidates are -0.5 [**15**]. This is because C_2 and C_3 always receive their lower quota, meaning that on average they receive 0.5 fewer delegates than their quota suggests they should receive. This implies that the asymptotic delegate bias for C_1 is 1.0; logically, the sum of the asymptotic delegate biases for the candidates is zero.

For $n > 3$, we use simulations to estimate the asymptotic delegate biases for C_1 and C_2. These simulations are again done using Python 3 code and are based on 1,000,000 runs; the results are shown in Table 4. Note that when $n \ge 4$, C_2 sometimes receives more than their lower quota as their delegate bias is larger than -0.5. Since C_2 almost never receives more than $\lceil q_2 \rceil$ delegates, the numbers in Table 4 indicate how often C_2 receives $\lceil q_2 \rceil$ rather than $\lfloor q_2 \rfloor$ delegates. In order for C_2 to receive $\lceil q_2 \rceil$ delegates more frequently than they receive $\lfloor q_2 \rfloor$, (and hence for the bias under ILQ to be positive, unlike under Hamilton's method), at least 12 candidates ($n = 12$) are required. Thus while the definition of ILQ suggests favoritism toward multiple top candidates, it appears that this bias only truly

TABLE 4. Asymptotic Delegate Biases for C_1 and C_2 for select values of n.

n	4	5	6	7	8	9	10	11	12
C_1	1.46	1.92	2.36	2.78	3.20	3.59	3.98	4.36	4.73
C_2	−0.46	−0.42	−0.37	−0.30	−0.23	−0.16	−0.08	0.00	0.07

favors the single leading candidate, at least for the small values of n that occur in a typical primary. And this bias is at the expense of all the other candidates.

In [**15**], we consider other notions of bias for the ILQ method and other methods used in Republican state primaries including Hamilton's method, by analyzing the differences between the allocation for C_1 and the allocation for C_n in close elections where each candidate receives about $1/n$ of the vote. Under most of the measures of bias we examine, the ILQ method shows the greatest amount of bias in favor of the leading candidate among all the delegate apportionment methods.

7. The ILQ method and the number of rounds

The ILQ method's allocation of delegates according to lower quotas using rounds (with an automatic allocation to the leading candidate(s) in the last round) makes it unique among apportionment methods. The rounds used by the ILQ method are different than the rounds used in the sequential allocation of cabinet minister positions by divisor (apportionment) methods. The application of apportionment methods to sequentially allocate cabinet ministers in Northern Ireland and Denmark has been analyzed by Brams and Kaplan [**6**] and O'Leary, Grofman and Elkit [**22**]. The sequential aspect in these applications of divisor methods is a priority notion to determine which cabinet minister positions a party fills, not the number of positions. Instead, the use of rounds is familiar in other fair division contexts, such as in cake cutting. For example, around 1960, Selfridge and Conway independently discovered a three-player, envy-free cake-cutting procedure in which a remainder is created when cutting a cake and then a procedure is used to divide the remainder after initial pieces are allocated (see Brams and Taylor [**8**] for a description). Brams and Taylor [**7**] provided an envy-free protocol for *any* number of players, guaranteeing that the steps would terminate in finite time. However, they could not find a bound on the number of steps as the number of steps depended on the players' utilities for the cake. More recently, bounded envy-free protocols for both four and any number of players have been constructed (see Aziz and Mackenzie [**2, 3**]). A natural question is then to ask how many rounds may be necessary to allocate all of the delegates under the ILQ method, and how likely is that maximum to occur.

We also note that when a rounds-based method is used in a political context, the complexity of the process can affect how voters view the method. For example, when jurisdictions choose to use instant runoff voting for local elections, a common argument against this rounds-based voting method is that it is too complicated for an average voter to fully understand, which may impede public support for the method (such an argument is articulated in [**12**], for example). For this reason, it is also important to investigate how the algorithm used in the ILQ method plays out. In this section, we determine formulas for the maximum number of rounds for the ILQ method and we explore the distribution of the number of rounds based on a uniform distribution of possible quotas.

Recall that when there are n candidates, $R_1 \leq n-1$. Intuition suggests that the maximum number of rounds occurs when R_1 is as large as possible, $R_1 = n-1$, and when all these remaining delegates are allotted to the strongest candidate, C_1, one delegate in each subsequent round. We show that this intuition is correct in the following theorem. Let $\Gamma(n, D)$ be the maximum number of possible rounds when D delegates are allocated to n candidates, and let $\Gamma(n) = \max_{D \in \mathbb{N}} \Gamma(n, D)$. That is,

$\Gamma(n)$ is the maximum number of rounds when allocating delegates to n candidates, where D is allowed to vary. Again, without loss of generality, assume C_1 receives the most votes.

THEOREM 3. *For each* $n \geq 3$ *and* D, $\Gamma(n, D) \leq \lfloor n/2 \rfloor + 2$. *For each* $n \geq 3$, $\Gamma(n) = \lfloor n/2 \rfloor + 2$.

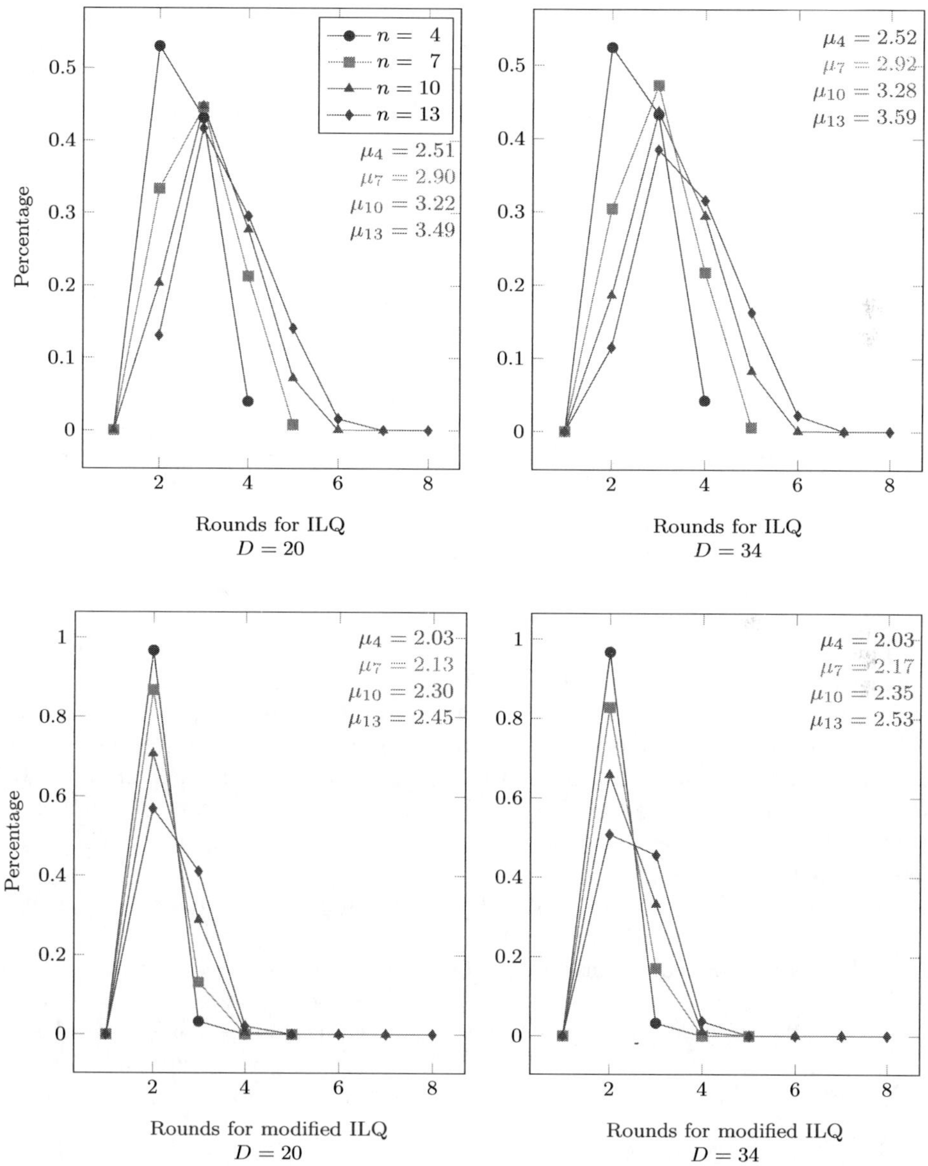

FIGURE 6. Distributions of number of rounds from 1,000,000 simulations for $n = 4, 7, 10$ and 13 for $D = 20$ and $D = 34$ for both the ILQ and modified ILQ methods. The averages μ_n are given for each distribution and for each n.

The proof of Theorem 3 appears in the Appendix. The extensive number of inequality conditions in the proof—which restrict the area in the simplex where the maximum number of rounds occurs—suggests that it is probably difficult to achieve the maximum number of rounds in practice. To get a better understanding of the typical number of rounds, we use simulations to determine the delegate apportionment and number of rounds under ILQ for a large number of vote distributions chosen uniformly over the simplex. The code for the simulations was written in Python 3, and was implemented as follows: for each run, choose a percentage distribution $\mathbf{p}$ at random, according to a uniform distribution over the simplex $\{(p_1, p_2, \ldots, p_n) : p_i \geq 0$ for all i and $\sum_{i=1}^{n} p_i = 1\}$. Simple algorithms exist to generate percentage distributions in this fashion; we use the so-called "broken-stick" technique [**18**]. We then determine the number of rounds under ILQ for $\mathbf{p}$, and repeat for a total of 1,000,000 runs. (We are happy to share the code upon request.)

Figure 6 (Top) illustrates the resulting distributions of rounds from our simulations for $n = 4, 7, 10$ and 13 for $D = 20$ and $D = 34$; recall that Georgia has 34 delegates in the Republican primary. For D larger than 34, distributions are similarly shaped. As can be seen, likelihood of the ILQ method requiring close to the maximum number of rounds is quite small.

Additionally, for large n, the last few rounds may consist only of C_1 receiving one or more delegates repeatedly, while the other candidates' apportionments remain fixed. If C_2 is assumed to have the second highest vote total, then once $R_j p_2 < 1$ for some j, the final apportionment is uniquely determined with C_1 receiving all of the remaining delegates. If the ILQ method is used until $R_j p_2 < 1$ and all R_j delegates are given to C_1, then we call this the *modified ILQ method.* The outcomes from the simulation for the modified ILQ method appear in Figure 6 (Bottom).

The average number of rounds μ_n for the ILQ and modified ILQ methods are given in Figure 6 for $n = 4, 7, 10$ and 13. As n increases, the average number of rounds increases for both the ILQ and modified ILQ methods, but the averages (and the distributions themselves) do not change greatly with increased D. By definition, for fixed n and D, the average number of rounds for the modified ILQ method is smaller than for the ILQ method, as detailed in Figure 6. Further, for fixed D, the difference between the averages of the ILQ and modified ILQ methods increases as n increases.

We conclude this section by considering the maximum number of rounds for the GA method. Under the GA method, the use of cutoffs decreases the number of candidates and hence the maximum rounds possible for small values of n. (For large values of n, the maximum number of rounds occurs when no player receives more than 10% of the vote, so the GA method reduces to the ILQ method.) Let $\Gamma_{\mathrm{GA}}(n)$ represent the maximum number of rounds that can occur under the GA method for n candidates, where D is allowed to vary.

TABLE 5. Achieving $\Gamma_{\text{GA}}(n)$ for $4 \leq n \leq 24$: C_1 receives v_1 votes (Column 3), other candidates' votes that exceed (resp. don't exceed) the cutoff are in Column 4 (resp. 5). For $14 \leq n \leq 24$, $f(n) = 9{,}999n^2 - 29{,}998n + 19{,}999$, $g(n) = 90{,}001n - 80{,}001$, and $h(n) = 10n^2 - 20n + 9$.

n	$\Gamma(n)$	v_1	v_2	$v_3 - v_n$	D
$4-8$	3	38	21	1	19
n	$\Gamma(n)$	v_1	$v_2 - v_4$	$v_5 - v_n$	D
$9-13$	4	200	100	1	35
n	$\Gamma(n)$	v_1	$v_2 - v_n$		D
$14-20$	$n-9$	$f(n)$	$g(n)$		$h(n)$
$21-24$	$\lfloor n/2 \rfloor + 2$	$f(n)$	$g(n)$		$h(n)$

THEOREM 4. *The maximum number of rounds that can occur under the GA method is*

$$\Gamma_{\text{GA}}(n) = \begin{cases} 3 & \text{if } 3 \leq n \leq 8, \\ 4 & \text{if } 9 \leq n \leq 13, \\ n-9 & \text{if } 14 \leq n \leq 20, \text{ and} \\ \lfloor n/2 \rfloor + 2 & \text{if } n \geq 21. \end{cases}$$

The proof depends on the following lemma; the proof of the Lemma appears in the Appendix.

LEMMA 1. *The maximum number of rounds that can occur under the GA method satisfies*

$$\Gamma_{\text{GA}}(n) \leq \begin{cases} 3 & \text{if the 20\% or 15\% cutoff applies,} \\ 2 & \text{if the 10\% cutoff applies and } n = 7, 8, \\ 4 & \text{if the 10\% cutoff applies and } n \geq 9, \\ n-9 & \text{if no cutoff holds and } 11 \leq n \leq 20, \text{ and} \\ \lfloor n/2 \rfloor + 2 & \text{if no cutoff holds and } n \geq 21. \end{cases}$$

PROOF OF THEOREM 4. If $n \leq 6$, either the 20% or 15% cutoff must hold so $\Gamma_{\text{GA}}(n) \leq 3$. If $7 \leq n \leq 10$, then one of cutoffs (20%, 15% or 10%) must hold so $\Gamma_{\text{GA}}(n) \leq \max\{2, 3\} = 3$ for $n = 7$ and 8 and $\Gamma_{\text{GA}}(n) \leq \max\{2, 3, 4\} = 4$ for $n = 9$ and $n = 10$. If $n \geq 11$, any one of the cutoffs could hold (leading to a maximum number of 4 rounds) or none of the cutoffs could hold. Since $n - 9 \leq 4$ if and only if $n \leq 13$, $\Gamma_{\text{GA}}(n) \leq 4$ for $11 \leq n \leq 13$, $\Gamma_{\text{GA}}(n) \leq n - 9$ for $14 \leq n \leq 20$ and $\Gamma_{\text{GA}}(n) \leq \lfloor n/2 \rfloor + 2$ for $n \geq 21$.

To conclude the proof, it suffices to show that there exist examples in which these upper bounds occur. When $n \geq 25$, the example given in the proof of Theorem 3 also demonstrates that the upper bounds occur for the GA method since no candidate receives more than 10% of the vote for n large. For $n \leq 24$, we give explicit examples in Table 5. The top vote-getting candidate receives v_1 votes, designated in the third column. Columns 4 and 5 indicate the number of votes earned by candidates who are above the cutoff and below the cutoff, respectively.

In Table 5, for $n = 4$ to 13, there are candidates of each type, while for $n = 14$ to 24, all candidates meet the cutoff. This completes the proof of Theorem 4. □

We close this section by noting that from a mathematical standpoint, the use of rounds by the ILQ and GA methods is a matter of interest, but is not *a priori* positive or negative. From a practical standpoint, however, the use of rounds is significant because the iterative nature of the methods makes them more difficult to understand. If an apportionment method is perceived to be overly complicated, this may adversely affect voter buy-in, which in turn may lead to disenfranchisement and a sense of unfairness. A similar phenomenon has been suggested in voting theory in conjunction with instant runoff voting. Voters who view the use of rounds as confusing or overly complicated may distrust the outcome of an election.

8. Apportionment paradoxes

Both the ILQ and GA methods are susceptible to a number of apportionment paradoxes. Since both methods are based on quotas, they are subject to classical paradoxes such as the Alabama paradox and the population paradox (cf. [**4**]), as explained below. The methods are also susceptible to other paradoxes that are meaningful in the context of delegate apportionment.

8.1. Delegate monotonicity. Our first paradox is the *Alabama paradox*, whose traditional description we modify to fit the delegate context.

DEFINITION 7. An apportionment method F is *delegate monotone* if for every $\mathbf{v}$ and D, $\mathbf{d} \in F(\mathbf{v}, D)$ implies that $\mathbf{d}' \in F(\mathbf{v}, D+1)$ for some $\mathbf{d}' \geq \mathbf{d}$ (so that $d'_i \geq d_i$ for all i).

That is, a method is delegate monotone if, when keeping the vote totals constant, an increase in the overall number of delegates does not lower the delegate allocation for any candidate. In the context of apportioning representatives in a house of representative setting, this property is referred to as house monotonicity.

Historically, house monotonicity was a desirable feature of apportionment methods of U.S. House of Representatives' seats because the number of House seats increased as the country's population increased. After conducting the census, potential House sizes were considered in which, paradoxically, despite an increase in the House size, a state could lose a seat. Indeed, this is what happened to Alabama in 1880 when an increase of the House size from 299 to 300 would have cost Alabama a seat [**4**]. For this reason, a method that fails to satisfy house monotonicity is said to suffer from the Alabama Paradox.

Because the number of delegates is fixed before the primary election and does not change after the votes are tallied, it may seem that delegate monotonicity is less important when apportioning delegates. But an apportionment method's failure of monotonicity shows sensitivity to the number of delegates awarded. Sensitivity of initial conditions is a common feature of chaotic systems; Saari [**28**] considers apportionment methods and how "rounding off in higher dimensions approximates chaos". More importantly, a method in which a candidate receives fewer delegates when more delegates are available would not instill confidence in the electoral process. Neither the ILQ or the GA method is delegate monotone as illustrated in the following example.

EXAMPLE 5. (The ILQ and GA methods are susceptible to the Alabama paradox) Consider an election with $n = 3$ for $D = 27$ and $D = 28$ with vote totals given in Table 6. Since each candidate receives between 20% and 50% of the vote, no candidate would be eliminated under the GA method, and hence the allocations under the ILQ and GA methods agree. The apportionments for $D = 27$ and $D = 28$ are $(11, 8, 8)$ and $(10, 9, 9)$ respectively. Thus C_1 loses a delegate despite the increase in D. Note that although these methods are biased in favor of C_1, the strongest candidate can still be hurt by increasing the number of delegates.

TABLE 6. Example demonstrating that the ILQ and GA methods are susceptible to the Alabama Paradox.

Cand.	Vote Total	$D = 27$ quota	$D = 28$ quota
1	1000	9.015	9.349
2	998	8.997	9.330
3	997	8.988	9.321

8.2. Population monotonicity. Our second paradox is the *population paradox* (cf. [**4**]), where (as with the Alabama paradox) we consider the paradox in the context of delegate apportionment rather than house seat apportionment.

DEFINITION 8. An apportionment method F is *delegate monotone* if for any two vote vectors $\mathbf{v}$ and $\mathbf{v}'$ where $\mathbf{d} \in F(\mathbf{v}, D)$, $\mathbf{d}' \in F(\mathbf{v}', D)$, and $v_i'/v_j' \geq v_i/v_j$ implies that

$$\text{either } v_i' \geq v_i,$$
$$\text{or } v_i'/v_j' = v_i/v_j \text{ and there exists } \mathbf{d}^* \in F(\mathbf{v}, D)$$

where $\mathbf{d}^*$ is the vector $\mathbf{d}$ except d_i and d_j are replaced by d_i' and d_j'.

That is, a method is population monotone if whenever C_i's vote total increases at a faster rate than C_j's vote total, it is not the case that C_i loses a delegate and C_j gains a delegate. If an apportionment method is not population monotone, then we say it suffers from the population paradox.

An example of this paradox under the ILQ method can be seen in Figure 7 which depicts a small section of Figure 2. Here, a change in the quotas from $(3.15, 1.95, 2.9)$ to $(3.85, 2.05, 2.1)$ results in C_1 losing a delegate to C_2 despite C_1's quota growing at a faster rate than C_2's quota. Since the percentages of each candidate in both cases are between 20% and 50%, this example also demonstrates that the GA method is susceptible to the population paradox. In the context of delegate apportionment, the changes in the quotas/votes could occur as the data from more polling places becomes available. So, for example, if updated delegate allocations are made public as voting results are tabulated, it could be possible for C_1 to lose a delegate despite doing better than C_2 when the tallies from the last polling place are incorporated.

Lack of population monotonicity under the ILQ method explains how close elections can be more helpful to the leading candidate than less close elections under either method. We illustrate this observation for the ILQ method with the following example taken from the election results in the 2008 Republican Party

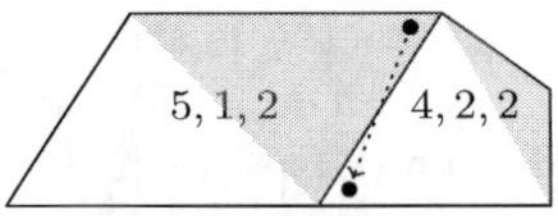

FIGURE 7. Illustrating the population paradox.

primary in Georgia. (At the time, the Georgia Republican Party used a winner-take-all system, which meant that Mike Huckabee was awarded all 34 delegates.)

EXAMPLE 6. (2008 GA primary) The vote totals for the top four candidates from the 2008 Georgia Republican primary are shown in Table 7. The vote totals for the top three candidates were relatively close, and had the ILQ method been used, Huckabee would have receive the 3 delegates remaining in Round 1 (column 3), resulting in an apportionment of $(14, 10, 10, 0, \ldots, 0)$.

However, had the vote been less close, then Huckabee would have received the remaining delegate in Round 1 (column 4) and the final apportionment would have been $(13, 11, 10, 0, \ldots, 0)$.

TABLE 7. 2008 at-large Apportionment in the Georgia Republican Presidential Primary

Candidate	Popular Vote	Quota	Alternate Vote	Quota	GA Quota
Huckabee	326,874	11.53	329,000	12.09	12.05
McCain	304,751	10.75	303,000	11.14	11.23
Romney	290,707	10.26	288,000	10.59	10.72
Paul	28,096	0.99	4,000	0.15	N/A
Others	13,113	0.46	1,000	0.04	N/A

This kind of "close election paradox" does not affect the leader under Hamilton's method. That is, using Hamilton's method the vote leader cannot be hurt when the leader's vote total increases and the vote totals of all other candidates decrease. Another way to frame this phenomenon is in terms of Huckabee's level of support. If Huckabee's support had been greater (had the election become "less close"), he would have fared less well, under ILQ. Thus, the additional support by voters worked to Huckabee's disadvantage, resulting in what is referred to as a *no show paradox*: Huckabee would have done better if some of his constituents stayed home instead of voting.

8.3. Elimination monotonicity. Our final paradox, the *elimination paradox*, was first explored by Jones, McCune, and Wilson in [**14**]. This paradox is similar to the population paradox except that it occurs when a candidate's apportionment decreases as a result of other candidates being eliminated. Because the the concept of the elimination paradox relies on candidates' being eliminated, this paradox does not arise in the context of apportioning House seats in proportion to population. Suppose that N is the set of candidates and $E \subset N$ is a subset of candidates that are eliminated.

DEFINITION 9. Given $\mathbf{v}$ and a subset $E \subset N$, let $\mathbf{v}_{N\setminus E}$ be the vector of vote totals for all candidates $i \notin E$. A method is *elimination monotone* if for all $E \subset N$, $\mathbf{d} \in F(\mathbf{v}, D)$ and $\mathbf{d}' \in F(\mathbf{v}_{N\setminus E}, D)$, if $d'_i < d_i$ for some candidate $i \notin E$ then there exists $\mathbf{d}^* \in F(\mathbf{v}, D)$ with $d^*_i = d'_i$.

An apportionment method that is not elimination monotone is susceptible to the *elimination paradox.*

Thus, an elimination paradox occurs if a non-eliminated candidate loses one or more delegates as a result of the elimination of other candidates, and the smaller delegate number was not a possible apportionment (as the result of a tie) had no candidates been eliminated. This behavior is paradoxical because the relationships among the surviving candidates haven't changed, and thus apportionment methods subject to the elimination paradox violate some notion of consistency. This problem is particularly acute if candidates are eliminated because of some cutoff value. Hamilton's method with a cutoff has been shown to succumb to the elimination paradox [**14**]. We would expect that the ILQ method would also be susceptible to the elimination paradox. In fact, the elimination paradox would have occurred had the 2008 Georgia Republican primary used the GA method.

EXAMPLE 7. (2008 GA primary) If the 2008 Georgia Republican primary had used the GA method (see data in Table 7), all but the top three candidates would have been eliminated, resulting in the quotas shown in column 6 of Table 7. Huckabee would have received the one remaining delegate in Round 2 and the final apportionment would be $(13, 11, 10)$. Thus, by eliminating all candidates receiving less than 20% of the vote, Huckabee would have lost a delegate to John McCain.

The incorporation of cutoffs with the ILQ method can also produce additional paradoxical behavior related to the population paradox. To illustrate, suppose $n = 5$, $D = 34$ and a 7% cutoff is used. If $\mathbf{v} = (450, 296, 296, 296, 100)$, then C_5 is eliminated leading to an apportionment of $(13, 7, 7, 7, 0)$. Suppose, under a recount, C_1 "gives" 5 votes to C_5 so that $v_1 = 445$ and $v_5 = 105$. Then C_5 is no longer eliminated and the new apportionment is $(14, 6, 6, 6, 2)$. Thus, because the cutoff is no longer enacted, C_1 gains a delegate even though v_1 has decreased while v_5 has increased: in this scenario, C_1 is better off losing some votes in a recount to an eliminated candidate so that the eliminated candidate can survive a cutoff. We discuss this version of the no show paradox for Hamilton's method in [**14**].

We conclude by noting that for $n = 3$, no candidate will lose delegates as a result of applying the elimination cutoffs of the GA method if D is even.

PROPOSITION 7. *If $n = 3$ and D is even, then no candidate can lose delegates under the ILQ method through the cutoffs of the GA method.*

PROOF. If $n = 3$ and no candidates are eliminated, then the GA method coincides with the ILQ method and C_1 receives the remaining delegates in Round 1 while C_2 and C_3 receive their lower quotas. Since the lower quotas can only increase for all candidates surviving elimination, the only candidate to potentially suffer from a loss of delegates is C_1.

Suppose that the cutoffs are enacted but C_1 receives less than 50% of the vote so that $\lfloor Dp_1 \rfloor \leq D/2$. Then C_1's apportionment will only decrease if: (1) prior to elimination $R_1 = 2$; and (2) $\lfloor D\hat{p}_1 \rfloor = \lfloor Dp_1 \rfloor$. If $\lfloor Dp_1 \rfloor < D/2$ then, since D is even and at most 1 additional candidate remains, $\lfloor D\hat{p}_1 \rfloor > \lfloor D/2 \rfloor \geq D/2 + 1$. If $\lfloor Dp_1 \rfloor = D/2$, then $R_1 = 1$. Thus an elimination paradox cannot occur. $\square$

9. Assessing the ILQ method and its implementation in the Georgia Republican primary

The Georgia Republican Party adopted the GA method for their 2016 primary. In 2020, because of the Covid pandemic and because President Trump was running unopposed, the state primary was delayed and the Georgia Republican primary used a winner-take-all election with only President Trump's name on the ballot. Thus, the GA method has been used to apportion delegates only once as of this writing. In this conclusion, we analyze the results from the 2016 Georgia Republican Primary. We also apply the previous theorems to give a partial assessment of the ILQ and GA methods.

The 2016 primary was held on March 1, with 13 candidates running. Statewide, only 3 candidates, Donald Trump, Marco Rubio and Ted Cruz received more than 20% of the vote (the next candidate received only 6%). They were also the only candidates to receive delegates in the 14 congressional districts, where the delegates were split $(2, 1, 0)$ in each district for different candidates, resulting in total district delegate counts of 26, 7 and 9 for Trump, Rubio and Cruz, respectively. For the $D = 34$ at-large delegates, after the other candidates were eliminated, the Round 1 apportionment was $(15, 9, 9)$ with the remaining delegate going to Trump in Round 2 for a final result of $(16, 9, 9)$ (see Table 8 columns 5-7).

For comparison, had the Georgia Republican Party dropped its cutoffs but continued to use the ILQ method, five candidates would have received delegates resulting in the apportionment in Table 8 (columns 2-4). Finally, had the Georgia Republican Party used Hamilton's method (with no cutoffs), the final two delegates would have gone to Rubio and Kasich, with the largest decimal remainders, yielding the apportionment in Table 8 (column 8).

TABLE 8. Allocation of the 34 at-large delegates in the 2016 GA Republican primary under ILQ, GA and Hamilton's method.

		ILQ		adj.	GA		
Candidate	quota	Initial	Final	quota	Initial	Final	Hamilton
Trump	13.20	13	15	15.19	15	16	13
Rubio	8.31	8	8	9.57	9	9	9
Cruz	8.02	8	8	9.24	9	9	8
Carson	2.12	2	2		-	-	2
Kasich	1.90	1	1		-	-	2
Total		32	34		32	34	34

The differences between these last results are not surprising. From Theorem 2, we know that the probability of the two methods agreeing decreases to a limiting value of $1/n!$, assuming every vote distribution is equally likely. Even had there been only three candidates running initially, Theorem 1 indicates that the probability of the methods agreeing would have been 17.65%, assuming a uniform distribution across the simplex.

In this instance, the leading candidate did not receive a large number of additional delegates after Round 1 under either the ILQ or GA methods, and the ILQ algorithm only took 2 rounds. Under ILQ, Trump would have received only one delegate above his upper quota; all the other candidates received their lower quota. Given the large number of candidates running in the 2016 GA primary, it is possible that the ILQ method might have produced results that differed greatly from a proportional allocation and taken a large number of rounds. Indeed from Table 4, with $n = 12$ candidates, the asymptotic delegate bias of C_1, based on a uniform distribution across the simplex, is 4.73. The fact that Trump only received 0.81 delegates more than his quota suggests that Trump was somewhat "unlucky". However, his 16 delegates under the GA method, and what would have been 15 under the ILQ method, are still more than he would have received under Hamilton's method.

With such a large number of candidates, the opportunity for the leading candidate to receive a majority of delegates also increases. From Proposition 5, to possibly receive $k = 18$ delegates—a bare majority—Trump would have to have at least a fraction of at least $I_{18} = \frac{18-12}{34} = 0.18$, not too far from his 13.20% of the vote. To be guaranteed a majority of the delegates, he would have needed $S_{17} = \frac{1}{2}$ of the vote. Such a wide range illustrates the sensitivity of the ILQ method to the particular vote distribution.

We next consider the maximum number of rounds that could occur in using the ILQ method in the Georgia Republican primary. As described above, the 2016 Georgia Republican primary took only two rounds, because only three candidates remained after elimination and because the distribution of votes to these three candidates were such that there was only one delegate remaining after Round 1. However, by Theorem 4, the number of rounds could have been as high as 4, given that there were originally 13 candidates (or 8 rounds, by Theorem 3, if the ILQ method had been used). In practice, due to where the Georgia Republican primary is in the primary calendar, it seems unlikely that the number of candidates running would be greater than 13, and hence the number of rounds is not likely to exceed 3 or 4.

Finally, we consider the susceptibility of the ILQ and GA methods to a number of paradoxes. Some of the more obvious, such as lack of delegate monotonicity, may not be too serious in the context of delegate apportionment, since it is unlikely that the number of delegates would be increased while vote totals remain constant. More troubling are the violations of population monotonicity and the related effects of eliminating a candidate. Recall Example 6: if Huckabee had received fewer votes he might have received more delegates. Indeed, as the discussion about the no show paradox indicates, under some circumstances, it might be better for the leading candidate to actively campaign for an opponent! Such behavior seems too irrational to be encouraged by an apportionment method, and hence violations of population monotonicity are significant. Furthermore, since the ILQ and GA methods are designed to favor top vote-getting candidates, it is concerning that top vote-getting candidates can be affected negatively by the population and elimination paradoxes. Since this method has only been used once, there is scant empirical data to suggest how often this may occur in practice. We suspect that the frequencies with which these paradoxes affect the vote leader is relatively high, which would lead to a somewhat unusual hypothesis about these paradoxes: the more an apportionment method favors the top vote-getting candidate, the more vulnerable this candidate is to such paradoxes.

To conclude the assessment of the ILQ and GA methods, we address the question of bias. While bias is undesirable in the context of political representation, when delegates are apportioned to candidates based on vote totals, bias in favor of top candidates is arguably less of a problem: indeed it may be an advantage. By giving more delegates to the top vote-getting candidate, the ILQ and GA methods help consolidate support for the leader. This may be particularly appropriate during later stages in the primary season as momentum builds to the national convention. Conversely, in the early stages, it may be advantageous to use a method, such as Hamilton's method, that allows more candidates to be competitive. Ultimately, whether or not an apportionment method should be biased in favor of top (or bottom) candidates depends on the purpose of the apportionment and one's notion of fairness in a given context.

The ILQ and GA methods hold a natural interest because of their use in the 2016 Georgia Republican presidential primary. However, their novelty is also compelling within the broader landscape of apportionment methods. These methods transform the apportionment problem from a matter of rounding, to one where apportionment is conducted through a series of rounds. The question of how "good" these methods are is ultimately one of political theory; this article articulates some of the mathematics needed to provide an answer.

Appendix

PROOF OF THEOREM 1. We consider the case $D \equiv 1 \bmod 6$; the other five cases are analogous. We calculate the area of the regions in Σ_2^D where the Hamilton and ILQ methods overlap, in units equal to the area of a full hexagon, divided by the total number of full hexagons in the simplex which is $D^2/2$.

Let $R(\mathbf{d}) \subset \Sigma_2^D$ be the region corresponding to apportionment $\mathbf{d}$ under Hamilton's method. Since $\mathbf{d} \in R(\mathbf{d})$, there is one region for each lattice point in Σ_2^D. Moreover, the $R(\mathbf{d})$ are space-filling (they do not leave any gaps and they overlap only on the boundaries); hence each $R(\mathbf{d})$ lying in the interior of Σ_2^D has area equal to 1 hexagon. There are $\binom{D-1}{2} = \frac{(D-1)(D-2)}{2}$ of these interior hexagons. In addition, there are $3(D-1)$ half-hexagons along the edges and 3 one-sixth hexagons at the vertices.

Straightforward calculations show that each full hexagon can be divided into six equally sized pie-shaped subregions of area $1/6$. Three subregions correspond to $R_1 = 2$ and three to $R_1 = 1$. Of the latter, exactly one of these subregions corresponds to the top candidate having the largest remainder. Each half-hexagon and one-sixth hexagon also contains one of these subregions. Thus, each $R(\mathbf{d}) \subset \Sigma_2^D$ contains a subregion of area equal to $1/6$ of a full hexagon in which the Hamilton and ILQ methods agree.

As noted previously, every $R(\mathbf{d}) \subset \Sigma_2^D$ has a corresponding region (quadrilateral or triangle) of positive area in the ILQ simplex with the exception of hexagons corresponding to apportionments with a tie involving the top candidate. We claim there are $\lfloor D/2 \rfloor$ of these exceptions.

To see this, let $D = 6k+1$ and $\mathcal{T}_k$ be the set of apportionments with a tie involving the top candidate. If $k = 1$ there are exactly $\lfloor 7/2 \rfloor = 3$ such hexagons (corresponding to $(3,3,1)$, $(3,1,3)$ and $(1,3,3)$). Assume $|\mathcal{T}_k| = \lfloor D/2 \rfloor = 3k$. Since $6(k+1)+1 = 6k+7$, it follows that $\mathcal{T}_{k+1}$ is the union of two kinds of apportionment: those of the form (d_1+2, d_2+2, d_3+2) for $(d_1, d_2, d_3) \in \mathcal{T}_k$, and the three additional apportionments $(3k+3, 3k+3, 1)$, $(3k+3, 1, 3k+3)$ and $(1, 3k+3, 3k+3)$. Thus, $|\mathcal{T}_{k+1}| = |\mathcal{T}_k| + 3 = 3k+3 = \lfloor (6k+7)/2 \rfloor$, which proves the claim.

Therefore the total area in which the two methods agree, measured in units equal to the area of a full hexagon, is

$$\frac{1}{6}\left(\frac{(D-1)(D-2)}{2}+3(D-1)-\lfloor D/2\rfloor+3\right)=\frac{1}{12}(D^2+3D+2-2\lfloor D/2\rfloor).$$

Dividing by $D^2/2$ yields the result. □

PROOF OF LEMMA 1. Let m be the number of candidates remaining after a cutoff is applied and $\hat{p}_i$ be the adjusted percentage of C_i. Then for each $1 \leq i \leq m$, $\hat{p}_i = \frac{v_i}{v_1+\cdots+v_m} \geq \frac{v_i}{v_1+\cdots+v_n}$ where the inequality is tight if and only if $m = n$, so that no candidates are eliminated.

Suppose the 20% cutoff holds. If C_i is not eliminated then $\hat{p}_i \geq p_i \geq 0.20$; hence $m \leq 5$, because at most five candidates can have $p_i \geq 0.20$. In order for the apportionment to take more than 3 rounds, it must be that $m = 4$ or $m = 5$ because the iterative process ends after at most three rounds if $m \leq 3$ by Theorem 3. If exactly 5 candidates remain, then $n = m = 5$, which implies that no candidates were eliminated, and $\hat{p}_i = p_i = 0.20$ for each candidate. Since $R_1 \leq 4$, this implies $R_1 p_1 < 1$; hence there can be no more than 2 rounds. Alternatively, if $m = 4$, then $\hat{p}_1 < 0.40$ (because $\hat{p}_j \geq 0.20$ for $j = 2, 3, 4$), and so $2\hat{p}_1 \leq 1$. Since $R_1 \leq 3$, this implies a maximum of 3 rounds.

Suppose the 15% cutoff holds. Similarly to the previous case, if C_i is not eliminated then $\hat{p}_i \geq p_i \geq 0.15$ so $m \leq 6$. Since no candidate must have received more than 20% of the vote, $p_i < 0.20$ for all i, so $\hat{p}_1 = \frac{p_1}{\sum_{i=1}^m p_i} < \frac{0.20}{0.15m} = \frac{4}{3m}$. But $(m-2)\frac{4}{3m} < 1$ for $m = 3, \ldots, 6$, which implies a maximum of 3 rounds.

Suppose the 10% cutoff holds. If C_i is not eliminated then $\hat{p}_i \geq p_i \geq 0.10$, so $m \leq 10$. If there are exactly 10 candidates remaining then $n = m = 10$, no candidates were eliminated and $\hat{p}_i = p_i = 0.10$. Since $R_1 \leq 9$, this implies $R_1 p_1 < 1$, and so there are no more than 2 rounds.

Alternatively, suppose $m \leq 9$ candidates remain. Since each of the $n - m$ candidates eliminated satisfies $p_j < 0.10$, $\hat{p}_1 \leq \frac{p_1}{1-\sum_{j=m+1}^n p_j} < \frac{0.15}{1-0.10(n-m)} = \frac{3}{20-2n+2m}$. But if $n = 7$ or 8, $(m-1)\frac{3}{20-2n+2m} < 1$ for all $2 \leq m < n$; hence in this case, there is a maximum of 2 rounds. If $n \geq 9$, then $\hat{p}_1 = \frac{p_1}{\sum_{i=1}^m p_i} < \frac{0.15}{0.1m} = \frac{3}{2m}$. Since $(m-3)\frac{3}{2m} < 1$ for $2 \leq m \leq 9$; this implies a maximum of 4 rounds.

Finally, suppose no candidate receives at least 10% of the vote and hence no cutoff holds. Since $n - 9 < \lfloor n/2 \rfloor + 2$ for $11 \leq n \leq 20$, it suffices, by Theorem 3 to show that $\Gamma_{\text{GA}}(n) \leq n - 9$ for $11 \leq n \leq 20$. To see this, note that $p_1 < 0.10$ implies $R_{\tau-2} \geq 11$. So the maximum number of rounds will occur when $R_1 = n - 1$ and $R_i = R_{i-1} - 1$ for $i = 1, \ldots, \tau - 2$. This corresponds to a maximum number of rounds of at most $n - 9$. This completes the proof of the lemma. □

PROOF OF THEOREM 3. Suppose $n \geq 3$ and D are fixed and that for some $\mathbf{v}$, the ILQ algorithm takes τ rounds. Let a_i denote the number of delegates allocated in round i. The (a_i) form a non-increasing sequence of positive integers, except possibly for the last term $a_\tau = R_{\tau-1}$ which is equal to the total number of remaining delegates.

Suppose $\tau > \lfloor n/2 \rfloor + 2$. If $a_i \geq 2$ for all $i = 2, \ldots, \tau - 1$, then the number of delegates allocated after Round 1 must be at least $2(\tau-1)+1 > 2(\lfloor n/2 \rfloor + 1) + 1 > n-1$, which is impossible. Thus, $a_i = 1$ for some $i = 2, \ldots, \tau - 1$. Let a_k be the first element in the sequence such that $a_k = 1$, which implies $a_i = 1$ for $k \leq i \leq \tau - 1$.

Then C_1 receives all the delegates starting in round k. Moreover, the sequence R_k, R_{k+1}, $\ldots$, $R_{\tau-1}$ decreases by 1, and so there are $R_{k-1} - R_{\tau-1}$ rounds in which 1 delegate is allocated, and then a final round in which $R_{\tau-1}$ delegates are allocated.

We claim $R_{\tau-1} \geq \lfloor R_{k-1}/2 \rfloor$. To see this, note that since C_1 receives 1 delegate in rounds k and $\tau-1$, $R_{k-1}p_1 < 2$ and $R_{\tau-2}p_1 > 1$ where $p_1 = v_1/V$. Rearranging, gives $1/R_{\tau-2} < p_1 < 2/R_{k-1}$ and so $R_{\tau-2} > R_{k-1}/2$. Thus, $R_{\tau-1} = R_{\tau-2} - 1 > R_{k-1}/2 - 1$. This implies $R_{\tau-1} \geq \lfloor R_{k-1}/2 \rfloor$, which proves the claim.

Thus, the number of rounds between k and $\tau - 1$, which is equal to the number of delegates allocated in rounds k through $\tau - 1$, is at most $R_{k-1} - R_{\tau-1} \leq R_{k-1} - \lfloor R_{k-1}/2 \rfloor \leq \lceil R_{k-1}/2 \rceil$. This implies, the number of rounds between rounds 2 and $k-1$ is at least $\tau - 2 - \lceil R_{k-1}/2 \rceil > (\lfloor n/2 \rfloor + 2) - 2 - \lceil R_{k-1}/2 \rceil = \lfloor n/2 \rfloor - \lceil R_{k-1}/2 \rceil$.

Since at least 2 delegates are allocated in each of these rounds, the number of delegates allocated in rounds 2 through $k-1$ is at least $2(\lfloor n/2 \rfloor - \lceil R_{k-1}/2 \rceil + 1) \geq (n-1) - (R_{k-1}+1) + 2 = n - R_{k-1}$. This implies the number of delegates allocated in rounds 2 through τ is at least $n - R_{k-1} + R_{k-1} = n$, which is impossible. Hence $\tau \leq \lfloor n/2 \rfloor + 2$. Since $\mathbf{v}$ was arbitrary, $\Gamma(n, D) \leq \lfloor n/2 \rfloor + 2$, which implies $\Gamma(n) \leq \lfloor n/2 \rfloor + 2$.

To show the other direction, it is sufficient to find a D and $\mathbf{v}$ for which the number of rounds is equal to $\lfloor n/2 \rfloor + 2$. For $4 \leq n < 12$, it is straightforward to construct specific examples for the number of rounds $\Gamma(n) = \lfloor n/2 \rfloor + 2$. For instance, for $n = 10$, and $D = 96$, the vote distribution in which C_1 receives 2400 votes and the remaining nine candidates receive 788 votes each results in 9 delegates remaining after Round 1, with C_1 receiving the remaining delegates in the sequence 2, 1, 1, 1, 1, 3 in successive rounds.

For $n \geq 12$, we provide a general example for n even; the case for n odd is similar. Let $D = (n-1)(n-4)(n-5) - 1$, $p_1 = \frac{2n-9}{(n-5)(n-4)}$, and $p_i = \frac{1-p_1}{n-1} = \frac{n^2-11n+29}{(n-5)(n-4)(n-1)}$ for $2 \leq i \leq n$. We claim first that there are $n-1$ delegates remaining after Round 1. To see this, note that

$$Dp_1 = (2n-9)(n-1) - \frac{2n-9}{(n-5)(n-4)}$$

and

$$Dp_i = n^2 - 11n + 29 - \frac{n^2 - 11n + 29}{(n-1)(n-4)(n-5)}.$$

Since $\frac{2n-9}{(n-5)(n-4)}$ and $\frac{n^2-11n+29}{(n-1)(n-4)(n-5)}$ are both less than 1 for $n \geq 8$,

$$Dp_1 - \lfloor Dp_1 \rfloor = 1 - \frac{2n-9}{(n-5)(n-4)} \quad \text{and} \quad Dp_i - \lfloor Dp_i \rfloor = 1 - \frac{n^2 - 11n + 29}{(n-1)(n-4)(n-5)}.$$

Thus the remainders sum to

$$1 - \frac{2n-9}{(n-5)(n-4)} + (n-1)\left(1 - \frac{n^2 - 11n + 29}{(n-1)(n-4)(n-5)}\right),$$

which simplifies to $n - 1$. This proves the claim.

Next, we claim that C_1's sequence of allocations after Round 1 is equal to 2, 2, 1, $\ldots$, 1, $n/2 - 3$, where there are $n/2 - 2$ rounds in which C_1 receives 1 delegate.

In order for this to be true, the following inequalities must hold:

$$\begin{aligned}
\frac{2}{n-1} &\le p_1 < \frac{3}{n-1},\\
\frac{2}{n-3} &\le p_1 < \frac{3}{n-3},\\
\frac{1}{n-5} &\le p_1 < \frac{2}{n-5},\\
\frac{1}{n-6} &\le p_1 < \frac{2}{n-6},\\
&\vdots\\
\frac{1}{n/2-1} &\le p_1 < \frac{2}{n/2-1},\\
\frac{1}{n/2-2} &\le p_1 < \frac{2}{n/2-2}, \text{ and}\\
&\quad p_1 < \frac{1}{n/2-3} = \frac{2}{n-6}.
\end{aligned}$$

It is easy to see that $p_1 = \frac{2n-9}{(n-5)(n-4)}$ satisfies all these inequalities. In addition, $p_i = \frac{n^2-11n+29}{(n-5)(n-4)(n-1)} < \frac{1}{n-1}$, so no other candidate receives additional delegates from any of the R_j. This proves the second claim. Thus the apportionment occurs in $\Gamma(n) = \lfloor n/2 \rfloor + 2$ rounds. □

Acknowledgment

Thank you to Katie Bird for her assistance with the simulation code.

References

[1] Stephen Ansolabehere and Gary King, *Measuring in presidential nominations the consequences of delegate selection rules*, J. Politics **52** (1990), no. 2, 609–621.

[2] Haris Aziz and Simon Mackenzie, *A discrete and bounded envy-free cake cutting protocol for any number of agents*, 57th Annual IEEE Symposium on Foundations of Computer Science—FOCS 2016, IEEE Computer Soc., Los Alamitos, CA, 2016, pp. 416–427, DOI 10.1109/FOCS.2016.52. MR3631004

[3] Haris Aziz and Simon Mackenzie, *A discrete and bounded envy-free cake cutting protocol for four agents*, Proceedings of the forty-eighth annual ACM symposium on Theory of Computing, 2016, pp. 454–464.

[4] Michel L. Balinski and H. Peyton Young, *Fair representation: Meeting the ideal of one man, one vote*, Yale University Press, New Haven, Conn., 1982. MR649246

[5] Brent A. Bradberry, *A geometric view of some apportionment paradoxes*, Math. Mag. **65** (1992), no. 1, 3–17, DOI 10.2307/2691355. MR1146589

[6] Steven J. Brams and Todd R. Kaplan, *Dividing the indivisible: Procedures for allocating cabinet ministries to political parties in a parliamentary system*, Journal of Theoretical Politics **16** (2004), no. 2, 143–173.

[7] Steven J. Brams and Alan D. Taylor, *An envy-free cake division protocol*, Amer. Math. Monthly **102** (1995), no. 1, 9–18, DOI 10.2307/2974850. MR1321451

[8] Steven J. Brams and Alan D. Taylor, *Fair division*, Cambridge University Press, Cambridge, 1996. From cake-cutting to dispute resolution, DOI 10.1017/CBO9780511598975. MR1381896

[9] Kevin J Coleman, *Presidential nominating process: Current issues*, Congressional Research Service, 2011.

[10] Mathias Drton and Udo Schwingenschlögl, *Asymptotic seat bias formulas*, Metrika **62** (2005), no. 1, 23–31, DOI 10.1007/s001840400352. MR2236294

[11] Kristi Ann Geist, Michael A. Jones, and Jennifer M. Wilson, *Apportionment in the Democratic primary process*, Math. Teacher **104** (2010), 214–220.

[12] Sonja Hutson, *Is ranked-choice voting too complicated? utah lawmakers may let cities experiment with approval voting, too*, `https://www.kuer.org/politics-government/2021-10-12/is-ranked-choice-voting-too-complicated-utah-lawmakers-may-let-cities-experiment-with-approval-voting-too`, 2021, Accessed: 2022-08-02.

[13] Caitlin E Jewitt, *The Republican party and the unsuccessful 2012 presidential nomination reforms*, Presented at State of the Parties Conference, Akron, OH, 2013.

[14] Michael A. Jones, David McCune, and Jennifer M. Wilson, *The elimination paradox: Apportionment in the Democratic party*, Public Choice **178** (2019), no. 1, 53–65.

[15] Michael A. Jones, David McCune, and Jennifer M. Wilson, *New quota-based apportionment methods: the allocation of delegates in the Republican presidential primary*, Math. Social Sci. **108** (2020), 122–137, DOI 10.1016/j.mathsocsci.2020.05.001. MR4172893

[16] Michael A. Jones, David McCune, and Jennifer M. Wilson, *Delegate apportionment in the US presidential primaries—a mathematical analysis*, Studies in Choice and Welfare, Springer, Cham, [2023] ©2023, DOI 10.1007/978-3-031-24954-9. MR4561873

[17] Michael A. Jones and Jennifer M. Wilson, *Evaluation of thresholds for power mean-based and other divisor methods of apportionment*, Math. Social Sci. **59** (2010), no. 3, 343–348, DOI 10.1016/j.mathsocsci.2010.02.002. MR2662119

[18] Dominique Lepelley, Ahmed Louichi, and Fabrice Valognes, *Computer simulations of voting systems*, Advances in Complex Systems **3** (2000), no. 1, 181–194.

[19] Arend Lijphart and Robert W. Gibberd, *Thresholds and payoffs in list systems of proportional representation*, European Journal of Political Research **5** (1977), no. 3, 219–244.

[20] William F. Lucas, *The apportionment problem*, Political and Related Models (Steven J. Brams, William F. Lucas, and Philip J. Straffin, Jr., eds.), Modules in Applied Mathematics, Springer-Verlag, New York, 1983, pp. 358–396.

[21] Scott R. Meinke, Jeffrey K. Staton, and Steven T. Wuhs, *State delegate selection rules for presidential nominations, 1972 - 2000*, J. Politics **68** (1990), no. 1, 180–193.

[22] Brendan O'Leary, Bernard Grofman, and Jørgen Elklit, *Divisor methods for sequential portfolio allocation in multi-party executive bodies: Evidence from Northern Ireland and Denmark*, American Journal of Political Science **49** (2005), no. 1, 198–211.

[23] Georgia Republican Party, *Rules of the Georgia Republican Party, Inc.*, `https://gagop.org/wp-content/uploads/2019/07/GAGOP-State-Party-Rules-Adoped-6-16-19.pdf`, 2019, Accessed: 2022-08-15.

[24] Friedrich Pukelsheim, *Proportional representation*, Springer, Cham, 2017. Apportionment methods and their applications; Second edition of [MR3156971]; With a foreword by Andrew Duff. MR3751357

[25] Joshua T. Putnam, *The impact of rules changes on the 2012 Republican presidential primary process*, Society **49** (2012), no. 5, 400–404.

[26] Douglas Rae, Victor Hanby, and John Loosemore, *Thresholds of representation and thresholds of exclusion*, Comparative Political Studies **3** (1971), no. 4, 479–488.

[27] Stein Rokkan, *Elections: electoral systems*, Macmillan Company, 1968.

[28] Donald G. Saari, *Chaos, but in voting and apportionments?*, Proc. Natl. Acad. Sci. USA **96** (1999), no. 19, 10568–10571, DOI 10.1073/pnas.96.19.10568. MR1712530

[29] Karsten Schuster, Friedrich Pukelsheim, Mathias Drton, and Norman R. Draper, *Seat biases of apportionment methods for proportional representation*, Electoral Studies **22** (2003), no. 4, 651–676.

Mathematical Reviews, 416 Fourth St., Ann Arbor, Michigan 48103
Email address: maj@ams.org

Department of Mathematics and Data Science, William Jewell College, Liberty, Missouri 64068
Email address: mccuned@william.jewell.edu

Department of Natural Sciences and Mathematics, Eugene Lang College, The New School, New York City, New York 100011
Email address: wilsonj@newschool.edu

Contemporary Mathematics
Volume **795**, 2024
https://doi.org/10.1090/conm/795/15971

Double moves by each player in chess openings make the game fairer

Steven J. Brams and Mehmet S. Ismail

ABSTRACT. Unlike tic-tac-toe or checkers, in which optimal play leads to a draw, it is not known whether optimal play in chess ends in a win for White, a win for Black, or a draw. But after White moves first in chess, if Black has a double move followed by a double move of White and then alternating play, play is more balanced because White does not always tie or lead in moves. Symbolically, *Balanced Alternation* gives the following move sequence: After White's (W) initial move, first Black (B) and then White each have two moves in a row (BBWW), followed by the alternating sequence, beginning with W, which altogether can be written as WB/BW/WB/WB/WB... (the slashes separate alternating pairs of moves). Except for reversal of the 3rd and 4th moves from WB to BW (underscored), this is the standard chess sequence. Because Balanced Alternation lies between the standard sequence, which favors White, and a comparable sequence that favors Black, it is highly likely to produce a draw with optimal play, rendering chess fairer.

> "I believe there will be one day in the future that all games between computers will finish in a draw, as the approximately +0.2 advantage White has after moving first is not enough to win the game.... [Under Balanced Alternation], I believe that the +0.2 advantage... has now disappeared and the initial evaluation is closer to 0.0."
> Veselin Topalov, Grand Master and FIDE World Chess Champion, 2005–2006.[1]

1. Introduction

The rules of chess have evolved over the past 1,500 years, but beginning in the 19th century they were standardized to facilitate national and international competition. While the most sophisticated chess-playing computer programs are now able to defeat the best human players, we still seem no closer to answering the question of whether chess is *fair*: When the players make optimal choices, neither

2020 *Mathematics Subject Classification.* Primary 91A46; Secondary 91A18.
Key words and phrases. combinatorial games, chess, fairness, sequencing.
This paper is adapted with minor changes from Brams and Ismail [**2**]. It is published with the permission of the IEEE.

[1] E-mail communication from Topalov to Ismail (03/18/2022).

White, who moves first, nor Black, who moves second, can force a win, rendering the outcome a draw.

Although most chess experts believe that a draw is the product of optimal play, there is no proof of this. Because of the astronomical number of choices in chess, no brute-force check of all possible moves, even with the fastest computers, can verify this. Accordingly, we take an indirect approach by showing that a small variation in the standard rules of chess—giving Black and White double moves after White's initial move, which reverses the order of play of the 3rd and 4th moves from White-Black to Black-White—balances the opportunities of White and Black to win with optimal play.[2] If neither player has a decisive advantage, it is highly unlikely that either player can force a win for itself (and loss for its opponent), making the outcome of optimal play a draw.

By *optimal play*, we mean that (i) if one player defeats its opponent, its opponent could not have done better—drawn or won—by making different moves; (ii) if the outcome is a draw, neither player could have won by making different moves. If both these conditions are met, the optimal strategies of the players constitute a Nash equilibrium in a 2-person zero-sum game of perfect information (like chess), so neither player would depart from its optimal strategy (if known). Zermelo [**10**] showed in chess that either (i) one player can force a win or (ii) both players can force a draw with optimal play (for a translation of this paper and discussion of its results, see [**8**]).

In principle, backward induction, by working backwards from a final move when checkmate or a draw (e.g., from a rule that specifies a draw after a three-fold repetition of moves) occurs, can be used to find optimal moves of the players in a finite game such as chess. In checkers, this approach was applied by Schaeffer et al. [**7**] to prove, using multiple computers making calculations over almost two decades, that optimal play always ends in a draw. But because chess is a far more complex game than checkers, a comparable calculation to ascertain the outcome of optimal play of chess appears beyond the capability of computers for the foreseeable future.

Accordingly, we take a different approach by making a sequencing argument for the variation in chess we mentioned earlier—that it gives Black more opportunity to win and thereby creates more of a balance between Black and White. It also has a major practical advantage, obviating the need of contestants to play both Black and White in a tournament, for reasons we give in the concluding section. In section 4, we supplement our sequencing argument with an analysis of four well-known first moves by White in standard chess and suggest how they would play out under Balanced Alternation. In section 5, we conclude that Balanced Alternation would render chess fairer by creating more even-handedness between Black and White, which in section 3 we show is decidedly biased in favor of White.

2. Statistics on chess

As background, we begin with a brief review of the Elo rating system, which was developed by Elo [**4**] to evaluate chess players according to their skill. (Zermelo [**11**] devised an earlier rating system in 1929; see [**12**].) The greater a player's Elo rating,

[2]According to the FIDE (International Chess Federation) Laws of Chess, the term "move" refers to a single action performed by a player during his or her turn. This corresponds to a "ply" in computer science.

which takes into account the results of previous games, the greater his or her chances of winning a game (see https://en.wikipedia.org/wiki/Elo_rating_system). In 2014, the reigning world chess champion, Magnus Carlsen, attained the highest Elo rating in history with a score of 2882. Most Grand Masters in chess often have a rating of 2500 or above.

In tournament games that have a winner in chess (*decisive games*), White on average beats Black in 55 percent of them (Elo ratings of 2100 or above), but for elite players (Elo ratings of 2700 or above), the winning percentage is 64 percent. The proportion of draws also increases with skill from 35 percent for nonelite players to 58 percent for elite players [**1**, p. 68]. These statistics have not changed much in recent years: White on average enjoys about a 2:1 advantage over Black in decisive games at the highest (human) level of play, but at this level most games end in a draw. (These statistics are based on chess games played under classical time controls. For more information, see https://www.chessgames.com/chessstats.html.)

González-Díaz and Palacios-Huerta [**5**] report White's advantage in all expert matches (players with an ELO rating above 2500) between 1970 and 2010. They found the winning percentage of White to be 64%. Moreover, although each player plays an equal number of games as White and Black in a match, they found that the winning percentage of the player who plays the first game as White to be 57%, which rises to 62% when only elite players (with an ELO rating above 2600) are considered. These findings illustrate the advantage enjoyed by White when each player plays an equal number of games as White and Black.

Statistics from computer play of chess by the strongest programs substantially amplify the advantage of White at the same time that they increase the proportion of draws. In 2020, when the reputedly most powerful chess engine in the world, AlphaZero, played against itself in 10,000 games, taking one minute per move, White won in 86 percent of the decisive games, but these games constituted only 2 percent of the total—98 percent were draws [**9**].[3]

Other expert programs, including Leela Chess Zero and Stockfish, when pitted against each other in the superfinal of the unofficial world computer chess championship (TCEC), give White even greater odds of winning, but the outcome is still a draw in the large majority of games (see https://tcec-chess.com). Despite the fact that computer programs start play from 50 preselected opening positions in the TCEC superfinal (once as White and once as Black), it is remarkable that Black has not won a single game in the last two TCEC superfinals (seasons 19 and 20). All 49 decisive games were won by White, which was either Stockfish or Leela Chess Zero in the Superfinal tournament.

Although these statistics indicate that chess is biased against Black in decisive games, Black is usually able to survive by drawing. It is, nevertheless, surprising that White, by making the first move, is able to achieve almost a 6:1 advantage of winning in decisive games, based on the aforementioned AlphaZero statistics. Whether White has an *inherent* advantage—can force a win when both players make optimal choices—remains an open question.

More light would be shed on this question if the two leading machine-learning chess programs, AlphaZero and Leela Chess Zero, were taught to play with our

[3]A chess engine is a computer program that analyses chess positions and recommends the "best" moves.

proposed change in the order of the 3rd and 4th moves from White-Black to Black-White. Still, even if this change enabled Black to win a greater proportion of decisive games, it would not prove that a draw is the inevitable product of optimal play, just as White's advantage in standard chess does not prove that it can always win. As advanced as machine-learning programs are today, they are not able to mimic perfectly all the moves prescribed by backward induction from every possible endpoint—a draw or a win for one player—in chess.

In section 3, we offer reasons why we think our proposed reversal of two opening moves would tend to equalize the probability that either Black or White can win and, consequently, make the game fairer. Other proposed rule changes that might render chess more balanced or provide other desirable changes (e.g., speed up play) are extensively analyzed in Tomašev, Paquet, Hassabis, and Kramnik (2020). None of this study's rule changes, however, such as the elimination of castling, is as simple as our reversal of two opening moves in chess or as likely to create balance between the two players and force a draw with optimal play.

3. Making chess fairer

Our argument for changing the order of two opening moves is theoretical: The new order lies between the present one that favors White and a comparable sequence, which we discuss below, that favors Black. The favoritism each player obtains from a sequence, we postulate, is mainly a function of being able to move earlier than its opponent or sometimes later, after observing its move.

Moving earlier gives a player a greater opportunity to "set the stage," whereas moving later enables a player to observe the move of its opponent and respond to it. These two factors are in conflict, creating a tension between moving earlier or later. There are occasions in chess in which responding to a move can put a player in a more advantageous position than moving first, which is called *Zugzwang* (see `https://en.wikipedia.org/wiki/Zugzwang`). However, these occasions almost always arise late in a game, when a player's king is in peril; we know of no instances in which Zugzwang can or has happened in the opening moves of chess, which is the kind of change we focus on here.

We assume, consistent with the statistics in section 2, that White has an advantage over Black with the standard chess sequence,

(1) WB/WB/WB. . . .

Black can counter this advantage if it has two moves in a row after White's initial move (WBB), followed by the alternating sequence beginning with W, WB/WB/WB. . . . The resulting sequence WBB/WB/WB/WB. . . , can be written as

(2) WB/BW/BW/BW. . . ,

where the slashes separate pairs of moves.

Observe that both (1) and (2) start with WB, but the alternation from the 3rd move on of (1) is WB/WB/WB. . . , whereas that for (2) is BW/BW/BW. . . . This makes (1) *White favorable*, because W precedes B for every pair from the 3rd move on. By comparison, with (2), B precedes W from the 3rd move on, so this sequence is *Black favorable*.

In effect, giving Black a double move after White's initial move swings the pendulum from favoring White with the standard chess sequence to favoring Black because of the potency of two moves in a row for Black. This raises the question of whether there is an intermediate sequence that creates a balance—between the White favorable sequence of (1) (standard chess) and the Black favorable sequence of (2)—that does not favor either side?

If we add a double move by White (WW) immediately following Black's double move (BB) in the Black favorable sequence of (2), then Black (B) and White each have two moves in a row (BBWW), followed by the alternating sequence beginning with B on the 6th move (underscored). This gives W/BB/WW/$\underline{\text{B}}$W/BW/BW... , which can be written as

(3) WB/$\underline{\text{BW}}$/WB/WB/WB... ,

which we call the *balanced sequence*, or *Balanced Alternation.* Under Balanced Alternation, we assume that a player can check or capture only on the second move of a double move to ensure that the opponent can respond to it.

Except for reversal of the 3rd and 4th moves from WB to BW (underscored), the balanced sequence is the standard chess sequence. The balanced sequence is bracketed by the White favorable (standard) sequence of (1) and the Black favorable sequence of (2), making (3) a neutral alterative that favors neither White nor Black.

Notice that there is no difference in the first four moves of (2) and (3), which both start with WB/BW. The difference lies in the alternating sequences, beginning on the 6th move, with the alternation of (2), BW/BW/BW... , favoring Black and the alternation of (3), WB/WB/WB... , favoring White. The alternation of (3) offsets Black's early double move of (2) and provides—with White's later double move of (3)—the balance we seek.

This is not to say that we can prove that the Balanced Alternation of (3), which translates into a switch of the 3rd and 4th moves of standard chess, always leads to a draw with optimal play. But because Balanced Alternation neutralizes (i) the apparent advantage that White derives from (1)—moving first in standard chess—and (ii) the apparent advantage that Black derives from (2) with a single double move, it seems more likely to force a draw with optimal play than either (1) or (2). This does not rule out the possibility that (1), (2) or both, though biased, also force a draw with optimal play.

It is worth pointing out that the first four moves of (2) and (3), WBBW, are identical. They are also the same as those given by the Prouhet-Thue-Morse sequence (see `https://en.wikipedia.org/wiki/Thue-Morse_sequence`), which can be written as

(4) WBBW/BWWB/BWWB/WBBW... ,

and is what Brams and Taylor [**3**] also call "balanced alternation," as opposed to the "strict alternation" of WB/WB/WB... that is the standard chess sequence. Although the Prouhet-Thue-Morse sequence is arguably fair, it has the disadvantage of allowing more than two double moves, as in the first 16 moves of (4) which contain four double moves.

Another well-known sequence, known as Marseillais chess (see `https://en.wikipedia.org/wiki/Marseillais_chess`), uses only double moves,

(5) WW/BB/WW/BB/WW/BB... .

Like the Prouhet-Thue-Morse sequence, (5) critically alters the strategy and the tactics in chess.

In both (4) and (5), a player can make at most two moves in a row, whereas in "progressive chess" the number of moves a player makes on a turn increases progressively. The move order can be written as

(6) W/BB/WWW/BBBB/WWWWW/. . . .

Due to the rising number of legal movements every turn, the traditional chess engines do not perform well in this version. Hence, Janko and Guid [**6**] propose a new chess program for this variant.

To make a rule change acceptable to chess players, we think it should not have more than two double moves, as does (3). The fact that (3) requires only a reversal of the 3rd and 4th moves from WB to BW makes it even more palatable—it's only a minor revision in the order of play at the beginning of a game, not, as with (4) and (5), throughout. Besides being impartial, (3) seems to be the sequence most likely to force a draw with optimal play, which would make it fair as well.

4. Chess analysis

We next present evidence for the greater fairness of Balanced Alternation by analyzing how moves under it might play out following four well-known first moves by White: (i) 1.e4; (ii) 1.d4; (iii) 1.Nf3; and (iv) 1.c4. For our analysis we assume the reader is familiar with chess notation (see `https://www.chess.com/terms/chess-notation` for an introduction to this notation). From these opening moves, we evaluated the most plausible positions that would be reached under Balanced Alternation by applying some basic chess principles (e.g., winning a free pawn in the opening is generally beneficial) and standard chess reasoning.

To assess the positions that would be reached after the first five moves, W/BB/WW—once Black and White have each made their double moves—we used the Stockfish 13 NNUE chess engine on the Lichess website and on the Chessify server. Stockfish recently won the superfinals of the TCEC Season 20 against Leela Chess Zero and is currently considered one of the top chess engines. Although there are tens of thousands of distinct positions that that can occur after the first five moves, we focus only on the four that begin with White's aforementioned first moves and are followed by Black's and White's arguably strongest double moves in response:

(i) 1.e4: Black responds with d5 and then captures the e4 pawn by dxe4. White recaptures Black's e4 pawn by first choosing a move such as Nc3, d3, or f3. If White chooses to recapture the pawn with a knight, then the resulting position would be the one given by the following notation: 1.e4 d5 2.dxe4 (B) Nc3 (W) 3.Nxe4, where the only difference with respect to the standard chess notation is that the order of moves under (2) is reversed, as indicated by B and W in parentheses. This brings us to an already "novel" position that is unlikely to occur under a standard chess sequence when White plays 1.e4. Stockfish gives an even "0.0" evaluation to this position. (The depth is 46 in the "cloud," which indicates that the evaluation of this position was already computed and available.) However, if White captures Black's e4 pawn with another piece, then Stockfish gives Black a slight advantage, which would be even greater if White chose not to capture the pawn on e4. Overall, Black is better off under these lines than, for example, under Ruy Lopez (i.e., 1.e4 e5 2.Nf3 Nc6 3.Bb5), where Stockfish's evaluation is +0.2 (depth 56), favoring White by 2/10 of a pawn (positive values indicate a better position for White).

(ii) 1.d4: Black responds by playing d5 followed by either e6 or c6, which may lead to standard defenses such as the Queen's Gambit (Declined) and the Slav Defense. In addition, Black can create at least two novelties under Balanced Alternation, capturing White's d4 pawn with its c-pawn. Suppose that Black captures White's d4 pawn by playing c5, followed by cxd4, and White responds by playing c3 followed by cxd4, which leads to the position given by 1.d4 c5 2.cxd4 (B) c3 (W) 3.cxd4. In this position, Stockfish's evaluation is +0.1 (depth 51), which slightly favors White.

(iii) 1.Nf3: Black has solid and well-studied defense options against 1.Nf3, such as d5 and e6, or d5 and c6. White can then create two reasonable novel lines, 1.Nf3 d5 2.e6 (B) e4 (W) 3.exd5 and 1.Nf3 d5 2.e6 (B) e4 (W) 3.e5. The evaluations of both positions are 0.0 (depth 47 and 49, respectively). In addition, the position after 1.Nf3 d5 2.e6 (B) c4 (W) 3.d4 can be reached under both Balanced Alternation and the standard chess; its evaluation is 0.0 (depth 47).

(iv) 1.c4: Black plays a standard defense against the English opening. Alternatively, Black may create two novelties by capturing the c4 pawn with either its d-pawn or b-pawn; the latter choice is favored by Stockfish. White's most promising continuation is to capture Black's c4 pawn by playing b3 and bxc4, resulting in the position given by 1.c4 b5 2.bxc4 (B) b3 (W) 3.bxc4. In this position, Stockfish's evaluation toggles between 0.0 and +0.1 as the depth increases, and stabilizes at 0.0 from depth 41 up to and including 47. By contrast, White keeps at least a slight advantage in the English opening under standard chess. (At a lesser depth (up to 30), we have also run Leela Chess Zero evaluations on the positions we present, but we have not obtained any significant differences compared to Stockfish's evaluations.)

Needless to say, our reporting of the evaluations of the novel lines under Balanced Alternation should not be considered as final evidence that those lines are preferable for Black over more established defenses. Our aim in this section has been to illustrate that there are some new plausible lines of defense for Black that seem to create balanced (such as under 1.e4 and 1.c4) or near balanced (e.g., under 1.d4) positions—while ensuring that some of Black's standard defenses are available (e.g., in response to 1.Nf3).

In summary, Black does strictly better under Balanced Alternation than under standard chess for the first moves (i), (ii), and (iv). Only under (iii) are the evaluations the same as those for standard chess. All in all, we believe our analysis shows that Black will be able to do at least as well, and sometimes better, than it does under standard chess.

5. Conclusions

The Balanced Alternation of (3) is relatively easy to implement, reversing only the 3rd and 4th moves of standard chess. Our study indicates that Balanced Alternation is more likely to be fairer to Black than (1), and fairer to White than (2), and more likely than either to force a draw with optimal play.

Balanced Alternation makes tournament play more efficient. Presently, because of the bias in favor of White in standard chess, contestants in most tournaments must play White and Black an equal number of times to neutralize this bias.

But this would probably not be necessary with Balanced Alternation. In knockout tournaments, in particular, immediate elimination could occur without the need to play White and Black an equal number of times. Thereby tournaments could accommodate twice as many contestants, each playing as few as one game before elimination. To be sure, physical limitations are not a problem with computer tournaments, but we presume that the attraction of in-person play will resume after the Covid-19 pandemic has subsided.

Without strictly alternating moves from the beginning, there will certainly need to be an adjustment in players' thinking about optimal openings. White, for example, might make a different first move under Balanced Alternation than it would under strict alternation. And Black, able to make two moves in a row (BB) after White's initial move, might use each move for a different purpose—or use both moves to launch a broader attack or defense—as might White with its subsequent double move (WW).

We believe that our reforms are compelling for two reasons: (i) they require only a switch in the order of the 3rd and 4th moves; and (ii) they would not be difficult for chess players to learn and adapt to, after which the familiar alternating moves of WB/WB/WB... occur. Perhaps the main benefit of Balanced Alternation is that it almost surely will make chess fairer, putting Black on a par with White, even if we cannot guarantee that it forces a draw with optimal play.[4]

This is not to say that there may not be other sequences, such as giving only Black a double move but later than we assumed in (2), that makes chess fairer. What especially appeals to us about (3) is that the reversal occurs early and strikes an even balance between the White favorableness of (1) and the Black favorableness of (2).

We wish to make clear that we have not proved that Balanced Alternation equalizes the chances of White and Black winning, much less that it forces a draw with optimal play. By giving a boost to Black that puts it on a par with White, Balanced Alternation renders chess fairer than any other reform of which we are aware.

There is a catch, however, that some aficionados of chess worry about—namely, that it will make the game even more drawish than it presently is and hence more boring to watch, especially at the elite level. This is not an unfounded concern. Our view is that new excitement can be injected into the game by giving human players less time to make moves (and so be more prone to make mistakes)—without necessarily going to the extreme of blitz games—that has the advantage of making games shorter and more immediately engaging.

Another reasonable concern is whether Balanced Alternation reduces the variety of chess openings. In contrast, we anticipate that Balanced Alternation will increase the variety of openings. Indeed, many rare openings in standard chess become viable under Balanced Alternation. For instance, it is uncommon for White to begin the game with 1.Nc3 or 1.c3 under the standard chess, but both openings are comparable to 1.e4 under Balanced Alternation, along with 1.b3, 1.h3, and 1.a3.

In comparison, the so-called pie rule (see `https://en.wikipedia.org/wiki/Pie_rule`) also makes the game fairer by giving Black the option to switch colors

[4]The best-known variant for making chess fairer is Fischer random or Chess960 (see `https://en.wikipedia.org/wiki/Fischer_random_chess`), in which the initial placement of pieces behind the pawns, with minor restrictions, is randomized, giving 960 different starting positions and making it essentially impossible for players to memorize best openings. But this makes winning partially dependent on what realization was selected—some opening positions are even more unbalanced than that of standard chess—-which makes winning for both Black and White less comparable across games and more dependent on the luck of the draw.

after observing White's first move. Under this rule, however, we anticipate that the most popular chess openings, such as 1.e4, 1.d4, and 1.Nf3, will be played infrequently.

In conclusion, Balanced Alternation is likely to change what are regarded as "good" openings in the first five moves at the same time that it reduces, if not eliminates, the advantage that White now enjoys in standard chess.

References

[1] Andras Adorján, *Black is Still OK!*, Sterling Publishing Company, Inc., 2004.

[2] Steven J. Brams and Mehmet S. Ismail, *Fairer Chess: A Reversal of Two Opening Moves in Chess Creates Balance Between White and Black*, 2021 IEEE Conference on Games (CoG), 2021, DOI:10.1109/CoG52621.2021.9619066, pp. 1–4.

[3] Steven J Brams and Alan D Taylor, *The Win–Win Solution: Guaranteeing Fair Shares to Everybody*, WW Norton, 1999.

[4] Arpad E Elo, *The Rating of Chess Players, Past and Present*, New York: Arco Publishing, 1978.

[5] Julio González-Díaz and Ignacio Palacios-Huerta, *Cognitive performance in competitive environments: Evidence from a natural experiment*, Journal of Public Economics **139** (2016), 40–52.

[6] Vito Janko and Matej Guid, *A program for Progressive chess*, Theoret. Comput. Sci. **644** (2016), 76–91, DOI 10.1016/j.tcs.2016.06.028. MR3543071

[7] Jonathan Schaeffer, Neil Burch, Yngvi Björnsson, Akihiro Kishimoto, Martin Müller, Robert Lake, Paul Lu, and Steve Sutphen, *Checkers is solved*, Science **317** (2007), no. 5844, 1518–1522, DOI 10.1126/science.1144079. MR2348441

[8] Ulrich Schwalbe and Paul Walker, *Zermelo and the early history of game theory*, Games Econom. Behav. **34** (2001), no. 1, 123–137, DOI 10.1006/game.2000.0794. With an appendix by Ernst Zermelo, translated from German by the authors. MR1895174

[9] Nenad Tomašev, Ulrich Paquet, Demis Hassabis, and Vladimir Kramnik, *Assessing game balance with AlphaZero: Exploring alternative rule sets in chess*, arXiv preprint `arXiv:2009.04374` (2020).

[10] Ernst Zermelo, *Über eine Anwendung der Mengenlehre auf die Theorie des Schachspiels*, Proceedings of the Fifth International Congress of Mathematicians, vol. 2, Cambridge University Press: Cambridge, 1913, pp. 501–504.

[11] Ernst Zermelo, *Die Berechnung der Turnier-Ergebnisse als ein Maximumproblem der Wahrscheinlichkeitsrechnung*, Mathematische Zeitschrift **29** (1929), no. 1, 436–460.

[12] Ernst Zermelo, *The Evaluation of Tournament Results as a Maximization Problem in Probability Theory*, Annotated Readings in the History of Statistics, Springer, 2001, pp. 167–186.

Department of Politics, New York University, New York City, New York 10012
Email address: `steven.brams@nyu.edu`

Department of Political Economy, King's College London, London, United Kingdom
Email address: `mehmet.ismail@kcl.ac.uk`

SELECTED PUBLISHED TITLES IN THIS SERIES

795 **Michael A. Jones, David McCune, and Jennifer M. Wilson, Editors,** Mathematical Analyses of Decisions, Voting and Games, 2024

792 **Shaoming Guo, Zane Kun Li, and Brian Street, Editors,** Recent Developments in Harmonic Analysis and its Applications, 2024

791 **Mee Seong Im, Bach Nguyen, and Arik Wilbert, Editors,** Algebraic and Topological Aspects of Representation Theory, 2024

790 **Peter Crooks and Alexandru I. Suciu, Editors,** Compactifications, Configurations, and Cohomology, 2023

789 **I. S. Krasil′shchik, A. B. Sossinsky, and A. M. Verbovetsky, Editors,** The Diverse World of PDEs, 2023

788 **I. S. Krasil′shchik, A. B. Sossinsky, and A. M. Verbovetsky, Editors,** The Diverse World of PDEs, 2023

787 **Hemen Dutta, Editor,** Mathematical Modelling, 2023

786 **Hemen Dutta, Editor,** Mathematical Modelling, 2023

785 **A. Leroy and S. K. Jain, Editors,** Algebra and Coding Theory, 2023

784 **Dinh-Liem Nguyen, Loc Hoang Nguyen, and Thi-Phong Nguyen, Editors,** Advances in Inverse Problems for Partial Differential Equations, 2023

783 **Alla Detinko, Michael Kapovich, Alex Kontorovich, Peter Sarnak, and Richard Schwartz, Editors,** Computational Aspects of Discrete Subgroups of Lie Groups, 2023

782 **Galina Filipuk, Alberto Lastra, and Sławomir Michalik, Editors,** Recent Trends in Formal and Analytic Solutions of Diff. Equations, 2023

781 **Emanuel Indrei, Diego Marcon, and Levon Nurbekyan, Editors,** Geometric and Functional Inequalities and Recent Topics in Nonlinear PDEs, 2023

780 **Erik Koelink, Stefan Kolb, Nicolai Reshetikhin, and Bart Vlaar, Editors,** Hypergeometry, Integrability and Lie Theory, 2022

779 **Samuele Anni, Valentijn Karemaker, and Elisa Lorenzo García, Editors,** Arithmetic, Geometry, Cryptography, and Coding Theory 2021, 2022

778 **Carlos Galindo, Alejandro Melle Hernández, Julio José Moyano-Fernández, and Wilson A. Zúñiga-Galindo, Editors,** p-Adic Analysis, Arithmetic and Singularities, 2022

777 **Bang-Yen Chen, Nicholas D. Brubaker, Takashi Sakai, Bogdan D. Suceavă, Makiko Sumi Tanaka, Hiroshi Tamaru, and Mihaela B. Vajiac, Editors,** Differential Geometry and Global Analysis, 2022

776 **Aaron Wootton, S. Allen Broughton, and Jennifer Paulhus, Editors,** Automorphisms of Riemann Surfaces, Subgroups of Mapping Class Groups and Related Topics, 2022

775 **Fernando Galaz-García, Cecilia González-Tokman, and Juan Carlos Pardo Millán, Editors,** Mexican Mathematicians in the World, 2021

774 **Randall J. Swift, Alan Krinik, Jennifer M. Switkes, and Jason H. Park, Editors,** Stochastic Processes and Functional Analysis, 2021

773 **Nicholas R. Baeth, Thiago H. Freitas, Graham J. Leuschke, and Victor H. Jorge Pérez, Editors,** Commutative Algebra, 2021

772 **Anatoly M. Vershik, Victor M. Buchstaber, and Andrey V. Malyutin, Editors,** Topology, Geometry, and Dynamics, 2021

771 **Nicolás Andruskiewitsch, Gongxiang Liu, Susan Montgomery, and Yinhuo Zhang, Editors,** Hopf Algebras, Tensor Categories and Related Topics, 2021

770 **Stéphane Ballet, Gaetan Bisson, and Irene Bouw, Editors,** Arithmetic, Geometry, Cryptography and Coding Theory, 2021

For a complete list of titles in this series, visit the AMS Bookstore at **www.ams.org/bookstore/conmseries/**.